Sustainable Civil Infrastructures

Sustainable Infrastructure impacts our well-being and day-to-day lives. The infrastructures we are building today will shape our lives tomorrow. The complex and diverse nature of the impacts due to weather extremes on transportation and civil infrastructures can be seen in our roadways, bridges, and buildings. Extreme summer temperatures, droughts, flash floods, and rising numbers of freeze-thaw cycles pose challenges for civil infrastructure and can endanger public safety. We constantly hear how civil infrastructures need constant attention, preservation, and upgrading. Such improvements and developments would obviously benefit from our desired book series that provide sustainable engineering materials and designs. The economic impact is huge and much research has been conducted worldwide. The future holds many opportunities, not only for researchers in a given country, but also for the worldwide field engineers who apply and implement these technologies. We believe that no approach can succeed if it does not unite the efforts of various engineering disciplines from all over the world under one umbrella to offer a beacon of modern solutions to the global infrastructure. Experts from the various engineering disciplines around the globe will participate in this series, including: Geotechnical, Geological, Geoscience, Petroleum, Structural, Transportation, Bridge, Infrastructure, Energy, Architectural, Chemical and Materials, and other related Engineering disciplines.

More information about this series at http://www.springer.com/series/15140

Mohamed Shehata · Fernanda Rodrigues
Editors

Project Management and BIM for Sustainable Modern Cities

Proceedings of the 2nd GeoMEast International Congress and Exhibition on Sustainable Civil Infrastructures, Egypt 2018 – The Official International Congress of the Soil-Structure Interaction Group in Egypt (SSIGE)

Springer

Editors
Mohamed Shehata
EHE-Consulting Group in the Middle East
Cairo, Egypt

Fernanda Rodrigues
University of Aveiro
Aveiro, Portugal

ISSN 2366-3405 ISSN 2366-3413 (electronic)
Sustainable Civil Infrastructures
ISBN 978-3-030-01904-4 ISBN 978-3-030-01905-1 (eBook)
https://doi.org/10.1007/978-3-030-01905-1

Library of Congress Control Number: 2018957410

This Springer imprint is published by the registered company Springer Nature Switzerland AG
The registered company address is: Gewerbestrasse 11, 6330 Cham, Switzerland

Contents

About the Editors

Dr. Mohamed Farouk Shehata PhD, MBA, MSC, PMP, CLAQ, ASQ CEO and Founder of the EHE-Consulting Group in Middle East

Dr. Mohamed Farouk Shehata He has more than 25 years' experience of many mega, large, and small projects in the Middle East. He was the leader of the multidisciplinary engineering works, so he has gained experience in the architectural, master planning, urban planning, project management, project preparations, decision making, and value engineering of the projects. In addition to all previous manager works, he has a professional expertise in the geotechnical, structural, and bridge engineering.

Dr. Fernanda Rodrigues holds a Ph.D. in civil engineering with a specialization in BIM management from the University of Aveiro. She is the responsible lecturer of the curricular unities of Site Construction Management and Legal Requirements, Construction Management and Safety Coordination, and Construction Modeling (BIM) in the Civil Engineering Department of the University of Aveiro. Assessment of buildings' deterioration state, development of a methodology to evaluate the deterioration/maintenance condition of buildings, energy efficiency of buildings, facility management, and building information modeling are some of the fields of her research. She is the author and the co-author of various publications in national and international scientific journals and conferences.

Implementing an Occupancy Sensor Lighting Control System in a University Lab Classroom

Lonny Simonian(✉) and Mitchell Packard

California Polytechnic State University, San Luis Obispo, CA, USA
{lsimonia,mtpackar}@calpoly.edu

Abstract. Occupancy sensors are lighting control devices that automatically turn lights on when they detect motion, and off when motion is not detected for an allotted time period. These sensors can be applied anywhere, indoor or outdoor, and can provide substantial energy savings. One potential application where occupancy sensors are not often used is in classrooms. This paper will examine the operation and maintenance of the current lighting control system in a university lab classroom; the operating cost of the current lighting system, the specifics of a proposed occupancy sensor system, the payback period of the occupancy sensor system, conclusions of the study, and recommendations for implementing a wireless occupancy sensor system.

1 Introduction

Occupancy sensors are a simple, affordable method to reduce lighting energy consumption. These sensors may be mounted on ceilings or walls, and automatically turn the lights on when motion is detected. When the sensors do not detect motion for a preset amount of time, the lights are turned off, creating an opportunity for energy savings. A large number of occupancy sensors exist, each with different features that vary in price, levels of performance, and overall sustainability. Von Neida (2000) discusses the results of a case study where classrooms with manual controls were monitored for their lighting energy consumption, determining the amount of wasted energy when rooms were unoccupied. The study revealed that during 20% of daytime operations and 17% at night (with a composite rate of 19%), the lights were left on in an unoccupied room (Fig. 1). The authors conclude the installation of occupancy sensors would drastically reduce levels of energy waste. When compared to a baseline, daytime energy savings would be more than 50%, and over 80% at night (Fig. 2). As a result, occupancy sensors can substantially reduce electricity bills, while preserving fossil fuels and other energy sources.

The subject university for this case study hosts several classes for Construction Management (CM) students; and if these students are enrolled in a university lab course, they may access their lab classroom at any time. With constant access to labs, it is probable that an even greater percentage of energy is wasted in these lab classrooms than those studied by Von Neida. Currently, the lab classrooms utilize manual toggle switches that control specific lights in each lab. Due to the developing technology in the lighting controls industry, the current system is not only outdated, but relies too

M. Shehata and F. Rodrigues (Eds.): GeoMEast 2018, SUCI, pp. 1–9, 2019.
https://doi.org/10.1007/978-3-030-01905-1_1

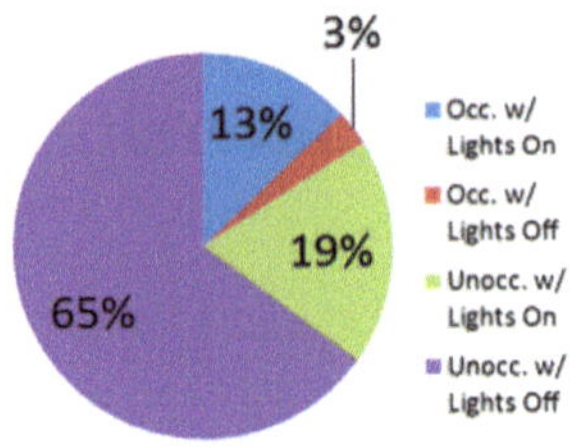

Fig. 1. Classroom lighting conditions (Von Neida)

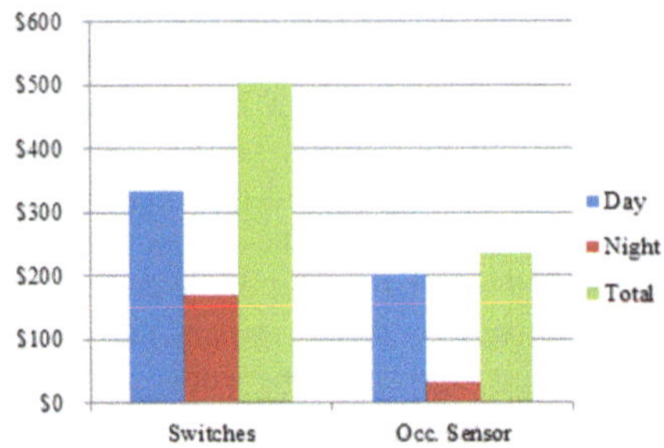

Fig. 2. Classroom lighting conditions (Von Neida)

heavily on human control. Personal experiences and observations have shown that these labs are often illuminated with no occupants during non-lecture hours, wasting a significant amount of energy. To improve the university carbon footprint, a new automated lighting control system, including occupancy sensors, is being considered.

2 Methodology

The objectives of this research were to:

- Identify the maintenance required for a current lighting control system
- Assess the need for a new lighting control system in these labs
- Calculate the operation cost of the current lighting system
- Determine a suitable occupancy sensor system for implementation
- Discover the payback period of the proposed occupancy sensor system

The methodology chosen for this study included fieldwork and qualitative research methods. A student survey was sent to all CM students to determine how well they manage the current lighting control system. The survey focused on how often they adjust the light level, if they find empty classrooms fully illuminated, and other questions to establish a basis for how much energy is being wasted with the current system. Cost data was gathered and the operating cost for the current lighting system calculated. Lastly, potential occupancy sensor systems were considered, along with associated components, features, cost, and the payback period.

Understanding the current lighting control system is essential. First, its proposed technical performance must be understood to determine whether it performs to current standards. As shown in Fig. 3, Partial Fixture Schedule for Light Fixtures in Lab Classrooms, classroom Fixture ‘G’ has one up-light and two down-lights for a row of fixtures (California Polytechnic State University 2006). In other words, each of two switches turns on a portion of a row of lights. There are six rows of lights and associated switches; therefore, when all switches are turned on, the room is illuminated at full brightness. This typically occurs during lecture hours. The next question regarding the existing lighting control system is whether students maintain the system properly, since switching provides the ability to reduce wasted energy. Lights may be turned off before leaving a classroom, or only turning on the necessary lights for a workspace, as opposed to switching on all the lights. However, this current system can also waste a substantial amount of energy. Simply observing these lab classrooms at various times during the day and night, lights are often turned on at, or near, full brightness when there are few or no people occupying the room. To better understand why this occurs, a student survey was designed.

G		4" SUSPENDED INDIRECT FIXTURE COMPLETE WITH ONE PIECE EXTRUDED ALUMINUM HOUSING, SPECULAR ALUMINUM REFLECTOR, HPF ELECTRONIC BALLAST, ICLS COMPATIBILITY AND MOUNTING ACCESSORIES AS REQUIRED. LAMP OPERATION: 2-LAMP GENERAL LIGHTING MODE AND 1-LAMP A/V LIGHTING MODE.	3	32	F32/T8	ELECTRONIC	1	277	FINELITE ALERA CORELITE	FORM 10 "IOC" SERIES "A1" SERIES	PENDANT

Fig. 3. Partial fixture schedule for light fixtures in lab classrooms

2.1 Composing Survey Questions

The online survey was sent directly to the university’s CM majors and minors. These individuals typically utilize lab classrooms on a daily basis due to the wide variety of construction-related classes that are offered. The opinions of the students was considered essential because they are responsible for managing the use of the current lighting system, and may not be aware of their role in conserving energy. To address the need for an occupancy sensor system, this survey was designed to determine if the current system is being utilized correctly. An introduction to the survey informed respondents that the focus was lab classrooms in particular, as well as the significance of a proposed occupancy sensor system. Also included was a brief, technical summary of the current lighting control system to raise awareness of how the toggle switches could be operated for specific uses, including conserving energy. While some of the survey questions were applicable to the current toggle-switch system, most were associated with its effectiveness. The remaining survey questions addressed the new occupancy sensor system and solicited opinions on its implementation.

2.2 Survey Results

The survey yielded a total of 39 responses from anonymous CM students, both majors and minors. Based on the results, the overall consensus amongst respondents favored the installation of occupancy sensors in the lab classrooms. Nonetheless, the responses indicated that a portion of the students adequately maintained the current system, conserving a moderate amount of energy. One factor to consider was that the ages of respondents varied, with some upperclassmen who were more familiar with the lighting system, and consequently, more mindful to operate it efficiently. Another item of consideration was that there was not a direct correlation between the non-lecture hours of classroom use and the individuals who do not manage the system properly. If this survey was not anonymous, then each respondent's answers could be compared throughout the survey, which would help identify the time intervals of zero occupancy with full lighting capacity. Overall, the survey results revealed that the lighting system was not managed properly and wasted energy, validating that an occupancy sensor system should be considered.

Of the 39 respondents, 15% never turned on lights in any of their lab classrooms, or at least do not remember this occurring (Fig. 4). Another 56% turn on lights only once every few weeks. This could be a coincidence if they simply entered the classroom shortly after another student entered it and turned the lights on, or exited but forgot to turn the lights off. Although this may be the case, it is assumed that these individuals consistently enter a fully-illuminated classroom with few-to-no occupants. To compound the situation, these students may be on the same weekly schedule, allowing energy waste to remain consistent.

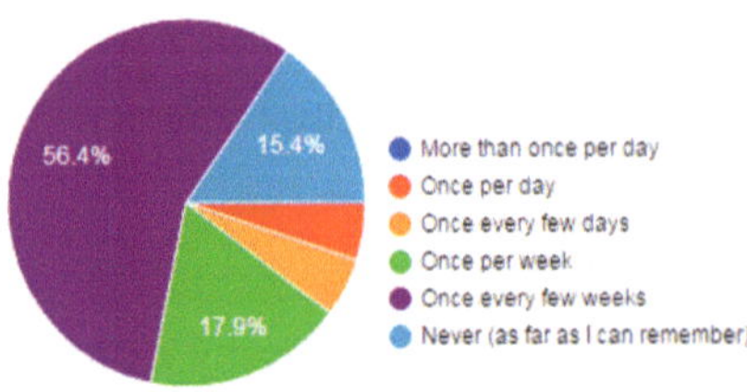

Fig. 4. Frequency that students need to turn lights on

Of these same students, 28% never turned the lights off, and 21% never considered turning the lights off (Fig. 5). Even so, there is the possibility that a person, or group of people, still occupied the room as these individuals left. However, this does not justify the 8% of individuals who claim to never turn the lights off, even if they were the last to leave. Other students turned off the lights sometimes, but not every time, even if they are the last to leave. This would be only a minor issue if it were a few students, but data shows 36% of respondents fit into this category. There is a compounding effect over time, as the sum of wasted energy increases.

Perhaps the most impactful survey question to address the amount of wasted energy in these classrooms was the number of students who enter empty classrooms to find the lights at full illumination (Fig. 6). While 26% of respondents experienced this occurrence only sometimes, 46% report that they walk into an unoccupied, fully-illuminated room often. Finally, 21% found that this situation occurred every single time they entered a lab classroom. Again, the amount of time these lights were left on was not identified, but even short periods of time waste a substantial amount of energy when considering how many students find this commonplace. The current lighting control

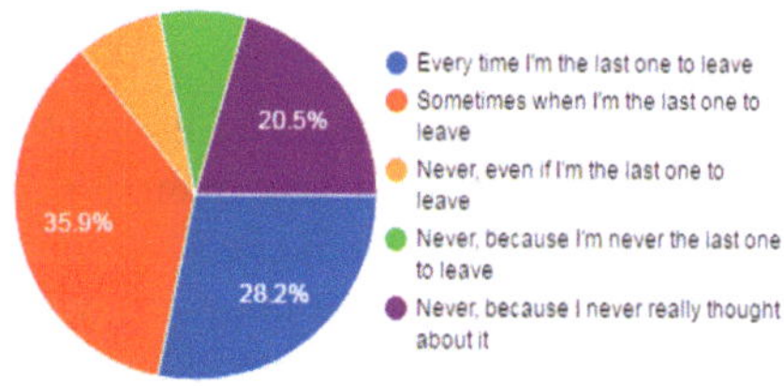

Fig. 5. Frequency that students turn lights off

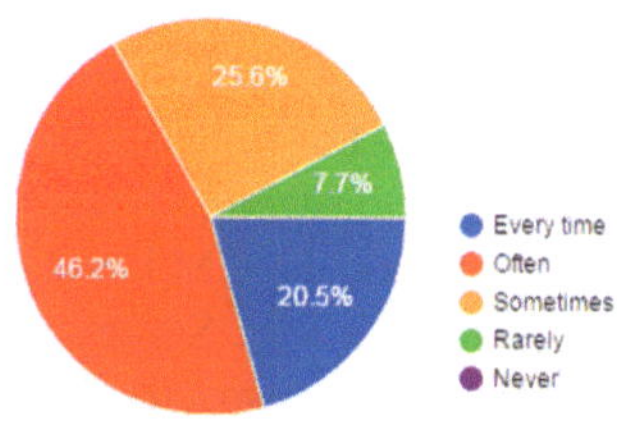

Fig. 6. Frequency that lights are on in unoccupied classrooms

system places the responsibility of turning lights off directly with CM students, who tend to forget to switch off the lights. As previously mentioned, a smaller percentage of respondents had never turned their lab classroom's lights off. This indicates that many of the students who failed to respond to the survey make up a large portion of those who often leave lights on in the labs, into which the survey respondents entered.

2.3 Current System Cost Analysis

To further understand the need for a lab classroom lighting occupancy sensor system, it is important to note the current electrical usage on campus. The university Facilities Department tracks consumption of campus resources, such as electricity, water, and gas. Electrical consumption is most relevant since lighting systems require electrical power to operate. Unfortunately, only campus-wide consumption levels were available.

Most, if not all campus classrooms use a similar toggle-switch system to those in the CM labs. Therefore, the campus consumption rates should correlate with the actual CM lab electrical use, due to the similar lighting control systems. The principal issue with this comparison is that all other classrooms will most likely differ in student-use during non-lecture hours. Also, some facilities operate their lights at full illumination throughout each day.

According to a Facilities Department Energy and Sustainability Analyst (Veium 2017), the university purchases primary power at 69 kV from the local utility at a blended energy rate of \$0.11/kWh Although this does not represent the specific lighting consumption cost of the lab classrooms, it is assumed that this rate can be applied to the building. This \$0.11/kWh is a key component in calculating the cost of the current system.

2.4 Cost Analysis

Since Cal Poly does not currently log the number of hours that the CM lab lights are turned on, an assumption must be made as to how many hours per day that the lights are illuminating each lab classroom. The current class schedule reveals that some labs are used at least 3 to 4 h per day during lecture hours. Other lab classrooms host two classes per day, which means the lights would be turned on for at least 6 to 8 h per day depending on the class subjects and allotted units. Lights are assumed to be fully illuminated during these lecture times, therefore, there is no opportunity to save energy during these times. However, based on personal experiences and observations, it was assumed that these classrooms' lights are turned on an additional 12 h per day, during non-lecture hours. In this 12-h span, students utilize their labs to study, work on schoolwork, and even just relax. Nonetheless, the survey results prove that students do not always occupy these rooms during non-lecture hours, but often leave the lights on. Considering lecture and non-lecture hours, the CM lab classroom lights are illuminated for approximately 16 to 18 total hours per day. Survey results support these numbers, as students indicated that they use their respective lab classrooms throughout all timeslots of a 24-hour period. In addition, 64% of students failed to turn off the lights. Whether the classrooms were occupied or unoccupied, the lights are turned on for about 12 extra hours each day, outside of lecture hours.

The next consideration was the amount of watts that are consumed per hour and the cost of that power. Per Fig. 1, each 4-foot (1.2 m) light fixture consumes 96 watts using a high power factor electronic ballast. There is a total of 30 of the 4-foot (1.2 m) light fixtures in each room, and this is multiplied by 96 watts per fixture to produce 2,880 total watts consumed per hour. Using the previously determined value of $0.11 per kilowatt-hour, the 2,880 total watts must be converted to kilowatts by dividing 1,000. Finally, this value is multiplied by $0.11 per kWh, equaling an operating cost of approximately $0.32 per hour for a single classroom's lights when they are turned on (1). When considering the 12 non-lecture hours of lighting consumption per day, this $0.32 per hour amounts to $1,170.89 per classroom in one school year of about 44 weeks, or 3,696 h (2). Assuming all lab classrooms consume the same amount, this value increases to $14,050.71 for all nine lab classrooms.

2.5 Proposed Occupancy Sensor System

The final step was to select an occupancy sensor system that provides the optimal level of performance, cost, and reliability. Lowering the consumption of light fixtures can provide significant energy savings, considering that lighting accounts for an average of 38 percent of electricity used in commercial buildings—more than any other building system (Lutron 2014). However, the system must meet the needs of these specific CM lab classrooms and the students' learning environment. If an appropriate occupancy sensor system is not implemented in a space, or if the chosen sensors are not adjusted properly, they may shut off the lights while people are still occupying the space. To prevent this, the sensors must be sufficiently sensitive to detect the slightest motion, while having a great range of motion detection.

Interestingly, the CM building project plans (California Polytechnic State University 2006) specified potential occupancy sensors for these lab classrooms, however, those proposed sensors were never installed. It is assumed this was most likely due to value engineering by the contractor during construction. In order for the university to maximize their energy savings, the labs should implement occupancy sensors and accompanying devices. This selected system features occupancy sensors with passive infrared (PIR) technology that detect even the slightest motion in a 1,500 square-foot area. Accompanying the sensors is a relay module that transmits the signal from the sensors to the lights, allowing them to turn on automatically. Also included in the system are wireless controls that turn the system on and off. These remotes are mounted on the wall replacing the CM labs' existing light switches. All components of the system should work seamlessly to conserve a minimum of 30 percent on lighting costs (Lutron 2014). According to the original project specifications (California Polytechnic State University 2006), this proposed system should comply with the original design requirements.

The original requirement state that sensors to be installed in the CM labs must be either PIR, ultrasonic, or dual-technology. As previously mentioned, the selected system uses PIR, thus satisfying this requirement. Furthermore, the detailed requirements for a PIR system state that it must cover an area of 1,000 sq. ft. (93 sq. m) when mounted on a 8-foot (2.4 m) high ceiling and detect occurrences of 6-inch (15 cm) minimum movement (California Polytechnic State University 2006). Although the lab classroom ceilings are higher than that stated, the selected sensors can be wall-mounted and satisfy this height requirement. Also, the state-of-the-art technology programmed into these sensors has no problem detecting even the slightest motion.

Due to its wireless nature, the system has other benefits. Typical hard-wire systems require a number of other materials. This wireless system eliminates the cost of excess wires, conduit, and other electrical commodities. With fewer materials there is a faster installation time, reducing the labor cost as well. With an easy setup, it is anticipated the systems should require approximately 1 h per classroom for installation time. Assuming an installation learning curve, all nine classrooms should have the occupancy sensors installed in 1 day. Once installed, the sensors can be moved as needed throughout the room, another benefit of the wireless system.

2.6 Operation

Once the system is operating, the sensors will begin detecting motion throughout the classroom. If all occupants have left the room and the programmed shut-off time has been exceeded, the sensors will turn off the lights. However, the occupancy sensor system continues to operate, even with the lights off. Subsequently, when other individuals enter, the lights will switch on. The system will remain in this state of operation until the 'off' button on a remote device is depressed. When this is done, the lights will no longer turn on when there is motion inside the lab classroom. For the system to turn back on, one must manually switch them back on by pressing the remote 'on' button.

2.7 Occupancy Sensor Payback Period

The assumed 12 non-lecture hours per day that the lights are turned on can assist in calculating the system's payback period. A payback period is the amount of time that it takes for the new occupancy sensor system's generated savings to pay for the total upfront cost of the system, including materials and labor. However, calculating this payback period is unique in that a desired payback period must be preselected. The final value that must be determined is the ratio of the total number of hours that must be conserved in an entire school year. To begin calculating the one-year payback period, the total cost of 532 (USD) per classroom is divided by the operating cost $0.32/hour per classroom (1). This equals approximately 1,680 h, therefore, the new system needs to conserve 1,680 h of consumption to pay for itself (2).

There are approximately 44 school weeks, or 308 days, per one calendar year when considering vacations and school breaks. These 308 days are then multiplied by the 12 non-lecture hours per day of light consumption, totaling 3,696 h (3). Finally, to find the ratio of conservation hours to consumed hours, 1,680 h is divided by 3,696 h, yielding 0.454. Therefore, the lights must be turned off 45.4% of the 12 non-lecture hours during a day, or 5.4 h, to achieve a one-year payback (4). When referring to the survey results as to how often the rooms are unoccupied with the lights on, this 5.5 h per day seems attenable. Payback period for other levels of conservation are shown in Fig. 7.

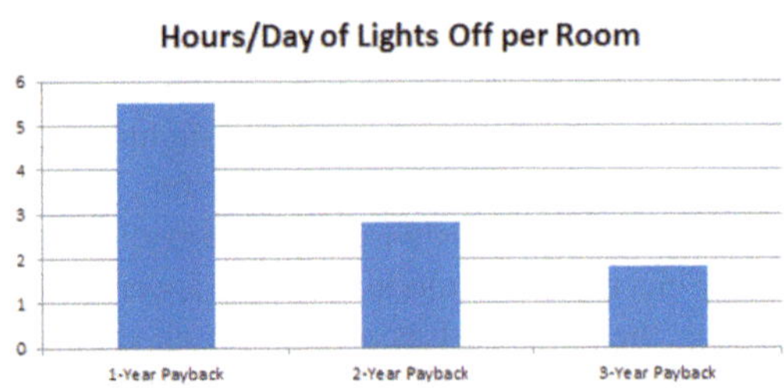

Fig. 7. Hours per day of lights off needed for payback

3 Conclusions and Future Research

The subject university labs currently utilize a manual switch lighting control system. Various studies, as well as survey results of Cal Poly CM students, reveal that these manual switch systems tend to waste energy, specifically when the rooms are unoccupied but lights are left on. The survey results further determined that over 64% of students do not always turn of the lights when they should, and over 92% sometimes, often, or always find unoccupied classrooms illuminated. Observations and survey results have concluded that there are about 12 h outside of lectures that the lights are turned on, which wastes a considerable amount of energy when considering how often students aren't occupying their rooms. This wasted energy costs over 14,000 (USD) per school year for the entire building. This supports the need to implement a new lighting occupancy control system, which has been proven to reduce energy consumption. The wireless occupancy sensor system appropriately addresses the size and motion-detection specifications required in the lab classrooms per the original design requirements. The system conserves enough energy to reasonably payback its initial cost within 1 year.

3.1 Future Research

Accurate data of lab lighting consumption in classrooms should be gathered. This would prevent the need to make assumptions, and would allow researchers to determine the actual cost of the current system, as well as the precise payback period of a proposed new system to implement. The consumption data can be obtained through the use of lighting data loggers, which analyze electrical consumption. Although this would present accurate data, it would create labor and materials costs, as well as time consumption to collect the data. Nonetheless, these steps are necessary in determining the most accurate data possible. Occupancy sensors are progressing in various applications, and their implementation in classrooms has steadily increased over time. With more efficient performance of occupancy sensors and the recent development of strict energy codes, they are becoming more appropriate to include in either new or existing classrooms. Future researchers should make sure these new sensors comply with California energy codes such as Title 24, and could choose exact locations to install the sensors in various classroom layouts to achieve maximum energy savings.

4 Calculations (All $ in USD)

$$96\,\text{w/fixture} \times 30\,\text{fixtures/room} \times 1\,\text{kw}/1,000\,\text{w} \times \$0.11/\text{kw-hr} \\ = \$0.317/\text{hour/room} \tag{1}$$

$$44\,\text{weeks/school year} \times 7\,\text{days/week} \times 12\,\text{hours lights are on/day} \\ = 3,696\,\text{hours/year} \tag{2}$$

$$\$532.00\,\text{system}/\$0.317/\text{hour/room} = 1,680\,\text{hours to payback system} \tag{3}$$

$$1,680\,\text{hours to payback system}/3,696\,\text{hours/year} \times 12\,\text{hours/day} = 5.5\,\text{hours/day} \tag{4}$$

References

California Polytechnic State University: New Construction – 100% C.D. Engineering/Architecture Renovation & Replacement, Phase IIB: California Polytechnic State University (2006)

Lutron: Occupancy/Vacancy Sensor Design and Application Guide. Lutron Electronics Co., Inc, Coopersburg (2014)

Veium, E.: Personal communication (2017)

Von Nieda, B., Manicria, D., Tweed, A.: An analysis of the energy and cost savings potential of occupancy sensors for commercial lighting systems. In: Proceedings of the Illuminating Engineering Society. Paper #43 (2000)

Urban Regeneration and Resilience: Evaluating the Impact of Regeneration Projects on Social Resilience in Glasgow's Sighthill

Yasser Majdi Khaldi(✉)

The University of Edinburgh, Edinburgh, UK
ymkhaldi@gmail.com

Abstract. The Concept of resilience has the capacity to describe issues holistically across different urban layers in diverse fields. Besides, Urban regeneration projects (URPs) result in shifts in the physical environment and perceptual landscape of its communities, where its interventions are explored carefully in this paper to evaluate which interventions stimulate or hinder resilience. For the purpose of understanding URPs impact on social resilience, Sighthill case-study was investigated as part of an ongoing transformational regeneration scheme across Glasgow using place-based approach. Emphasis on social dimensions is given, in addition to the extent of institutions preparedness and involvement to stimulate resilience. This research demonstrates that resilience isn't only a reactive approach that is confined to response to sudden shocks. However, it's more promising if considered as a proactive long-term approach, which gradually leads to socio-ecological and institutional transformation. Encouraging bottom-up approaches lead to socially innovative activities that are found to be one of the main stimulants of resilience across communities and stakeholders. Lastly, four strategies are derived from the analysis of findings as part of the recommendations including empowering locals; initiating place-based solutions; upgrading efforts for economic growth; and developing civic participation.

1 Introduction

1.1 Overview

Resilience is gradually found in academic discussions as a challenging notion, with a more extensive utilisation of the concept in planning and policy discourses (MacKinnon and Derickson 2013). Vale (2013) recognised this academic concern and claimed that unlike sustainable development that proposes moderately ambiguous promises for sustaining environments, the concept of resilience is exclusive for preparedness by adding a significant sense of security, safety, and well-being from potential vulnerabilities encountered. Indeed, this paper discusses how resilience involves more than absolute responsiveness or preparedness for emergencies. In planning, resilience approach has been upgraded to examine the associations between societies and their corresponding environment (Wilkinson 2011).

M. Shehata and F. Rodrigues (Eds.): GeoMEast 2018, SUCI, pp. 10–53, 2019.
https://doi.org/10.1007/978-3-030-01905-1_2

There are many definitions for the concept of resilience; but to grasp the term in an evolutionary manner, urban resilience refers to the capacity of societies, organisations and systems to cope, adapt, and progress against critical shocks, and enduring long-term pressures that weakness the urban fabric (100RC 2017). This definition considers resilience as a proactive approach to the planning, policy-making and strategic steering arena. It allows communities to shape their neighbourhoods on a resilient basis through their capacity for active learning, ability to innovate and adaptability to change. Urban areas adopting resilience structure are viewed as complicated yet flexible schemes, where their arrangement considers the two-way relationship between society and environment. Social interactions in this context have a significant role as much as the local build-up environment do (e.g. neighbourhood). To assess this socio-ecological relationship, governance effectiveness accompanied with public policy should be reviewed. Moreover, the capacity of society to adapt and transform, in addition to the benefits provided by nature are crucial as well (De Groot et al. 2002). Understanding the socio-ecological perspective can lead decision-makers to consider place-based implications of 'people, planet and profit' in both the short and long terms. Nowadays, an emphasis is given to regional competitiveness as the major stimulant of social and economic resilience (Uyarra and Flanagan 2012). Subsequently, added efforts are required to allow local groups in deciding the shape and identity of their regions.

Looking at urban regeneration projects as interventions that transform local places, it's valuable to understand how it affects the resilience of the associated community. Urban regeneration Projects (URPs) are comprehended as an inclusive vision, which aims at achieving permanent development through interventions on the social, economic and environmental aspects of the designated area (Nunes et al. 2013). Conventional URPs are considered to be inadequate due to the associated 'side effects', and inability to entirely attain sustainable-resilient development. Nonetheless, many theoretical methods such as Sustainable, Green and resilient Urbanism have emerged to accomplish better resilience performance (Farr 2008; Lehmann 2010; Applegath et al. 2017).

1.2 Purpose and Significance

This paper intends to understand the impact of Glasgow's investments regarding regeneration projects on the resilience of communities. The paper plans to set up the type and degree of these effects, to find out the relative viability of various regeneration actions and policies, to advise on existing policies and practices in the national context. Glasgow is obtaining substantial funding for rehabilitation programs to enhance and transform disadvantaged neighbourhoods and groups (Kearns 2012). Urban renewal is embedded conspicuously in social policy; yet, knowledge on the impact of resilience methodology on resilience is not enough because numerous regeneration schemes have been ineffectively examined (Davoudi et al. 2012, p. 310).

This research will investigate the biggest URP in Glasgow, which is Sighthill transformational regeneration area (TRA) that has significant location 1 min nearby the city centre. To ensure a holistic understanding is obtained, the study will engage with key actors from decision-makers and the community who are involved in the process to get an in-depth analysis of the context. As the regeneration project is still in its initial phases, there is an opportunity to influence upcoming stages. Thus, outcomes of this paper will be shared by the researcher among decision-makers for their consideration, which can affect the course of the development not only locally, but also across other regenerated areas in the city. This research will focus on social dimensions as it acknowledges the necessity of obtaining an in-depth understanding of the social structure in progressing resilience; in other words, understanding key aspects that support resilience capacities. The outcomes could assist decision-makers in considering particular interventions and avoiding others in regards to resilience strengthening of locally existing arrangements of social capital (Davoudi et al. 2012).

Sighthill is considered to be one of the most deprived areas in Scotland (Goodwin 2017). Glasgow Centre for Population Health's (2014) study suggests that since the people live in an environment with numerous challenges, aiming at solving issues of poverty, health inequalities and the accompanying deprivation persists to be ingrained and problematic to address. Increasingly, literature and studies encourage for attaining higher resilience levels concerning both people and places. Instead of solving problems by decision-makers and organisations, equipping communities and their corresponding environments would magnify efforts to handle confronts (GCPH 2014). Consequently, quality and quantity of local amenities are improved, in addition to community empowerment and participation in decision-making. This papers will aim at Identifying place-based reasons for why individuals and communities react differently to adverse change, to demonstrate that it can support communities and decision-makers in developing practical approaches to future stresses, hence build resilience.

2 Literature Review: Concept of Resilience

The notion of resilience is acknowledged universally in planning and policy making. It is regularly mentioned in discourses related to the idea of resisting variation and recovering to the preliminary condition (Davoudi et al. 2012). The concept varies in definition according to the speciality; however, in the field of planning such as Resilience Forums in the United Kingdom, it is usually constrained to emergency planning (Mehmood 2015), which is linked to safety channels and their preparedness towards distresses and hazards (e.g. flooding). Recently, attention towards long-term stresses is encompassed under the resilience umbrella, as stresses weaken the urban fabric on different levels and have direct impact on the overall resilience of individuals, communities, and institutions (OECD 2016).

On the other hand, there are different scenarios considering the notion of retrieving to a previous phase after disturbance. Examples can be drawn from the work of Jones and Mean (2010) concerning resilience thinking. The first set of evidence relates to the idea of equilibrium that accepts that people and ecosystems react in a chronological and anticipated fashion to instabilities (e.g. environmental changes). The second group of examples discusses the non-equilibrium method, which acknowledges outside factors as a significant role that yields in a less foreseeable response from both communities and the environment against adversities. Furthermore, a third observation reflects on the evolutionary and incorporating responsibility for people and environment in avoiding unfavourable fluctuations before natural limits are reached. Communities, in this case, are also accountable to establish, and preserve the resilience of these joint arrangements (Jones and Mean 2010).

Although discussions about resilience thinking in the planning arena aroused lately, the notion can be traced back to fields of ecological, engineering and even social sciences. Cities are constantly evolving, meaning that their dynamic state necessitates altering our thinking of only achieving a former equilibrium state. In engineering, resilience is described as the extent of disruption or amount of time leaping back into equilibrium after a system disruption (Gunderson 2009, p. 2). This is profoundly close to emergency planning discourses. However, Holling (1973) crafted a new understanding of resilience through an ecological perspective. He clarifies that there could be more than one state of equilibrium in a structure. In other words, a system could adjust to distribution by retrieving to a stable but dissimilar state. Resilience from both engineering and ecological perspectives has two unique features that aids in deciding whether an environment could cope with disruption until a previous state is reached, or transform into new distinctive but stable situation (Davoudi et al. 2013).

The nonlinear nature of resilience questioned the conventional equilibrium understanding adopted by other fields of knowledge (Vale 2013). Elements of dynamics, maximum limits and improbability illustrate the impact of slow-change phases alongside quick-change phases on spatial forms. Cities enclose complex interactions on different levels where change is seen as a new constant. Literature suggests that resisting change doesn't fulfil the requirements of cities, especially those vulnerable to their surrounding environment (Simmie and Martin 2010). Related issues entail active commitment from local groups in the development stage (Adger 2000, p. 347). Additionally, construction and regeneration projects have direct implications on the social fabric with little empirical research on the relationship between these schemes and social resilience (Davoudi et al. 2012).

Upon the rise of the concept, some critiques indicate that resilience is exploited to upsurge the growth of neoliberal policy models (Slater 2014). These recent observations rouse predominantly with rapid transfer of accountability discourses to local communities without suitable preparedness (Cretney 2014, p. 633). In other words, responsibility is imposed on the community to face challenges beyond their current capabilities; this relates to the lack of the appropriate training and resources to be prepared. There is little attention given to dimensions of social capital including its organisation, influence and politics. Social justice issues, in this case, are being ignored by this approach of utilising resilience (Cote and Nightingale 2011).

2.1 Social Resilience

According to Keck and Sakdapolrak (2013), there are numerous definitions for social resilience; yet, the notion has particular connections with the theoretical development of resilience as illustrated previously. It begins with the limited perspective of social resilience as the competence to react to disturbances only. Subsequently, the concept advances to incorporate progressive dimensions such as social learning and adaptation and then continues to recognise the significance of participation and politics with rising uncertainty (Keck and Sakdapolrak 2013, p. 8).

Social resilience definitions are concerned with capacities of individuals, communities, and/or organizations to persist, adapt, and grow to both environmental hazards and social stresses of various natures. Pressures in this manner are usually assumed to emerge from external sources, however they might rise from internal dynamics or from combination of both (Gallopín 2006).

The evaluated case studies on social resilience tackle a broad spectrum of pressures. According to Marshall (2007), most reviewed studies guide their research using this concept, where they highlight specific threats that can be categorised into three sets. The first emphasizes on the long-term impact of environmental volatility, resource scarcity and natural resource organization. The second group of studies focused on disaster risks and natural vulnerabilities. The third set of papers addresses several types of social and institutional transformation while assessing policy and growth strategies (Marshall 2007). Other tackled topics included: health threats, immigration, tourism, and socio-economic change.

To attain a deeper understanding of social resilience, scholars proposed three capacities as shown in Table 1.

Table 1. Social resilience capacities' categories and explanation. (Keck and Sakdapolrak 2013, p. 10)

	Coping capacities	Adaptive capacities	Transformative capacities
Response to risk	Ex-post	Ex-ante	Ex-ante
Temporal scope	Short-term	Long-term	Long-term
Degree of change	Low, status quo	Medium, incremental change	High, radical change
Outcome	Restoration of present level of well-being	Security of future well-being	Enhancement of present and future well-being

The first is coping capacities, which is concerned with the re-establishment of the current degree of welfare immediately after system disturbance (Béné et al. 2012). Though the proactive procedures are demonstrated in the second column, which is the adaptive capacity. This ability promotes people's dedication to learning from previous involvements, foresee imminent dangers, and hence modify their behaviours (Obrist et al. 2010). Lastly, transformative capacities, which comprise of people's aptitude for participation and access to services from the broader socio-political environment (such as local association and governmental institutions), to contribute to governing procedures. Moreover to establish bodies that enhance their wellbeing and promotes collective strengths regarding prospect disasters (Lorenz 2010).

2.2 Society in Resilient Cities

To understand the relationship between URPs and social resilience, it is essential to identify possible elements that could support communities to overcome disturbance. Neighbourhoods that are equipped with decent access to public amenities, education and jobs while demonstrating strong social networks with robust social infrastructure are predicted to behave better under shocks. Since society is considered to be one of the four drivers of resilience (others include environmental, economic, and institutional), three significant parameters were found in the reviewed studies that stimulate resilience in this context (OECD 2016).

1. Resilient places strengthen inclusiveness and social cohesion

The first parameter investigates the significance of social cohesion and inclusiveness in a resilient neighbourhood. The outcomes of community regeneration projects are presented in a positive context while neglecting the associated side effects. Although such developments could deliver prospect jobs, and enhance accessibility to services that can shape social capital, variations in the structure of the community might cause inequalities that are predominantly applicable in this context (Gray 2017).

If enhanced resilience is targeted as an outcome, decision-makers are ought to identify the sources of socio-economic inequality in a community in the process of regeneration planning. Moreover, other concerns should be considered relating to types of governance adopted and what interventions implemented in the new scheme to yield in an increased resilience among the community and mitigate actions that weaken the achievements of improved resilience. Parallel to this, social infrastructure of people, places and institutions shall be accounted for to strengthen the existing capacities and stimulate growth and outputs while maintaining community function (Pitrenaite-Zileniene and Torresi 2014). The following can display few signs regarding the community's skills required to endure unexpected threats and long-term stresses.

To understand what impacts the social cohesion and inclusiveness parameter, population, income and sense of safety should be examined as they are altered by regeneration (OECD 2016). Demographics variations affect collective inclusiveness; therefore, assessment for population outflow and inflow caused by Glasgow's TRAs is vital. The growing economy of Glasgow and the new appealing areas produced is attracting young people searching for education and jobs prospects whether nationally or internationally (GCC 2016b). The influx of higher income or young, or highly educated people might trigger gentrification leading to the displacement of original residents. This paradox might be due to a sudden increase in prices that some residents can't afford, or change in neighbourhood character and networks (Atkinson 2004, p. 117). From a resilience standpoint, this can be leveraged as new knowledge, values, and identities are being introduced which increases both diversity and flexibility; such movements across the city impact the social structure (OECD 2016). Likewise, if migration occurs at high rate involving large populace, it could erode existing social network among local groups.

To acquire an understanding of economic vitality as an outcome to regeneration, income can be a suitable measure as it offers a perception to consumption capacity. URPs should plan for direct and indirect methods of job provision (e.g. social enterprise), as a source of income is an essential component of people's lives (Pitrenaite-Zileniene and Torresi 2014). It allows residents to meet their fundamental needs; therefore, it implies welfare and social relations. The influence of employment resonates to the relationship among people, places and institutions (Brezzi 2014).

According to Choi and Matz-Costa (2015), safety is the pre-requisite of social cohesiveness in a neighbourhood. Their study suggests that safety encourages interaction among inhabitants, neighbourhoods and even local businesses, which is a critical characteristic of social resilience. Moreover, sense of safety can build or damage neighbourhood's reputation affecting its overall attractiveness. Accordingly, economic vitality is affected following the capability of the place to invite professionals and talented employees to work, live and play. Attention to crime rates in URPs should be highlighted (particularly crime policy) because of the direct influence on the general prosperity of the community (Gaigné and Zenou 2015). Empirical evidence illustrates how enhanced levels of education and schooling can drop crime rates; likewise with job availability. These aspects can be tackled in URPs if prioritised in the regeneration planning process (Lochner and Moretti 2004).

2. Resilient neighbourhoods encourage social networks and interactions

The second parameter emphasises the significance of active community networks in resilient societies. Certain characteristics should be evaluated to determine the degree of resilience in an area. These include but not restricted to: communities with high levels of interactions among each other; quality business and commercial life to initiate day-to-day activates; and local associations to support communities through empowerment and participation. These features are critical to establish stronger societies and encounter adverse changes (Bernier and Meinzen-Dick 2014).

On a wider lens, the current rapid urban growth is altering population densities, and accumulating pressure on societies. This can also impact the social infrastructure of communities, such as causing social isolation issues that would constitute major problems, which might lead to additional vulnerabilities against adversities. Conversely, neighbourhoods that establish reliable social networks, support sense of belonging, and obtain stronger social identity tend to overcome shocks, while adapting to enduring stresses and possibly leverage change to transform (Cretney and Bond 2016, p. 11).

Resilience capacities pre-requisites efficient social networks across the neighbourhood and city scale. According to OECD Resilient Cities report (2013, p. 43), active community interactions improve four main capacities. First, adaptive capabilities that revolve around social learning and taking actions based on previous knowledge. Second, inclusiveness throughout the presence of various collective opinions. Third, robustness, which relates to the strength of infrastructure and built-up environment to captivate shocks. Lastly, and integration among systems to expand boundaries.

3. Resilient neighbourhoods provide access to local amenities and services

The third parameter focuses on service accessibility, which translates to prospects for education, work, and access to communal facilities. This parameter supports the overall welfare, accompanied by socio-economic development, and hence social resilience. To evaluate the coverage of service accessibility in a given cases study, both physical and economic accessibility should be analysed (Brezzi 2014). Physical accessibility relates to the location of services and facilities relative to communities. On the other hand, economic accessibility concerns the affordability perspective relating to services pricing and its correlated costs. To further evaluate the level of social resilience, access to services should be assessed to obtain deeper understanding into the issue of well-being and inequalities across communities (OECD 2014).

Services whether basic or advanced differ from one area to another. When thinking about boosting resilience, the degree of planned interventions should be different depending on the needs of the community. One of the pillars of resilience is achieved social justice; this implies that underserviced areas should be prioritised first (Vale 2013). URPs in this context are held accountable for implementing schemes that allocate services equally.

2.3 Evolution of Resilient Thinking

When thinking about people and places, decision-makers are currently adopting resilience approach in the planning process. As the concept is still developing, it's vital to trace how it was embedded in societies, institutions and even strategies. Evaluating the growth period known as the urban renaissance—which begun in the late 20th century and hindered by the 2008 financial crisis—could lead to clues of how resilience thinking started in the UK urban planning (Regan 2000).

Rapid growth yielded in complicated developments as a process of increased national expenditure underneath the New Labour reforms. Notwithstanding, the urban renaissance phase was reinforced to show a profound change in the post-Thatcherism era (Thatcherism relates to the political approach of the British politician Margaret Thatcher) by merging notions of socialism and capitalism forming a new alternative pathway (Mehmood 2015). The vast amounts of URPs were commenced targeting cities with an industrial legacy such as Glasgow that resulted in revitalisation as well as gentrification (Cameron 2003). Various public squares, retail areas, high-end streets, and apartments were established comprising diverse income and ethnic groups in urban districts due to physical proximity (Jones and Mean 2010). Creativity and innovation (growth wise) were evident in this phase fundamentally from medium to small cities. Nevertheless, the succeeding declining phase from the financial crisis uncovered a set of new problems across UK cities from the rapid urbanisation (Martin 2011).

Although the phase of urban renewal took years, sudden realization of segregation from the crisis was highlighted, particularly in spaces designated for the public in addition to retail districts, schools and even housing (Mehmood 2015). In this context, the outcome of various revitalized city centres reduced attention to place-making characteristics of local character. Cities in general demonstrated comparison regarding physical appearance corresponding to urban uniqueness and historical identities (Simms et al. 2005). As shops, high street retailers were similar in city centres; this transformation resonated to shifts in urban policies making public spaces function as systems for added control from monitoring to management using tools such as CCTV and policing. On the institutional horizon, the Civil Contingencies Act in 2004 generated a governmental agenda that made all agencies on different scales in the UK to preserve crisis-planning procedures to be equipped to manage severe dangers for public security (Mehmood 2015). Accordingly, Resilience discussions emerged, and syndicates were initiated to guarantee an organized formal work. Nonetheless, the resilience approach that was adopted by most institutions through their plans was limited to preparing for possible emergency conditions, planning processes for incidents, reacting immediately and retrieving urban schemes to their stable state (Greater London Authority 2010). This tactic is critiqued for holding a linear approach while being less pre-emptive since the model's objective is to sustain the existing equilibrium condition (Davoudi et al. 2017).

2.4 Resilience in Institutions

Efficient leadership and governance have major significance in fully achieving urban resilience. To build resilience in cities, institutions and governance structures should entrench resilience in their thinking, policies and operations (OECD 2016). Trust is essential between communities and their organisations; meaning confidence in institutions' capabilities to react, and equip as well citizens with high-quality, affordable resources for both shocks and long-term stresses (Pelling et al. 2015, p. 33). To grow resilience on the neighbourhood level and across the city, this necessitates strong political will and dedicated leadership. This means having competence amid local organisations, assets flexibility, and pledge to clarity (Colombatto 2013). To grow resilience in communities, many dimensions should be attained. This includes policy modifiers shifting institutions approaches to embrace a holistic resilience approach in all management and governance operations; unified short to long-term resilience plan accompanied by assessment instruments; open data strategies and effective community engagement; and confirming applicable regulations and principles (OECD 2016, p. 52).

3 Methodology

a. Research Approach

For the research approach, qualitative deductive research is employed to understand the impact of regeneration projects on social resilience. This technique aids in recognising pragmatic stimulants to resilience through regeneration interventions. The adopted approach helps the researcher understand the complex socio-ecological relationships by investigating case studies in their locales and perceive their mechanisms in their circumstances (Flint and Golicic 2009, p. 846).

The qualitative method allows the researcher to find and analyse information grounded on individual knowledge. To collect the planned data, case study scheme is used (Creswell 2003). Moreover, this approach can be valuable when using a case study to establish an in-depth understanding of social subjects (Flint and Golicic 2009, p. 845). It contributes to answering questions when the researcher has no control of the occurring measures. This is the primary reason why interviews with decision-makers and residents where conducted.

b. Samples/Characteristics

This research engaged non-probability sampling for its methodology. It has the edge of grasping the case-study efficiently in this investigative study. Conversely, the results may not be general, as it doesn't embody of the entire population (Lund Research 2012). The experience of building social resilience can be conveyed. Selective sampling is used for this research, particularly expert sampling. To understand how regeneration interventions stimulate or dissuade social resilience in practical terms and influence planning policies, it is essential that the interviewed sample be directly engaged to deliver appropriate material. As the investigation was conducted throughout a case study (Sighthill, Glasgow), the sample was chosen from all stakeholders involved. The conducted interviews included decision-makers from Glasgow City Council (GCC), Glasgow Housing Association (GHA), and Gowell research program. Other interviewees included community artists (hired by GCC) to understand how they are building empowerment through community engagement. Moreover, as Sighthill is the biggest housing-led regeneration project, an expert overview from the director of Policy Scotland was obtained.

The researcher chose to investigate Sighthill case study mainly because of the various active interventions. It is a part of an ongoing regeneration program in Glasgow that involves demolitions, renovation works, the establishment of new housing, and supporting locals throughout the provision of facilities and services. The previously mentioned activities constitute a prospect to investigate their impact on social resilience by analysing changes from the local community and decision-makers' perspective. There are various materials from the project that makes Sighthill highly suitable for investigation particular due to its significance regarding scale, location, and the fact that outcomes of the study can alter the course of future phases (since Sighthill development is still ongoing until 2020 according to GCC). Moreover, the researcher attended a community meeting to engage with people from Sighthill and acquire a different angle about resident's experience; part of the community meeting was designated for the researcher to ask questions to the participants to get feedback on certain issues that will be presented in detail in the finds and discussion section. Due to restrictions to conduct one-to-one interviews with the community, Gowell research program provided the researcher with an internal document where locals are interviewed on individual and family basis. This document was essential as it makes the data more reliable and comprehensive. Details of the sample population that are used in the findings & discussion chapter are shown in the Table 2.

c. Data Collection Methods

In the data collection phase, the researcher used both primary and secondary data. Desk research is the main tool utilised for secondary data, which was derived from academic articles, books, reports and websites. Moreover, internal documents provided from Gowell, GCC, and GHA were analysed. These sources constituted the basis for literature review and interview questions.

Regarding the primary data, it was gathered from semi-structured interviews, in addition to the analysis of attended community meeting. The interviews as mentioned above were conducted mainly with decision-makers involved in Sighthill TRA. Semi-structured interviews are useful in this context to permit the interviewees to expand on

Table 2. Participants' details from the conducted interview. (Lawson and Kearns 2017, p. 16)

ID (gender)	Household composition (age)	Reason for moving	Size/type of new build	Other information
SH1 (M)	Couple (51, 55)	Voluntary, health issues	1-bed, ground floor flat, adapted	Male not working, multiple health issues; BAME migrant
SH2 (F)	Single woman (59)	Clearance	1-bed flat, 2nd floor	Not working, mental health problems
SH3 (M)	Single man (26)	Voluntary, from Maryhill	1-bed flat, top floor	Part-time job (waiting for college); BAME migrant
SH4 (F)	Woman (23) lives with parents	Clearance	2-bed ground floor flat, adapted	Family have fast-food business; BAME migrant
SH5 (M)	Single man (71)	Clearance	2-bed flat, ground floor	Retired
SH6 (F)	Woman (28) lives with partner and 3 children	Clearance	2-bed flat, 2nd floor	Not working; would like bigger flat
SH7 (F)	Couple (34, 37) and 14 year old son	Clearance	2-bed terraced house, adapted	Female not working, partner in work; BAME migrants
SH8 (F)	Couple (30 s) and 6 children	Clearance	5-bed house	Parents both not working
SH9 (F)	Couple (both 38) and 3 children	Clearance	3-bed house	Female not working, partner in work; BAME migrants
SH10 (F)	Woman (46) and 3 children	Clearance	3-bed	Part-time work; BAME migrants
SH11 (F)	Woman (50), daughter (24) and grandson (4)	Clearance	3-bed	Not working (older woman), daughter part-time work

particular topics and possibly offer an in-depth perspective on certain issues (Gill et al. 2008). Each Interview lasted approximately 1 h and was reordered.

d. Data Analysis

Interview questions were formed mainly from the literature review arguments and case study research. The questions tend to cover all aspects of social resilience concerning URPs, were thematic and content analysis was used to analyse the data. After the interviews were recorded, content analysis was used to code the raw material (after the researcher transcribed the answers) and yielded in categories and sub-categories that relate to the overall themes presented earlier in the literature review (Aronson 1995). The following data is then evaluated thematically, which aids in evolving a deeper consideration to the qualitative aspects (Marks and Yardley 2011). Furthermore, Thematic approach enhances the analysis by detecting relationships and linking categories. These classes are further developed and linked with aspects of both the literature review and case study to help determine gaps that are discussed in the recommendations section for future research.

e. Validity and Reliability

In qualitative research, the reliability of thematic analysis can be doubted, as it's built on the researcher arbitration and dependent on research documents. Nevertheless, in qualitative research, validity has more significance compared to reliability (Guest et al. 2012). Moreover, to guarantee the validity, careful transcribing was conducted for all the interviews coherently. The researcher collected and analysed material that supports or opposes answers derived from the interviews whether from the masterplan documents, circumstantial evidence, and other materials that would be shown in detail in the case study section. Additionally, the appendix section has the interview questions to confirm transparency.

f. Limitations of the Study

Due to the time span of the research, a longitudinal approach is hard to implement, meaning to study the population before and after the interventions to assess how social resilience is changed. Moreover, capacities and skills (e.g. social learning, and networks) lost or obtained throughout regeneration are difficult to measure. Another restriction is the inability to conduct one-to-one interviews, though Gowell research program compensated for this by providing their own conducted interviews. Some of the outcomes will not be generalized, as the case study was done for particular type of projects and particular population sample. However as mentioned earlier in the significance of the study, there are seven other regenerated sites across Glasgow with similar characteristics and conditions that can benefit from most of the outcomes. Even though recommendations from this study will be conveyed to decision-makers for future consideration in the upcoming phases of the development, Sighthill masterplan is not complete which indicates that some findings are susceptible to change. This could have implications in supporting or opposing particular arguments in the paper.

4 Findings and Discussion

This chapter mainly intends to provide an in-depth understanding of the impact of regenerations interventions on social resilience using Sighthill TRA as a case study. Therefore, this section will start contextualization and description of the nature, history and current situation of the Sighthill. Subsequently, central themes will be presented to provide a holistic understanding. Other questions that will be tackled include: how do neighbourhoods, its form and social networks play a role in affecting, alleviating and facilitating change mechanisms? And what role do decision makers and institutions have in stimulating resilience?

a. Case Study Context

Sighthill is one of the biggest regeneration schemes across Glasgow; the area is located north of River Clyde in the wider part of Springburn district as shown in Fig. 1. It has an advantage of being approximately few miles away from Glasgow city centre and to Glasgow's main highway and rail systems.

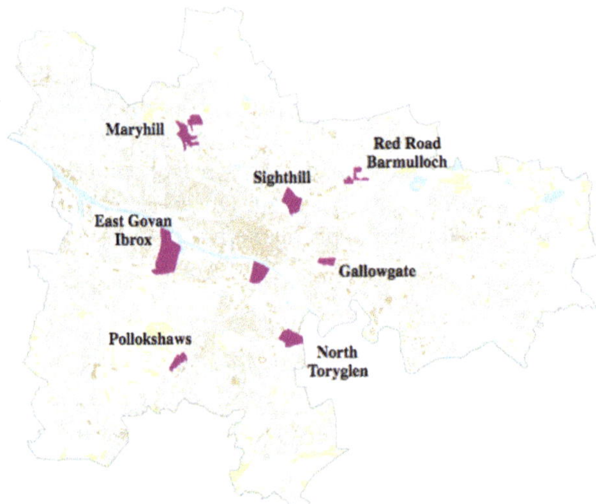

Fig. 1. Shows the eight sites where regeneration schemes is conducted (Graham 2017, p. 3)

Sighthill is part of The Transforming Communities Partnership, which was founded to supervise the provision of a substantial renewal and growth scheme throughout Glasgow. The partnership consists of Glasgow City Council, the Scottish Government, and Glasgow Housing Association. The program was initiated in 2005 and tackles eight sites known as Transformation Regeneration Areas (TRAs). Local Delivery Groups are created for every TRA as illustrated in Fig. 2, with inhabitant and neighbourhood representative.

The study site falls under the new regeneration approach that seeks to deliver sustainable housing with mixed communities and tenures accompanied with enhanced accessible amenities (GCC 2017). Furthermore, green space and commercial units are part of the project delivery. This regeneration projects which is also part of the complete Housing Regeneration Program for Glasgow aims at providing around 600 affordable social units, alongside with approximately 6500 homes for mid market rent (MMR) or reasonable sale (GCC 2017). Sighthill is one of the key infrastructure projects within Glasgow and Clyde Valley's City Deal. As illustrated in Fig. 3, The £250 million masterplan has Planning Permission in Principle (PPP) that comprises of the following: 141 units of social rented housing for Glasgow Housing Association (GHA); Around 650 houses for sale; 500 units designated for student accommodation; A renovated school estate; Commercial and local community services; A pedestrian bridge to connect Sighthill neighbourhood to the city centre; A road bridge to open a

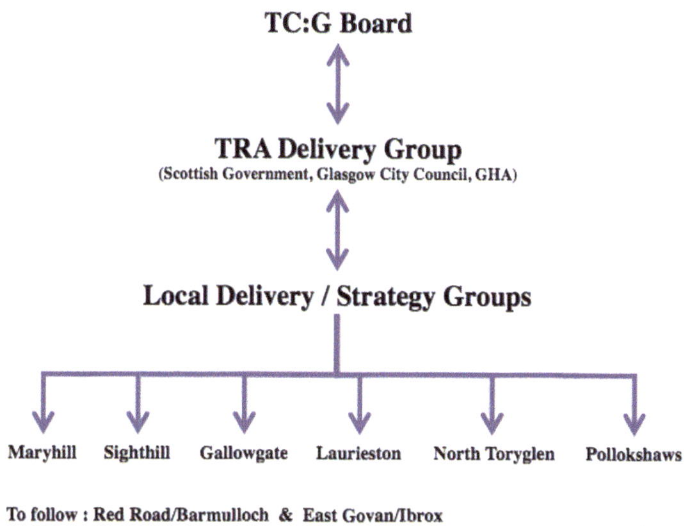

Fig. 2. Illustrates the TRAs partnership model (Graham 2017, p. 6)

route from Sighthill to both the north and west sides of the city; and a revived Park. Overall, The masterplan intends to build a lively pedestrian welcoming urban neighbourhood. The primary objective is to achieve sustainable living accompanied with several varieties of housing and an improved green system (GCC 2017).

i. History and Current Situation

In the past two centuries, Glasgow witnessed more change compared to its entire history. The city underwent major transformation due to the Industrial Revolutions, which triggered intense urban growth and rapid population upsurge. In the 1850s, Glaswegians suffered from low life expectancy that caused population loss and deindustrialisation in the succeeding years (GCC 2016c, p. 15). This was accompanied by a rise in unemployment rates mainly due to the disappearance of traditional industries. Numerous families relocated in the early 1950s from the packed city to adjacent emerging towns (GCC 2016c, p. 16). However, people who stayed were accommodated in new trends of high-rise apartments that escalated problems, particularly on the social aspect. High-rise buildings were thought to be the solutions, although they rapidly lost their primary desirability and became linked with bad reputation and crime. By 1990s, Glasgow's population has intensely diminished to half of its preceding rate (GCC 2016c, p. 16). The fabric of the city suffered from income, prospect and health inequalities. Although the city uplifted its status in the last decade, taking major steps in urban regeneration and development that positively impacted health, and wellbeing, it continues to have the poorest life expectancy in Scotland (GCC 2016c, p. 18).

Particularly for Sighthill, the former derelict land at a period hosted large chemical manufacturers functioning, where chemical waste was dumped until its closure in 1964. Therefore, poor soil quality existed that didn't allow residents in creating allotments or participating in community gardening events according to Coyle (2008, p. 1). Afterwards, Glasgow Corporation Housing Department delivered Sighthill project in 1964,

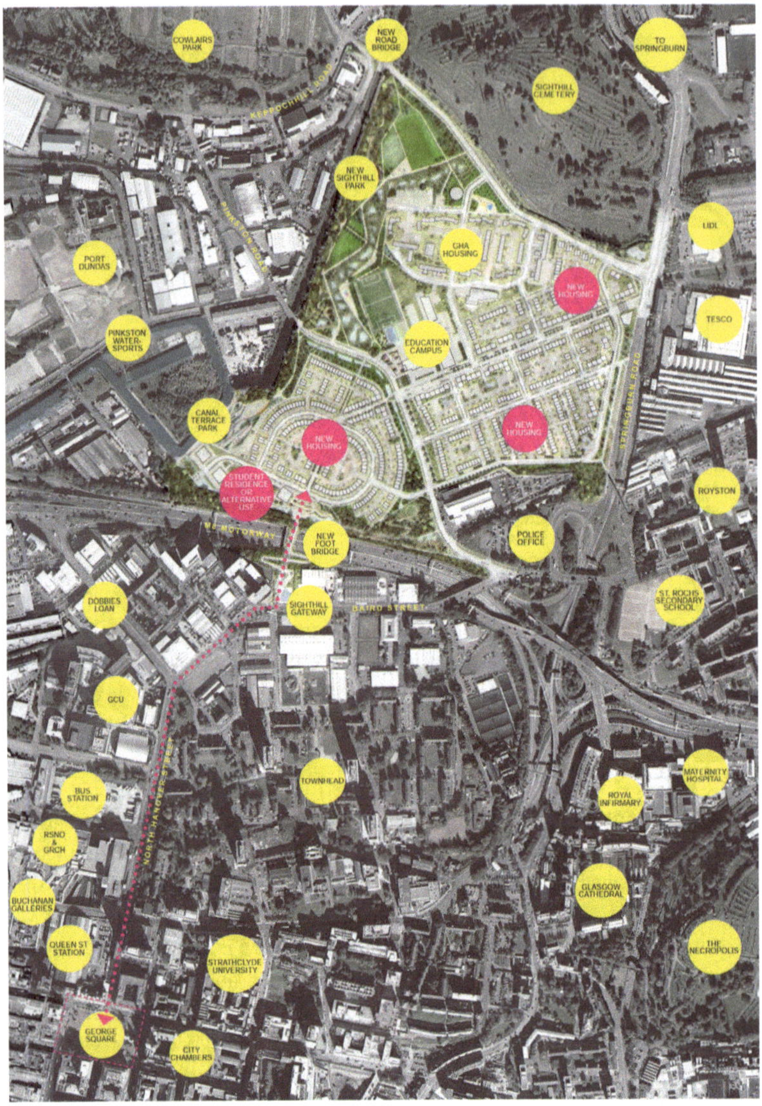

Fig. 3. Sighthill masterplan and its components. (GCC 2016d, p. 2)

which consisted of ten 20-storey high-rise structures, five-story maisonette buildings and five lines of tenements (UK Housing Wiki 2017). Following this, the area faced stagnation, particularly from living issues related to high-rise abundance. Notwithstanding its vicinity, the district was chosen in the 1990s as a temporary shelter site for immigrants and asylum seekers, as it has experienced low tenancy rate (UK Housing Wiki 2017). Therefore, Sighthill has high diverse ethnic population compared to other districts across Scotland. Leasing flats in 2005 were paused as Glasgow Housing Association determined a different future for the area. Glasgow City Council's bid for the 2018 Youth Olympics designated the site to be the Athletes village (UK Housing Wiki 2017). Although the proposal didn't proceed, the area's regeneration continued by modifying the masterplan (Fig. 4).

Fig. 4. Sighthill boundaries and GHA new built housing site in the north. (Lawson and Kearns 2017, p. 11)

Based on GHA presentation regarding TRAs scheme, before developing the masterplan, discussion among decision-makers, local groups and stakeholders took place (Graham 2017). Several transformative interventions were determined such as major high-rise demolitions plans that took place between 2009 until 2016. This process lead to displacement of people accompanied by drastic changes in Glasgow's skyline (since demolitions were tackling high-rise buildings). Although the council's intentions are clear as their primary objective is to regenerate deprived areas, (and this meant high-rise buildings weren't suitable for the new development), it caused controversies across the city concerning what guarantees that the new development is planned to last (Leslie 2015)?

Following demolitions, low-rise social housing was constructed as a replacement for the previous housing as shown in Fig. 5, in addition to hundreds of other units that are planned to finish within ten years as shown in Fig. 10. Before the demolitions process, the intervention gave local residents the choice of inhabiting the newly built housing or leaving to other neighbourhoods (Kearns and Darling 2013). Few families accepted taking new homes, while the majority of people living in active TRAs decided to live somewhere else for various reasons such as construction inconvenience, and doubts regarding the new developments. Currently, Sighthill is undergoing fundamental historical field decontamination process to establish the infrastructure (GCC 2016d). Parallel to this, the first phase of 141 units of social housing were delivered and occupied since 2015, containing a variety of designs and sizes (refer to Fig. 5).

Fig. 5. Transformation of tower blocks housing scheme into a new community. (Paterson 2016).

ii. Masterplan

Green areas are perhaps the most noticeable aspect of Sighthill masterplan as shown in Fig. 6, whether it's around the built-area, or embedded in between. The built form typologies in the masterplan can be mainly categorised into single 3–4 story buildings, and row low-rise buildings. Open and green spaces are designed to be available in the form of semi-public and public spaces.

Overall, the plan enhances connectivity to surrounding neighbourhood mainly the city centre by establishing the new pedestrian bridge, and the new road bridge. Internally, there are four zones as illustrated in Fig. 6 that are connected with a spine that cuts the masterplan from the southwest to the northeast. Although the project designates approximately 22% (141 out of 650 homes) for social housing, they are allocated in only one area, which is the GHA neighbourhood in the north. Since the community is isolated (acts as an island), it can be argued that this arrangement can cause segregation when the development is completed between the current residents and the new inhabitants of the other zones. The project manager of Sighthill commented on this claiming that such phenomena could occur; however due to future activities and arrangements that will take place, it will help merge all zones together and lessen social isolation for the social housing zone.

iii. Demographics

According to Gowell (2011), the population composition in Sighthill TRA is highly different and diverse when compared to Glasgow. Regarding ethnicity, the area contains primarily non-Scottish population including mostly asylum seekers, followed by black and other marginal groups. Overall, they have high child densities, and more people below the age of 18 than 25 (Gowell 2011), which could be associated with probable issues of informal social control.

The following household mix might generate particular challenges for both dwellers and planners. Concerns about the role of asylum seekers on the social fabric always existed whether expressed by the community, city council or even the media (Kearns 2012). During the community meeting attended by the researcher, people reflected on stories of displacement by saying how clearance for demolitions made the

Fig. 6. Sighthill Masterplan, services and facilities. (GCC 2017, p. 2)

people stick together regardless of the ethnic background. This intervention increased the cohesiveness of the society, which is a characteristic of resilient society as explained in literature review section (OECD 2016; Pitrenaite-Zileniene and Torresi 2014). Although measurements show the least scores regarding cohesiveness in TRAs, it is gradually increasing as shown in Fig. 7. Wilton (1998) explains how subjects of context, belonging and identity starts to rise when the communities decide to draw restrictions to maintain control by eliminating or including particular groups.

iv. Clearance Process

Most of Sighthill population was subjected to clearance from their high-rise flats due to announced demolitions (Lawson and Kearns 2017). As mentioned earlier, residents were given the option of leaving to different neighbourhoods across the city or take place in the new development. Those who decided to stay were moved to several flats before their final move to the new estate. A study conducted by Jones (2015) shows

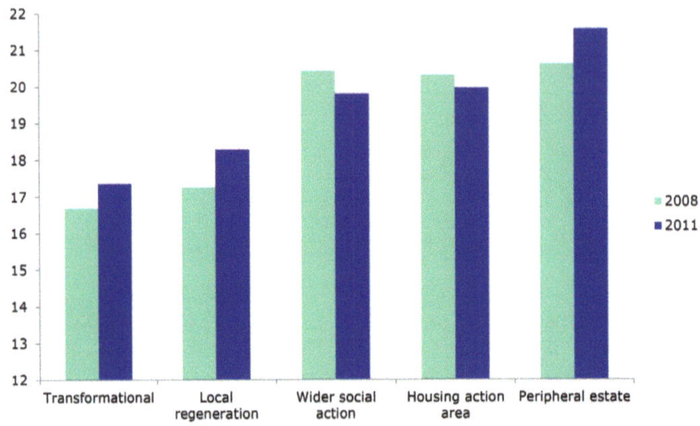

Fig. 7. Social cohesion score by regeneration type. (Gowell 2011, p. 13)

that families who experience high housing mobility report poor well-being and require health assistance more often. Frequent movement interrupts social connections and acquainted neighbourhood structure, leading to losing of local institutions support (e.g. churches, community councils). Notwithstanding, some prospects are accompanying residential mobility such as accessing enhanced local amenities and getting support from various institutions, which is the case in Sighthill context (Fig. 8).

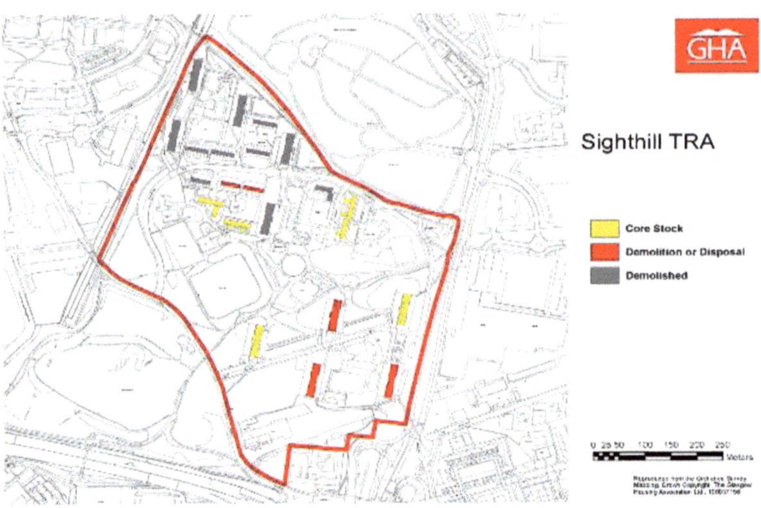

Fig. 8. Sighthill demolitions map. (Fletcher 2011, p. 19).

b. Resilience in Practice

In the data collection phase, the researcher observed that resilience definition varies from one person to another, which is evident in literature as well. Engineers explain resilience by the force a material can withstand before it deforms (Cretney and Bond 2016, p. 11). In practice, this indicates that resilient people ought to attain intellectual or inner strength. Such characteristics though useful are challenging to grow and measure in people. Besides, this approach regarding resilience overlooks the core reasons of obstacles that impact people. What about communities who encounter more pressures than others? Is it possible to blame them for being less capable of enduring? This was the case in Sighthill; their unique circumstances determined their resilient capacities. Therefore, understanding these patterns is essential for decision-makers to account for the differences.

Throughout the conducted research on the concept of resilience from available studies as well as fieldwork, it can be noted that it's an advantage to think about resilience holistically by grasping the different definitions relating to different fields. This way resilience can be thought of as a quality and a capacity that is not only embedded in communities but also shared across different systems of various scales. In other words, to analyse resilience, it's valuable to look at it as a concept that is found as a characteristic among communities (social sciences), in the physical environment (engineering) and networks between groups' interactions (urban geography). Last, beyond the people where decision-makers play a significant role in stimulating it. The below subsections will elaborate on the effectiveness of this inclusive approach to resilience.

i. Community Resilience

Resilience as communal quality varies depending on experiences, skills, memories and other related themes. Hence, it's valuable to analyse the experience of dwellers in disadvantaged circumstances and how they overcome their encountered pressures. Families who showed assurance in managing their conditions exhibited better resilience capacities despite their situation. In a similar context, some people reported their ability to handle themselves by resolving their problems throughout using their skills when encountered by issues either externally (e.g. construction inconvenience) or internally (e.g. house maintenance). Furthermore, some inhabitants mentioned their role in the future within the bigger development looking positively at aspects that they can benefit from while communicating potential outcomes of their decision.

The analysis of Sighthill community shows that having at least one of the following values -confidence, foresight thinking, intelligence, and problem-solving - indicates having higher resilience capacities. Though it's vital to note that these are specific unique features, researchers discovered that growing resilience relies on the efficiency and accessibility of social connections (Scheffran et al. 2012, P. 16). Due to the development procedures, Sighthill lacks a local voluntary sector. This is a vital feature of community resilience as it empowers residents to exploit local assets to prepare and adapt to long-term societal and ecological changes.

1. Living Changes

By summer 2016, the 141 units of housing estate established by GHA were occupied. Figure 9 shows the drastic improvement of the physical conditions of the houses and the surrounding environment. The development dedicated additional attention to the built-up environment (e.g. materials, streets, and green spaces) because of its benefits to the community. The new housing is set to be energy-efficient with modern design (Booker 2017).

Fig. 9. Sighthill row housing changes. (Lawson and Kearns 2017, p. 11)

A report by Policy Exchange (2014, p. 15) implies that demolishing tower blocks and constructing low-rise flats as replacement significantly progress the quality of life. This is because poor health, education and safety levels could be encountered as an outcome. The Create Streets Study (2013) suggests that high-rise construction and maintenance costs are higher compared to other low-rise schemes; moreover, high-rise living doesn't suit families best because of the confined spaces (e.g. no balcony or open spaces).

Participants in the community meeting expressed their content with their move mainly because of increased perceived safety, and health, especially families with children. One of the unexpected feedbacks came from a participant (SH6) who mentioned that the new housing replaced painful memories that she passed through with her family in her previous flat. This was concerning the constant risk of full dampness during winter that had implications on the quality of living. Another participant (SH5) who is an old retired man revealed how the new scheme gave him the option to practice his hobbies and interests, as it was easier to be outside than before. Regarding his living space, he had an extra room where he can host people proudly. Feelings of pride were mentioned several times throughout the discussion. More than one participant stated the fact that they can host a gathering, and invite people generally to their houses because of its "beautiful and well-sustained" nature, compared to their older flats that had maintenance and preservation issues. One of the participants (SH10) who is a single mother who owns a home-based job, can find more time for work and achieve the work-life balance as she can let her children play while monitoring them without consuming the time of having to be with them at the same time. One couple (SH7) that lived in high-rise flats used to be heavy smokers even indoors. Since they moved to the new estate, they changed their habits and reduced their consumption.

Various participants conveyed improvements in their overall wellbeing after moving, compared to their previous behaviours and health complaints. On the other hand, some of the participants (such as SH2) spoke on their behalf about the persistent state of despair they felt before and after the intervention. Although it has changed after moving into the new housing, their health issues are still unstable, and they cannot blend in with the community as they lost most of the people they used to know in this process. One of the participants (SH4) felt the need to change this place to somewhere safer; as she is still sceptical with the new development especially that this is only the first phase and they still have to go through more trouble.

People's responses reflect how the new type of housing is changing people's behaviour, hence the outcomes. In this context, some people feel less isolated than before. Surprisingly, the researcher observed that people are still not spending time outside as expected. Living in high-rise flats requires a different life style. People adapted to these conditions until it becomes part their habits through their behaviours. The analysis of people's behaviours showed that when suddenly moved to a different context, their activities remained similar with few modifications. However, for some people, they voluntary modify their habits, while others, just having the option of being able to go outside makes a difference for them (Fig. 10).

Fig. 10. Site perspective of the current and future development in Sighthill. (Nicoll 2013).

ii. Resilience Around Us

Knowledge from the post-war New Towns to new housing projects has recurrently implied the significance of equipping communities (at an early stage) with resources and expertise from local facilities and infrastructure to be resilient against confronts. Similarly, communities that exhibit high social interaction, where individuals can trust and relay upon each other are more likely to be resilient to adverse change (OECD 2016; Bernier and Meinzen-Dick 2014; Choi and Matz-Costa 2015).

1. Local Amenities

The new neighbourhood was generally described with features of peaceful atmosphere and suitable location. Researcher's observations showed that residents are enjoying the open spaces as they have more room to move around their neighbourhood compared to their older flats, which included many on-going activists across the corridors that caused noises and distribution. Conversely, the ongoing construction activates are causing not only distribution concerning the noise levels, but restricting the accessibility and movements in certain parts. Most inhabitants of Sighthill spend a major amount of time in the neighbourhood (Harvey 2017). This could be owed to having high unemployment levels, and child densities. Thus, the absence of essential services is giving people a feeling of remoteness. The recognised influence of access to local amenities on the community constitutes a substantial reason for the uncertainty and inequality that were part of Sighthill deprivation.

Currently, there is only one small shop and one pharmacist in the whole area. Provisional shops including Lidl, Tesco, and Costco are available across the main road (refer to Fig. 3), which is on the edges of the site. Other services that disappeared with demolitions and construction process as displayed in Fig. 11 including the post office, library, and playground for children. Moreover, there are no local health services provided in the area.

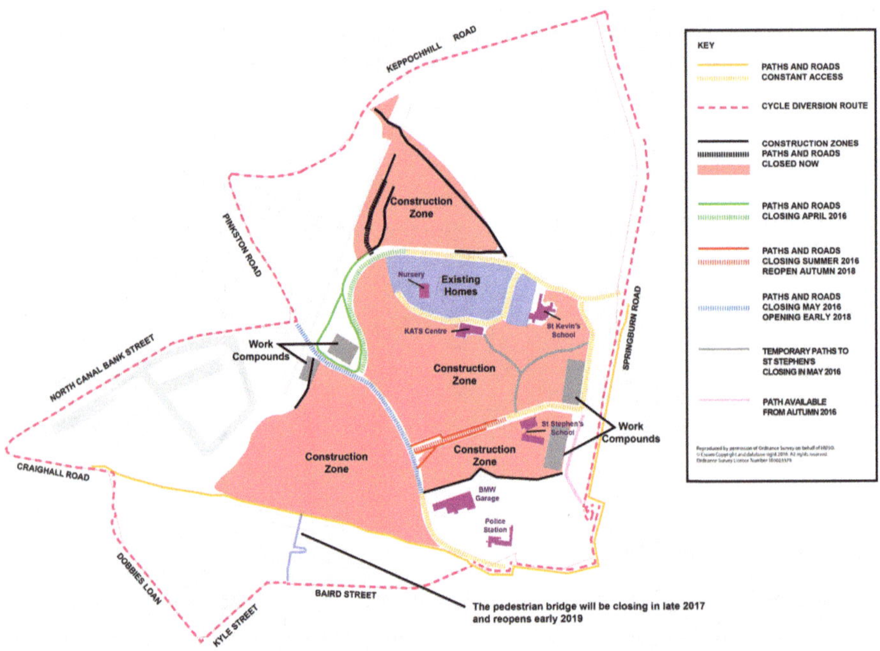

Fig. 11. Sighthill current local facilities, accessibility and construction zones. (GCC 2016a, p. 2)

The investigation showed two buses lines used to pass by the area, where they stopped suddenly to construction works and closure of different streets. People initiated a petition demanding buses to function again. At first, no response was given, and this is due to other operations on the site such as decontamination and construction works. However recently, a bus was provided for the residents for Sighthill with no fees to use, mainly to compensate for the disturbance, and to encourage people to move around the city and other neighbourhoods to reconnect with their people (Harvey 2017). These events make the people feel heard, hence giving them the chance to engage even more; in addition to not only feel they are in control but taking actual initiatives.

2. Ecosystem and Biodiversity

Sighthill masterplan promotes various elements of sustainable development. For instance, planners reallocated Sighthill Park for the advancement of the area, where they have concentrated on offering high-quality green spaces to accommodate for strong ties for the future social fabric according to the Graham (2017). Other initiatives include green infrastructure schemes (green and blue spine) such as planning a strategic urban drainage system (SUDS) strategy to allow for the efficient provision of amenity spaces (GCC 2017). Additionally, The program involves tree planting that will enhance and replace the current trees and woodland at Sighthill. Although poor quality biodiversity prevailed, future aims at transforming the landscape while improving accessibility to account for recreational activates for the local community and natural systems that increase biodiversity and enhance the overall physical integrity. Such interventions improve the physical resilience of the area to mitigate risks such as floods but also comprise of contributing to restoration experiences, which increase the health and wellbeing of the dwellers (Forest Research 2010). Yet, all of the mentioned plans to improve biodiversity will depend on the quality of the delivery that can only be assessed once completed (Fig. 12).

Fig. 12. Section of the park's path to illustrate the intended Biodiversity. (GCC 2016d, p. 10)

Deprived setting with a poor physical environment (e.g. unmaintained green space) is evidently recognised to adversely affect inhabitants' physical and psychological health (NCCRS 2001). This explains the severe health conditions residents suffered before the intervention. Conversely, the use and accessibility of quality green spaces including parks, and recreational areas have a positive influence on well-being as stated by Forest Research (2010). After carefully reviewing the biodiversity plans, and analysing the community and interviewees answers, it's clear that there is an increased active interest to change the entire neighbourhood environment. This involves developing the area to become more accessible and allow for new experiences where people (from diverse ages and abilities) can interact. All efforts put into increasing green spaces has the potential to transform the area and has a positive influence on other aspects related to health and well-being (Forest Research 2010). Again, the quality of maintenance and delivery will confirm the mentioned expectations.

The researcher noted that there is a high demand for allotment and gardening that exists compared to low supply in general. This was one of the activities conveyed, as the people showed a keen interest in gardening. Therefore, the council decided to provide the tools and education for this practice, because of its various benefits in different aspects (Harvey 2017). This includes but not restricted to reaching higher levels concerning people's empowerment, greater social interaction, and social belonging. However, during this time it is recognised, there will be limited access to green spaces due to construction procedures.

3. Safety

One of Sighthill's primary reasons for deprivation is the lack of security and justice as expressed earlier. Families lived in prevailing high crime rates, with poor access to resources for protection. These conditions are vital to identify to comprehend this history, which is full of psychological and trust insecurities affecting the current social cohesion and inclusiveness. Notwithstanding, Sighthill redevelopment reduced crime rate significantly which was agreed upon by many of the interviewees. There is an apparent transformation in resident's perception of safety, in addition to added state of openness and autonomy. The area drastically changed from being insecure and anti-social to being sociable with welcoming community, which adds to individual and communal well-being.

To ensure security and justice are prevailing, Glasgow city established the Violence Reduction Unit (VRU), which aims at diminishing crime levels and its related behaviours throughout collaborating with other institutions such as health and education (VRU 2005). Glasgow's approach also focused on strengthening social agencies through governance. The balanced delivery of safety assists in creating the confidence required to stand against uncertainty in poor contexts (Gaigné and Zenou 2015; Choi and Matz-Costa 2015). Building trust and offering amenities in local groups donate to societal and organisational makeover while contributing to the overall national growth (World Bank 2011).

Reflecting on the final physical outcome of Sighthill's masterplan, planning on reducing crime rates could lead to an undesirable influence on the residents if the neighbourhood is over-policed. This approach is comparable to the Crime Prevention Through Environmental Design (CPTED) that gradually gaining support as a crime avoidance scheme. The debate whether design involvements affect crime levels is complex, yet CPTED approach remains an appealing choice for communities, and planners (Cozens and Love 2015). The analysis shows little consideration given to the impact of spaces with a high level of surveillance on the community. This could end up in upsurging the problem instead of improving it, as over-policing might lead to social isolation and other implications. According to the Young Foundation (2009), many academics and urban thinkers such as William, H Whyte and Jane Jacobs have debated that communities can achieve safety throughout self-policing with a certain degree of assistance from the built environment. They argue that communities can handle themselves if they have an adequate degree of permeability to allow for active street movement, sociability, and civic surveillance. Notwithstanding, this can be achieved by rethinking planned spaces before the final execution while equipping the people to avoid over policing and encourage self-regulation, which is essential for reinforcing the resilience capacities of the community.

4. Community Groups and Social Opportunities

Researchers found that resilience also depends upon how a person accesses a network of support, or how a person relates to others (Bernier and Meinzen-Dick 2014; Cretney and Bond 2016). This encompasses values such as empathy, sociability, expressiveness and warmth. These social characteristics indicate that the core of resilience stretches past the individual. Resilience is about establishing enhanced relationships between community members that can be relied upon during times of emergency. The primary foundation of families' resilience is found to be the neighbourhood they belong to. Forrest and Kearns (2001) suggests explanations on the clear contradiction concerning the rise in communal movement and decreasing levels of confidence and trust. Sighthill is receptive to this aspect as the development is expanding at high rates.

Citizen's status in the social fabric reveals several prospects for creating unique connections and offers various methods for leveraging those opportunities (Forrest and Kearns 2001, p. 2136). At the current moment, there are two major centres for community activates and social opportunities. The first is KATS (refer to Fig. 11), which is the principal community centre that provides prospects for people to host events and establish social groups. A women's community group takes place, in addition to English classes as well as kid's summer clubs. Other services in Sighthill includes computers which are essential at this stage for the people as the government is encouraging them to use them to be able to access a whole new range of services whether its employment-related or even participate in the social media world (if the person doesn't own a smartphone). Although it was expected that many people would use the centre, as it was one of the initial community requests to have a place for gatherings, the majority don't benefit from it. People are calming that it doesn't provide what they need, in addition to offering only exclusive groups that not everyone can join (e.g. women's group) (Table 3).

Another location where services are offered is St Rollox Church that holds meetings occasionally, besides having a café and a 'bring-buy' shop. Currently, there are few things to do in the area, as there are no restaurants, pubs, or 'proper social clubs'. This is a major concern for the peoples since these places are the locations where friends could meet as they used to before the intervention. In his debate concerning local characteristics, Castells (1997) clarifies the motives for particular opinions on social transformation. People communicate in their neighbourhood and eventually establish systems of networks. Likewise, neighbourhood's significance develops to be the leading platform for recreational activities; the local area somehow progresses to be an extension of people's houses for community drives. Consequently, it is highly critical, as the neighbourhood becomes part of our identity.

Table 3. Community groups in Sighthill

Group
Sighthill Community Council
International Women's
St Rollox Church
St Stephens Primary
Pinkston Flats
Youth Group
St Kevins Primary
Sighthill Nursery
Meet the Builders Night

5. Social Networks

The previous sub-sections argued the significance of neighbourhoods in citizens' lives regarding meaning, identity, and provision of life-opportunities. According to Young and Willmott work in 1957, it is essential to consider the community's voices, as they portray overlooked environments engaged in social interactions, and communal support systems that constituted the core of inhabitants' resilience against uncertainties such as unemployment (Bunting 2017). Community voices depict the level of social capital while grasping how neighbourhoods operate. But why social networks and interactions are crucial to social capital? The contemporary policy discourses argue that social cohesion is the outcome of local social networks' arrangements, which is viewed as a bottom-up process. This model is similar to Jacobs' description of 'civilized self-government' (Forrest and Kearns 2001, p. 2137), which relates to the various social networks of urban localities in poor circumstances.

In deprived areas, it's perhaps the value of neighbouring that is crucial in the community's capacity to survive a declining physical setting (Forrest and Kearns 1999). Social aspects practically neighbouring hold high impact for the underprivileged and the ageing population, whereas most of the citizens acquire further spatially dispersed links relatively easier (Guest and Wierzbicki 1999). These patterns justify why the topic of social networks was mentioned numerous times in Sighthill community meeting attended by the researcher. People merely value and depend on relationships that they either have or lost. Consequently, if individuals' network still exists in the area, they tend to appreciate it. On the other hand, others who lost their relations tend to either build new ones or even relocate from the area searching for them. These movements explain why some people stay and why other leave, and most importantly highlight the importance of networks in people's life.

6. Sense of Belonging Familiarity and Pride

People agreed on having an increased sense of community describing their society as relatively supportive. Since most people either lived in Sighthill for very long time or even born in the area, there is a high level of sense of awareness in terms of knowing other families and residents, including those who moved in and quickly blended in the population. One participant (SH2) expressed the opposite by saying the area is tranquil as she hardly sees anybody outside their homes. She added, although she is not used to spending time outside, her old flat view outlook the M8 motorway. This gave her a sense of life knowing that city is consistently functioning.

Sighthill has been associated with a negative reputation in the past decade from local media in Glasgow. This negative image is also prevailing on the mass populace who are exposed to these media sources that eventually establish a mutual detrimental image of the neighbourhood (Kearns 2012). The communities living under such circumstances find it problematic to alter the imposed negative image, which is the case for Sighthill community. The analysis showed that this has implication on the sense of belonging, as people are no longer able to relate to place labelled with negative reputations, hence reduces the overall resilience (Table 4).

Table 4. Article coverage for Sighthill from 1998 to 2011. (Kearns 2012, p. 6)

	Sighthill (%)
Positive articles	9
Mixed articles	19
Negative articles	49
Neutral articles	23

According to Kearns (2012), the media exploited the influx of asylum-seekers and refuges to Sighthill in the previous decade, which drastically affected the area's reputation even more. The flow offered the media a dense material to create stories with dominated themes of violence, discrimination and social isolation for deprived people inhabiting poor living circumstances. People are eagerly accustomed to receiving reactions from external sources regarding their level of belongings. These structures (that extends from unofficial systems to politics) provide feedback to rationalise how communities relate to each other (sense of belonging) whether diverse or homogenous (Pelling et al. 2015, p. 25). Accordingly, media sources cannot be trusted or given the accountability alone towards development and growth of certain locality. Instead, institutions involved in neighbourhoods like Sighthill have to consider the media's impact as an added layer in the regeneration scheme to achieve successful redevelopment.

iii. Resilience of Institutions and systems

The analysis of organisations' involvement proved to alter how a community function in and perceive their neighbourhood. Beyond housing provision, GHA provided several schemes to support Sighthill community. Better Future Bursaries is one of these programs, which allows residents to accomplish and maintain their education. Other services available include obtaining fuel advice to reduce bills and welfare counsel. During the community meeting, many residents declared leveraging these schemes and services. Graham (2017), the GHA officer, claimed, "people are confident in contacting GHA, as their requests would be considered and sorted out". When the researcher asked the participants regarding this point, it was conveyed that at this stage, GHA including other institutions are very responsive and willing to offer help. Being able to extend your networks to organisations is essential for social transformation and resilience capacity (Bernier and Meinzen-Dick 2014).

Another party that is empowering the community using engagement practices is Gowell research program. Their primary objective is to evaluate the influence of Glasgow's redevelopment investments on the health and well-being of citizens (Kearns 2012). According to the CE Manager at Gowell, Community engagement is the primary tool employed by the Sighthill community because of two things "first, it can combine expertise with pragmatic opportunities to shape empowerment. Secondly, it can be creative and provide entertainment while achieving some pretty serious results in terms of what we learn and what individuals and organisations get out of community engagement"(Tabbner 2017). Leveraging learning and research, the community-based organisation is empowering people through interventions; likewise, they are acquiring a better understanding of the impact of these interventions on the social fabric.

The creative consultancy Pidgin Perfect's workshops conducted in Sighthill constituted a significant proof of the importance of community engagement to empowerment (Pidgin Perfect 2017). They were assigned to support the community in learning a new set of skills and capacity building in general. Moreover, they involved the residents in the process of providing an outdoor open space for community interaction until the construction phase is done. Furthermore, Community artists Rachel Mimiec and Karen Shaw were also appointed by GCC to create activates through organised, creative workshops mainly to assist the people in visualising their future living in the transformed neighbourhood as shown in Fig. 13. The inhabitants in Sighthill were keen on establishing a community area at the centre of the estate. The desired community space, as required, would be for community interaction, urban agriculture, and safe playing arena for children.

Community artists will continue to conduct workshops until the development span is completed. The workshops provide the opportunity for all residents to gather and participate in activates regardless of their age and social status. The content will focus on the mobile public open space and the permanent communal gardening space. Workshop participants were encouraged to propose ideas and consider their implication on the individual and collective perspective. Such initiatives pave the road for creating a sense of community while achieving actual targets of enhancing social resilience that leads eventually to higher social capital, and larger voluntary sector.

To broaden the research horizon, it can be settled through examining the Scottish Government resilience model, that their divisions, partnerships and policies adopt a "prepare-respond" model towards emergencies. For example, The Civil Contingencies Act 2004, which establishes the legislative foundation for crisis readiness in Scotland is concerned with defining and tackling emergency regulations (Mehmood 2015). Even though a Resilience Division exists in the government structure (office are in Glasgow, Edinburgh, Perth), their primary objective is to prepare individual institutions to provide emergency planning. This approach is also prevailing on the Scottish Guidance on Community Resilience established in 2013 that is targeting communities and voluntary segment associations to enhance their resiliency. The following resilience model mainly focuses on emergency planning while giving little attention to the significance of long-term resilience strategies that pave the road to social innovation, adaptive governance and its implications.

Fig. 13. Community workshops activities. (Mimiec 2017)

As for Glasgow, it was named in 2014 as a member of the Rockefeller Foundation's 100 Resilient Cities Network (GCC 2016c). Glasgow's membership in the network represents a significant step towards creating a strategic, multi-agency approach to govern the inevitable challenges, shocks proactively and stresses the city face. This membership entitled Glasgow to develop a resilience strategy (published in 2016) that will act as a roadmap to greater resilience. The strategy has four essential dimensions of urban resilience: health and well-being, economy and society, infrastructure and environment, and leadership and strategy (GCC 2016c, p. 7). Before 2014, the Resilience Unit in the Council accountable directly to the Chief Executive governed emergency-planning framework. This is still the case, however, under the resilience strategy, GCC brought emergency responders together with long-term planners through the resilience Steering Group as shown in Fig. 14 (Barrett 2017). The resilience work adopted a similar approach, albeit incorporating a broader range of stakeholders to represent community interests and social resilience (i.e. GCPH, housing associations, and community groups). GCC is now at a stage where the Steering Group has served its original purpose in supporting the development of the strategy. Therefore, it seems that the next phase of the resilience strategy work will be governed through the Community Planning Partnership which aims at bringing key partners together to deliver better services the public (GCC 2016c, p. 38). This will help to both elevate the agenda and give stronger links to the work being carried in and with communities through existing partnerships (Barrett 2017).

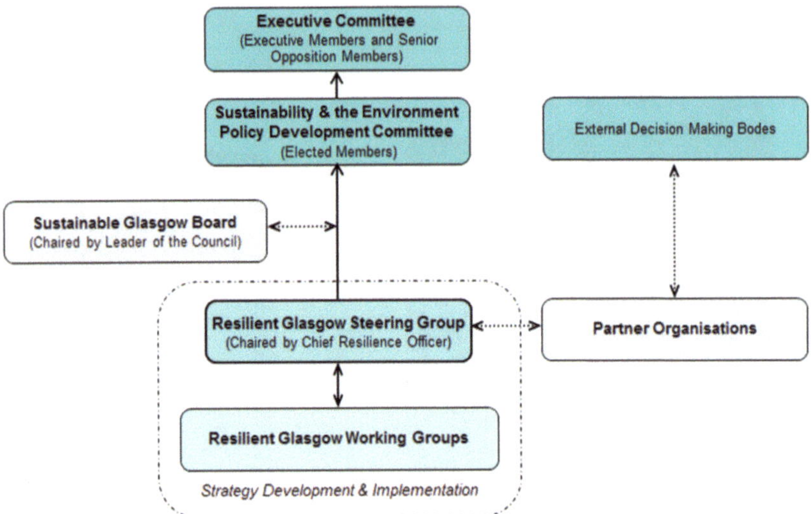

Fig. 14. The model illustrates the council governance structure for the Glasgow resilient strategy. (GCC 2016a, b, c, d).

c. Assessment

The case study discussed the regeneration interventions, and the different key players involved to understand how resilience is affected. The analysis of the masterplan showed that the regeneration scheme employed resilient urban design principles of having a low-rise, compact built form, diversity while using different forms of active travel and prioritising pedestrians. The project aims to offer comprehensive services to ensure complete, yet engaged communities in an assimilated natural system ecosystem. Nonetheless, procedure, strategies, and timing of the activities govern the social structure and how the community interrelate with their area.

The current population has diverse ethnicities, where social cohesion in this context can be tricky to measure, as it can be about exclusion of minority groups by majority's values. However, Sighthill showed the opposite because of demolitions, clearance and movement that caused the people to support each other regardless of the differences. The Large influx of people in Sighthill (delivery of new housing) might change population demographics. Therefore, the prevailing state of cohesiveness is subject to modification in the future.

The ties among demolitions and clearance, about revitalisation, have been fragile for many reasons. Firstly, the intention of the regeneration endeavour as it was intended to be the 2018 Olympics Athletes Village. Secondly, the initial absence of understanding for the masterplan between key stakeholders and the affected communities.

Thirdly, the aftermath of the global financial crisis and related funding consequence for social housing and infrastructure. Lastly, the shortage of private sector participation (Kearns and Darling 2013, p. 56). Demolitions and its associated activities caused the most damage to the community; whether it's the related health problems or deteriorated social networks. As demolitions are a citywide involvement, it embraces complex implications to be managed by a single association (GHA).

The scheme dedicates considerable efforts to upsurge biodiversity. In addition to developing infrastructure flexibility, and preparing structures to endure threats, various studies emphasise on the healing outcomes from landscape schemes in regeneration projects, primarily that interaction with nature could offer cognitive and restoration experiences for its users. Yet, planning didn't account for the absence of these spaces during construction works. Although most of the residents are not new, their environment entirely transformed. Hence, extra support that should be there is not fulfilled at the current moment. Other less-apparent services that benefit the people are similarly essential as they gradually develop distinctive local character and social affiliation. Examples of such services involve volunteers and social workers that could play a fundamental role in integrating residents throughout communal initiates. The analysis of the findings shows that absence of social infrastructure to assist inhabitants (particularly new ones) hinders the creation of strong collective fabric and possibly generates enduring issues that impact social welfare and prospects of new residents.

One of the practical ways to overcome the problem of confined open spaces is the executed workshop. They offer opportunities for community members to meet each other learn new skills and shape the future of Sighthill's spaces. Initiating consultation platforms intended to establish reachable prospects for Sighthill community, assists in producing of the community spaces to allow the people to interact, engage, and give feedback regarding the project at different phases. The participants in the community meeting expressed their content with the workshops considering them successful. This is possibly due to the diverse types of workshops held that ranges from community gardening, arts and envisioning events, to crafting lessons, and communal dinners. The researcher observed that community engagement initiatives are favoured and demanded by the community. Gowell research proved that empowerment is an outcome of engagement; this is due to the more residents conveying having enhanced well-being conditions (Tabbner 2017). Sighthill case study proved how leadership and institution's level of commitment has a profound influence on social resilience. Both by preparing strategies on the executive level and directly engaging with communities to generate essential trends that facilitate a growing sense of affiliation. Community development programs offered by decision-makers are assisting people to inhabit their new homes, build connections, and organise community events to form distinctive area character. This is stimulating empowerment capabilities and social affiliation (Silverman et al. 2005).

To assess Sighthill's community participation level, it's valuable to use Arnstein (1969) typology of eight levels of participation shown in Fig. 15 as it provides a detailed categorisation of the issue. Every level in the ladder resembles the degree of residents' authority in influencing decisions. From the case-study analysis, it can be determined that the community is shifting from the non-participation levels to degrees of tokenism with the new development. In other words, they are shifting from decision-makers entire involvement in making decisions on behalf of the community, to being heard. Activities (e.g. workshops) initiated by decision-makers before the development weren't empowering citizens to participate in planning issues, but to permit them to "educate" or "cure" residents (Arnstein 1969, p. 217). Nonetheless, with the new regeneration scheme, residents are given more space for participation that might indeed offer the chance for them to be heard. This level of involvement lacks the reign to confirm decision-makers will heed citizens' voices. Therefore, this restraining participation guarantees no power, thus no binding agreement of altering the prevailing conditions.

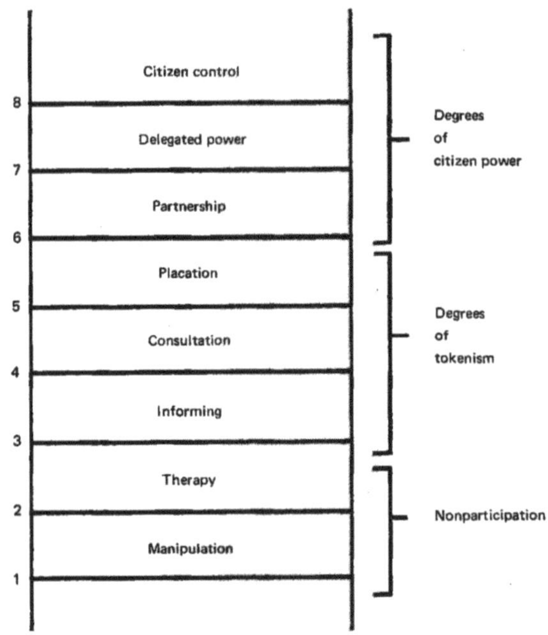

Fig. 15. Ladder is showing the people's participation level. (Arnstein 1969, p. 217)

As processed throughout the case-study evaluation, social resilience isn't limited to the capacity to preserve or recover to preceding state; it's also comprised of adjusting and accumulating knowledge to live with continues change and a certain degree of risk. In this context, and as clarified in previous sections, three categorises of capacities are crucial in assisting communities throughout this journey as shown in Fig. 16. The diagram is unique as it illustrates capacities in a spectrum, which is helpful to understand them from a different perspective. Sighthill community can be mapped close to the middle with a relative degree of flexibility. This is drawn from their ability to adapt to the new surrounding environment by accepting it and exhibiting a willingness to learn skills to enhance their participation further.

The resilience structure of GCC as explained earlier, embrace a comprehensive resilience approach of preparing for emergency planning, as well as long-term chronic stress. This is also reflected in their Resilience Strategy that identifies pillars, goals and

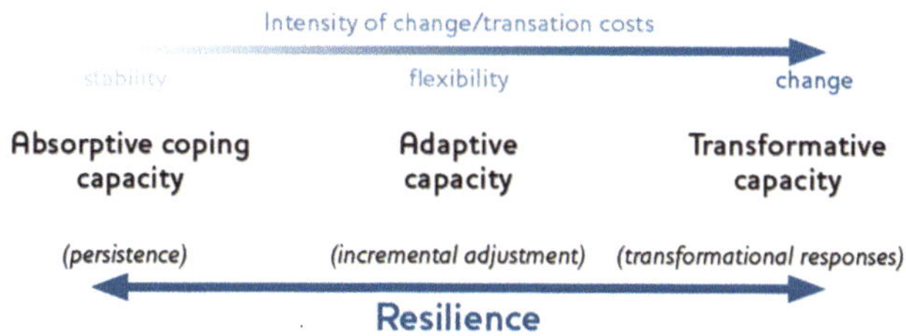

Fig. 16. Resilience capacities spectrum. (Béné et al. 2013, p. 1).

actions to achieve greater urban resilience among communities, and institutions (GCC 2016c). Although other involved parties such GHA implement some elements of resilient thinking in their operations, they still don't have a reliable resilience approach, which indications the existence of a gap among institutions.

5 Conclusion and Recommendations

a. Conclusion

The paper's objective is to deliver an in-depth understanding of how social resilience in practice can be stimulated or hindered throughout regeneration interventions. The use of the case-study assisted in framing the problem in pragmatic terms to illustrate the impact in particular context. URPs transform the physical features of the targeted area, as well as the social fabric by inserting new knowledge, values, and identities. This aper investigated how this transformation stimulates communities to cope, adapt and transform against adversity. The analysis showed how resilience capacities vary among the communities, yet, to have them is essential.

The study illustrated how resilience usage in URPs is valuable. Rather than examining elements in an isolated sectorial manner, resilience necessitates its users to think inclusively to embrace inter-disciplinary strategies, hence obtain a holistic outlook. Furthermore, resilience thinking contributes to several related features for regeneration schemes. Firstly, various sudden threats are gradually distressing entire communities, however, due to its focus on all-inclusive and cross-sectorial tactics, resilience offer the opportunity to comprehend the impact of these hazards on communities and how to react as an agency. Secondly, the procedures of regeneration ought to understand implications -such as market instability, population shifts and ecological fluctuations- at various levels and across diverse scales. Therefore, policy-makers can use resilience discourses as it aids in recognising how involvements at certain levels influence others; meaning that actions implied on a particular part of the system can progress or obstruct others. Thirdly, resilience as a concept and practice is instinctively appealing. This is due to present volatile environments that are always changing; growing resilience capacities, therefore, is a rational objective to counter the multiple stresses that hinder regeneration endeavours. Resilience from this perspective is evolving as a policy plot, which is employed to create partnerships across diverse stakeholders and from various fields below one methodology.

Regarding the question of how the surrounding environment plays a role in shaping and mediating change mechanism, findings showed that the nature of the place is fundamental to foster resilient capacities. However, if altered due to regeneration schemes with the intentions of enhancing socio-ecological conditions and tackling adverse environments, can develop a new but stable state with new set of skills, and capacities for resilience. Similar to the relational theory of place investigated in Massey's work in 1991, 2004, and 2004 that concerns change from globalisation as a hazard to perceptions of belonging (Cretney and Bond 2016), change of neighbourhood features affects social belonging and relations. This paper supports the argument that resilience development is not only achieved through maintaining place identities, but regeneration can equip areas to develop resilient capacities such as flexibility, and robustness to allow for better exposure, adaptive governance, and system sensitivity when encountered by immediate or enduring shocks.

The conducted research demonstrates how resilience capacities can gradually create resilient places despite the conditions of the surrounding environment with limited resources, and how change mechanism can be mediated. This is achieved by comprehending that resilience is applied to economy and infrastructure, in addition to society and its related dimensions. The notion isn't exclusive to readiness from sudden changes; it's more useful to long-term tactics to alleviate and adjust to economic, environmental and social hurdles. Instead of being passive about change mechanisms, adopting a proactive approach - by intentionally steering policy and development - can predict, produce, and determine change. Hence, social innovation is encouraged, where it can recognise methods to contribute to social transformation, by considering conditions that deliver solutions to complicated social-ecological problems. This approach is valuable as it extends our thinking to find what stimulates stability and change, especially at the social level. Meaning that in what circumstances change effectively develop and disperse, evolving social networks, and empowering inhabitants to meet their demands.

Decision-makers can leverage regeneration schemes as an opportunity to restructure themselves first, and to equip the people as a seconds step. The plans, activities and degree of involvement of decision-makers and institutions in URPs can stimulate or deter social resilience. Having a committed leadership to prepare and respond to crises is not enough. Long-term strategies are essential as well to account for sudden shocks and long-term stresses, which is the core of urban resilience. Such attitude entitles embedding resilience in all characteristics of policy and planning to ensure its affectivity. To open the room for social and institutional advancement, community engagement and trust should be achieved; these elements constitute the foundation of the community and the institution's relationship. Findings indicated that empowerment is the result of institutional engagement with the community; it is essential in this case and time of the project as the people are pursuing further prospects to launch or maintain their businesses and social groups. The number of community groups is increasing, which is serving in creating more social networks and uplifting the level of community's participation. The conducted analysis illustrates the rising degree of sense of ownership that is impacting individuals and the entire community. This is crucial to the resident's capabilities of taking additional control over their health and safety and gradually adapt to their new environment. Institutions' participation is making people gain more control as they have a responsibility towards not only inside their homes but also in the surrounding environment since its part of their neighbourhood.

The process of case-study evaluation illustrated the importance of understanding the different types of resilience, to provide the various effective forms of support. Resilience is about how individuals respond to different kinds of challenges. However, community support is critical in the face of these challenges (100RC 2017). The first type of resilience is 'getting back on your feet' (GCPH 2014), which relates to preserving equilibrium state after a disaster. The second category is associated with leveraging the sate of variation to change to be effective in the new conditions. This state requires having reliable networks; the community also necessitates help from the broader network who are aware of new prospects, and how to approach them. Networks in this case are indeed vital, where three forms of associations occur. The first includes help from people who have similar values, comprehend the same perspective and offer assistance through difficult times. The second class is connected to those who hold dissimilar principles, and can access different prospects. But what if all the people are subject to the same crisis? Here the third types of networks play a significant role, which is concerned about connections beyond us. This is correlated mainly to people who create decisions concerning the community to aid the community be heard.

b. Recommendations

There are four main pillars of recommendations that the researcher identified from the conducted research and they include: (1) Empowering locals, (2) Initiate place-based Solutions, (3) Upgrade efforts for economic growth, and (4) Develop civic participation.

1- Empowering Locals

To fully utilise the legislation passed on the issue of empowerment, its advised to explore the potential of The Community Empowerment Act that "will help to empower community bodies through the ownership of land and buildings, and by strengthening their voices in the decisions that matter to them" (Scottish Government 2015). This could imply community planning and participation levels. One of the main pillars of empowering the community is electing a community leader who will be the connecting link between institutions, and the local society. Therefore, a definition of what constitutes a good community leader should be reached with the involvement of different sectors using research and experimentations methods. Empowered community leaders allow for information transfer, which enhances system feedback and grows trust. This gradually establishes a more communicated long-term ability and development for resilience. Developing a local volunteering sector that is connected to the citywide volunteering network is fundamental quality for social resilience; it boosts community attachment and exploits social capital (Butland 2016).

To account for integrated planning, a local resilience strategy shall be created for critical services with regards to long-term stresses as part of the recreation scheme. Then, the strategy would be conveyed and adopted by different stakeholders including the community. Findings showed how construction period restricts the use of green and open spaces. Therefore, for future development phases in Sighthill and other regeneration schemes, a full plan should be prepared to consider construction inconvenience.

In other words, particular spaces should be available during construction for the residents to access them due to their impact on the short and long-term basis. A gap analysis should be conducted to ensure available resources are optimised to reach desired performance. Mapping services for the community in nearby locations to compensate for their absence during regeneration phase can be another method. This will minimise social isolation, and provide spaces for social interactions.

2- Initiate Place-Based Solutions

Implement and support place-based solutions to improve, community and economic resilience. Additional attention should be given to Placemaking, which aims at promoting public health and well-being as clarified in Our Resilient Glasgow Report (2016). Beyond what the regeneration program is delivering, inclusive Placemaking elements are vital to implement, which comprises of inclusive methods that account for the local context. Placemaking could aid in progressing resilience in places as it's grounded on accepting a variety of interests to yield in interrelated prospects across common practices (GCC 2016a, b, c, d, p. 50).

3- Upgrade Efforts For Economic Growth

Regarding economic opportunities, concrete plan for economic growth should be developed to generate meaningful opportunities for the community. This includes supporting the existing businesses and encouraging new ones. Workshops are useful in this context as shown in Sighthill case-study. Seminars in this context refer to the training services that teach the local community skills to allow for innovation practices. This comprises tackling deprivation, poverty, and progressing social innovation by persistently developing skills among the working populace.

Engaging educational institutions with the local community is recommended to amplify their responsibility in generating knowledge to stimulate economic growth. Moreover, planners should include the local industry in the cycle to learn from their experience, while focusing on skills compulsory for a better resilient economy with the current and future resources.

4- Develop Civic Participation

Additional work should be devoted to encouraging community contribution, confidence and resilience values among residents of Sighthill and the local associations that operate near them. This is vital in progressing capacities and new skills among the community members and decision-makers to qualify them to leverage resilience thinking.

Information sharing and knowledge are crucial features that are necessary for this pillar. This comprises of offering education to all Sighthill's community members and equipping them with suitable tools to entirely recognise the encountered challenges. Various stakeholders have in-depth competence in this field (refer to Fig. 17); yet to confirm the scheme is cooperative and efficient, Sighthill community ought to be further engaged with new endeavours to connect stakeholders with each other.

To uplift the efforts of above recommendations and apply them across Glasgow regeneration areas and possibly influence the resilience policy agenda, exploiting potential resilience partnerships is advised as shown in Fig. 17. Many of the

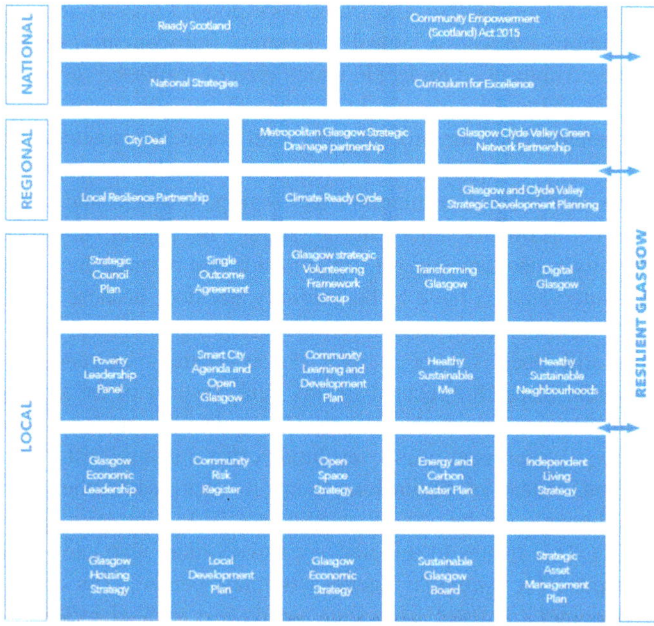

Fig. 17. Potential local, regional, and national resilience partnership in Glasgow. (GCC 2016c, p. 25)

organisations displayed above function as strategic policy mediums to cover gaps including topics within the resilience field; this is done through identifying tools to tackle them efficiently. These syndicates have the ability to assist the community by informing them on suitable practices based on their expertise. Moreover, their initiatives could inform the Scottish Government schemes particular on social resilience.

a. Future Studies

The question of evaluating social resilience with respect to URPs interventions is explored on one case-study (Sighthill); future studies should aim to investigate other regeneration sites in Glasgow and possibly in other cities to draw conclusions that could enhance our understanding of resilience in pragmatic terms. Qualitative methods should be implemented in the research (e.g. resilience indexes surveys) to further solidify outcomes from the qualitative assessment. Furthermore, future explorations should aim to understand the impact of redevelopment interventions on the political, environmental, and economic resilience, to complete the resilience objective of comprehensives.

Although resilience benefits are clear to regeneration policies and activities, certain restrictions are essential to be reviewed. Resilience as a critical agenda accepts systems to recover to a previous state or change to a new but stable state against socks and stresses (Vale 2013). Nonetheless, the framework doesn't specify that improving or maintaining conditions are safe or actually harming. Subsequently, more research in this field should be commenced to decide which type of resilience should be built, and who will benefit from it.

Resilience thinking continues to struggle with some fundamentals of the social dimensions. This has consequences as certain groups' resilience are advanced at the expense of others in the same community and region. Planners and decision-makers need to realise that resilience interventions should be applied at different extents according to the city's status to achieve social justice and balance the overall situation. Thus, resilience interventions have to be carefully decided, because if misemployed, can skim across essential elements of power. Adopting resilience in the decision-making process still doesn't explain why power-relations embedded in enduring poverty, due to little research executed in this field. Instead of associating resilience with concepts of power (Béné et al. 2013, p. 3), acquiring an understanding of power relations is necessary to prevent policy resolutions that unconsciously ingrain poverty and social isolation, instead of diminishing them. As expressed earlier in the literature review that resilience can be co-opted (Slater 2014; Cretney 2014; Cote and Nightingale 2011). This relates to the different stakeholders who can support resilient interventions to push for particular agendas of their benefit. The concern lies that sustainable development suffered from such trends that yielded in keeping the same status quo in some cases; so could resilience agendas experience the same? This risk will indeed exist, such as labelling certain developments with the notion of resilience without actually accomplishing its goals. To mitigate implications in this context, resilience should be redefined, accompanied with policy agendas that apply these considerations to magnify efforts and minimise risks for specific plans.

References

100 Resilient Cities: Catalyzing the urban resilience market. New York: 100 Resilient Cities, pp. 1–23 (2017). http://action.100resilientcities.org/page/-/100rc/pdfs/100RC%20-%20Catalyzing%20the%20Urban%20Resilience%20Market.pdf. Accessed 7 July 2017

Adger, W.: Social and ecological resilience: are they related? Prog. Hum. Geogr. **24**(3), 347–364 (2000)

Applegath, C., Arney, E., Fitzpatrick, D., Gómez-Palacio, A., Howard, P., Vomberg, A., Watkinson, W., Yazer, J.: ResilientCity | ResilientCity Home. Resilientcity.org (2017). http://www.resilientcity.org/. Accessed 1 Aug 2017

Aronson, J.: A Pragmatic View of Thematic Analysis. Article 3 [online]. The Qualitative Report, pp. 2–5 (1995). http://nsuworks.nova.edu/cgi/viewcontent.cgi?article=2069&context=tqr. Accessed 9 July 2017

Arnstein, S.: A ladder of citizen participation. J. Am. Inst. Plan. **35**(4), 216–224 (1969)

Atkinson, R.: The evidence on the impact of gentrification: new lessons for the urban renaissance? Int. J. Hous. Policy **4**(1), 107–131 (2004)

Béné, C., Newsham, A., Davies, M.: Making the Most of Resilience. IDS (2013). http://www.ids.ac.uk/publication/making-the-most-of-resilience. Accessed 22 July 2017

Béné, C., Wood, R., Newsham, A., Davies, M.: Resilience: new Utopia or New Tyranny? Reflection about the Potentials and Limits of the Concept of Resilience in Relation to Vulnerability Reduction Programmes. IDS Working Papers, 2012(405), pp. 1–61 (2012)

Bernier, Q., Meinzen-Dick, R.: Networks for resilience the role of social capital. International Food Policy Research Institute, Washington (2014). http://ebrary.ifpri.org/utils/getfile/collection/p15738coll2/id/128489/filename/128700.pdf. Accessed 28 June 2017

Brezzi, M.: How's life in your region? Measuring regional and local well-being for policy making. Global Forum UN-GGIM. OCDE, pp. 1–12 (2014). http://ggim.un.org/docs/meetings/Global%20Forum/OECDRegionalwellbeing_UNGGIM%20August%202014.pdf. Accessed 26 June 2017

Bunting, M.: Kin outrage. The Guardian (2017). https://www.theguardian.com/society/2007/apr/25/communities.britishidentityandsociety. Accessed 16 July 2017

Butland, S.: What is community resilience and how do you measure it? - Social Enterprise Mark. Social Enterprise Mark (2016). https://www.socialenterprisemark.org.uk/what-is-community-resilience-and-how-do-you-measure-it/. Accessed 13 July 2017

Cameron, S.: Gentrification, housing redifferentiation and urban regeneration: 'going for growth' in Newcastle upon Tyne. Urban Stud. **40**(12), 2367–2382 (2003)

Choi, Y., Matz-Costa, C.: Neighborhood safety, social cohesion, and well-being of older adults. Gerontologist **55**(2), 775 (2015)

Colombatto, E.: A theory of institutional resilience. Research Handbook on Comparative Law and Economics. Edward Elgar, Cheltenham, pp. 1–6 (2013). http://kie.vse.cz/wp-content/uploads/Theory-of-institutional-resilience.pdf. Accessed 4 July 2017

Cote, M., Nightingale, A.: Resilience thinking meets social theory. Prog. Hum. Geogr. **36**(4), 475–489 (2011)

Coyle, J.: Sighthill area information. GoWell, Glasgow, pp. 1–3 (2008). http://www.gowellonline.com/assets/0000/0794/Sighthill_Area_Information.pdf. Accessed 9 July 2017

Cozens, P., Love, T.: A review and current status of crime prevention through environmental design (CPTED). CPL Bibliogr. **30**(4), 393–412 (2015)

Creswell, J.: Research Design: Qualitative, Quantitative, and Mixed Methods Approaches, 2nd edn. Sage, Thousand Oaks (2003)

Cretney, R.: Resilience for whom? Emerging critical geographies of socio-ecological resilience. Geogr. Compass **8**(9), 627–640 (2014)

Cretney, R., Bond, S.: Shifting relationships to place: a relational place-based perspective on SES resilience. Urban Geogr. **38**(1), 8–24 (2016)

Davoudi, S., Brooks, E., Mehmood, A.: Evolutionary resilience and strategies for climate adaptation. Plan. Pract. Res. **28**(3), 307–322 (2013)

Davoudi, S., Mehmood, A., Brooks, L.: The London climate change adaptation strategy: gap analysis. ARCADIA Project. Global Urban Research Unit, Newcastle University, Newcastle, pp. 18–56 (2017). http://orca.cf.ac.uk/53010/1/ewp44.pdf. Accessed 13 July 2017

Davoudi, S., Shaw, K., Haider, L., Quinlan, A., Peterson, G., Wilkinson, C., Fünfgeld, H., McEvoy, D., Porter, L.: Resilience: a bridging concept or a dead end? "Reframing" resilience: challenges for planning theory and practice interacting traps: resilience assessment of a pasture management system in Northern Afghanistan urban resilience: what does it mean in planning practice? Resilience as a useful concept for climate change adaptation? The politics of resilience for planning: a cautionary note. Plan. Theory Pract. **13**(2), 299–333 (2012)

De Groot, R., Wilson, M., Boumans, R.: A typology for the classification, description and valuation of ecosystem functions, goods and services. Ecol. Econ. **41**(3), 393–408 (2002)

Farr, D.: Sustainable Urbanism, pp. 20–53. Wiley, Hoboken (2008)

Forrest, R., Kearns, A.: Social cohesion, social capital and the neighbourhood. Urban Stud. **38** (12), 2125–2143 (2001)

Forest Research: Benefits of green infrastructure. Farnham (2010). https://www.forestry.gov.uk/pdf/urgp_benefits_of_green_infrastructure.pdf/$file/urgp_benefits_of_green_infrastructure.pdf. Accessed 22 July 2017
Fletcher, D.: Transformational regeneration areas. Glasgow Housing Association, Glasgow (2011). https://www.glasgow.gov.uk/CHttpHandler.ashx?id=6055&p=0. Accessed 5 July 2017
Flint, D., Golicic, S.: Searching for competitive advantage through sustainability. Int. J. Phys. Distrib. Logist. Manag. **39**(10), 841–860 (2009)
Gaigné, C., Zenou, Y.: Agglomeration, city size and crime. Eur. Econ. Rev. **80**, 62–82 (2015)
Gallopín, G.: Linkages between vulnerability, resilience, and adaptive capacity. Glob. Environ. Change **16**(3), 293–303 (2006)
Gill, P., Stewart, K., Treasure, E., Chadwick, B.: Methods of data collection in qualitative research: interviews and focus groups. BDJ **204**(6), 291–295 (2008)
Glasgow Center for Population Health (GCPH): Resilience for public health: supporting transformation in people and communities. 12 Concept Series. Glasgow Center for Population Health, Glasgow, pp. 2–17 (2014). http://www.gcph.co.uk/assets/0000/4197/Resilience_Briefing_Paper_Concepts_Series_12.pdf. Accessed 5 July 2017
Glasgow City Council: Access and closure notification roads and paths. Glasgow City Council, Glasgow (2016a). https://www.glasgow.gov.uk/CHttpHandler.ashx?id=33448&p=0. Accessed 4 July 2017
Glasgow City Council: Glasgow economics strategy 2016–2023. Glasgow City Council, Glasgow, pp. 1–45 (2016b). https://www.glasgow.gov.uk/CHttpHandler.ashx?id=36137&p=0. Accessed 11 July 2017
Glasgow City Council: Our resilient glasgow: a city strategy. 100 Resilient Cities, Glasgow (2016c). http://100resilientcities.org/wp-content/uploads/2017/07/Glasgow-Strategy-PDF.pdf. Accessed 16 July 2017
Glasgow City Council: Sighthill cycling village. Glasgow City Council, Glasgow (2016d). https://www.glasgow.gov.uk/CHttpHandler.ashx?id=33446&p=0. Accessed 2 July 2017
Glasgow City Council: Sighthill Masterplan Update. Glasgow City Council, Glasgow, pp. 1–3. https://www.glasgow.gov.uk/CHttpHandler.ashx?id=33447&p=0. Accessed 9 July 2017
Goodwin, K.: The Glasgow effect: new study reveals causes of city's high mortality rates. The Guardian (2017). https://www.theguardian.com/cities/2016/jun/10/glasgow-effect-die-young-high-risk-premature-death. Accessed 2 July 2017
GoWell: Social capital, community cohesion and health. GoWell, Glasgow (2011). http://www.gowellonline.com/assets/0000/3725/GoWell_annual_event_presentation_-_Anne_Ellaway.pdf. Accessed 4 July 2017
Graham, T.: Transforming Communities: Glasgow Overview (2017)
Gray, N.: Neil Gray on the Glasgow Commonwealth Games' urban regeneration. Dezeen (2017). https://www.dezeen.com/2014/07/26/opinion-glasgow-commonwealth-games-urban-regeneration-neil-gray/. Accessed 11 July 2017
Greater London Authority (GLA): The Draft Climate Change Adaptation Strategy for London. Public Consultation Draft. Greater London Authority, London (2010). http://www.deltacities.com/documents/London_Climiate_change_adaptation.pdf. Accessed 12 July 2017
Guest, A., Wierzbicki, S.: Social ties at the neighborhood level. Urban Aff. Rev. **35**(1), 92–111 (1999)
Guest, G., MacQueen, K., Namey, E.: Validity and reliability (credibility and dependability) in qualitative research and data analysis. In: Applied Thematic Analysis, pp. 79–106. SAGE (2012). http://dx.doi.org/10.4135/9781483384436. Accessed 20 July 2017

Gunderson, L.: Comparing ecological and human community resilience. Research Report 5. Community Regional Resilience Institute (CARRI), Atlanta, pp. 1–13 (2009). http://www.resilientus.org/wp-content/uploads/2013/03/Final_Gunderson_1-12-09_1231774754.pdf. Accessed 7 July 2017

Holling, C.: Resilience and stability of ecological systems. Annu. Rev. Ecol. Syst. **4**(1), 1–23 (1973)

Jones, A.: Residential instability and obesity over time: the role of the social and built environment. Health Place **32**, 74–82 (2015)

Jones, S., Mean, M.: Resilient Places Character And Community in Everyday Heritage, 1st edn. [ebook]. Demos, London, pp. 29–47 (2010). https://www.demos.co.uk/files/Resilient_places_-_web.pdf. Accessed 14 July 2017

Kearns, A.: Area reputation: comparing newspaper coverage of the Sighthill and Red Road estates. Briefing Paper 19. GoWell, Glasgow, pp. 9–18 (2012). http://www.gowellonline.com/assets/0000/0561/Briefing_Paper_19.pdf. Accessed 9 July 2017

Kearns, A., Darling, L.: Giving the 'all clear': housing staff experience of the rehousing process in Transformational Regeneration Areas. Gowell, Glasgow (2013). http://www.gowellonline.com/assets/0000/0964/Clearance_Staff_Report_FINAL.pdf. Accessed 29 June 2017

Keck, M., Sakdapolrak, P.: What is social resilience? Lessons learned and ways forward. Erdkunde **67**(1), 5–19 (2013)

Lawson, L., Kearns, A.: Living in New Homes in Glasgow's Regeneration Areas: The Experience of Residents in the Pollokshaws and Sighthill Transformational Regeneration Areas. Gowell, Glasgow (2017)

Lehmann, S.: Green urbanism: formulating a series of holistic principles. S.A.P.I.EN.S. **3**(2) (2010). http://sapiens.revues.org/1057. Accessed 8 July 2017

Leslie, C.: Disappearing Glasgow: documenting the demolition of a city's troubled past. The Guardian (2015). https://www.theguardian.com/cities/2015/apr/22/disappearing-glasgow-documenting-demolition-city-troubled-past. Accessed 14 July 2017

Lochner, L., Moretti, E.: The effect of education on crime: evidence from prison inmates, arrests, and self-reports. Am. Econ. Rev. **94**(1), 155–189 (2004)

Lorenz, D.: The diversity of resilience: contributions from a social science perspective. Nat. Hazards **67**(1), 7–24 (2010)

Lund Research: Non-probability sampling|Lærd Dissertation (2012). http://dissertation.laerd.com/non-probability-sampling.php. Accessed 16 July 2017

Mackinnon, D., Derickson, K.: From resilience to resourcefulness: a critique of resilience policy and activism. Prog. Hum. Geogr. **37**(2), 253–270 (2013)

Marks, D., Yardley, L.: Content and thematic analysis. Research Methods for Clinical and Health Psychology, pp. 56–68. SAGE Publications, London (2011)

Marshall, N.: Can policy perception influence social resilience to policy change? Fish. Res. **86**(2–3), 216–227 (2007)

Martin, R.: The local geographies of the financial crisis: from the housing bubble to economic recession and beyond. J. Econ. Geogr. **11**(4), 587–618 (2011)

Mehmood, A.: Of resilient places: planning for urban resilience. Eur. Plan. Stud. **24**(2), 407–419 (2015)

National Centre for Culture and Recreation Statistics (NCCRS): The social impacts of sport and physical recreation (2001). https://www.ausport.gov.au/__data/assets/pdf_file/0006/276927/ABS-Social_impacts_of_sport.pdf. Accessed 13 Aug 2017

Nicoll, V.: Exclusive first look at the future of Sighthill. Evening Times (2013). http://www.eveningtimes.co.uk/news/13259545.Exclusive_first_look_at_the_future_of_Sighthill/. Accessed 20 July 2017

Nunes, D., Tomé, A., Pinheiro, M.: Urban regeneration - developing strong sustainable urban design perspectives. In: Portugal SB13: Contribution of Sustainable Building to meet EU 20-20-20 Targets. Guimarães, pp. 719–726 (2013). https://www.researchgate.net/publication/281586851_Urban_Regeneration_-_Developing_strong_sustainable_urban_design_perspectives. Accessed 10 July 2017
Obrist, B., Pfeiffer, C., Henley, R.: Multi-layered social resilience. Progr. Dev. Stud. **10**(4), 283–293 (2010)
OECD: Overview paper on resilient economies and societies. Meeting of the OECD Council at Ministerial Level. OECD, Paris (2014). https://www.oecd.org/mcm/C-MIN(2014)7-ENG.pdf. Accessed 28 June 2017
OECD: Resilient Cities. OECD, pp. 16–74 (2016). https://www.oecd.org/cfe/regional-policy/resilient-cities-report-preliminary-version.pdf. Accessed 17 July 2017
Paterson, S.: Before and after: the radical transformation of one of Glasgow's housing schemes. Evening Times (2016). http://www.eveningtimes.co.uk/news/14179616.Before_and_after__The_radical_transformation_of_one_of_Glasgow_s_housing_schemes/. Accessed 20 July 2017
Pelling, M., Sharpe, J., Pearson, L., Abeling, T., Swartling, Å., Forrester, J., Deeming, H.: Social learning and resilience building in the emBRACE framework. Deliverable 4.3. Newcastle upon Tyne, pp. 13–40 (2015). https://www.sei-international.org/mediamanager/documents/Publications/SEI-2015-TechnicalReport-SocialLearning-emBRACE.pdf. Accessed 3 July 2017
Pidgin Perfect: Sighthill Community Space|Pidgin Perfect (2017). http://pidginperfect.com/sighthill-community-space/. Accessed 29 June 2017
Pitrenaite-Zileniene, B., Torresi, F.: Integrated approach to a resilient city: associating social, environmental and infrastructure resilience in its whole. Eur. J. Interdiscip. Stud. **6**(2), 1–13 (2014). http://www.ejist.ro/files/pdf/383.pdf. Accessed 8 July 2017
Regan, S.: Towards an urban renaissance: the final report of the urban task force. Polit. Q. **71**(1), 115–118 (2000)
Scheffran, J., Marmer, E., Sow, P.: Migration as a contribution to resilience and innovation in climate adaptation: Social networks and co-development in Northwest Africa. Appl. Geogr. **33**, 119–127 (2012)
Scottish Government: Community Empowerment (Scotland) Act (2015). http://www.gov.scot/Topics/People/engage/CommEmpowerBill. Accessed 15 July 2017
Silverman, E., Lupton, R., Fenton, A.: A good place for children? Attracting and retaining families in inner urban mixed income communities. The Chartered Institute of Housing, London, pp. 22–58 (2005). https://www.jrf.org.uk/sites/default/files/jrf/migrated/files/1905018126.pdf. Accessed 20 July 2017
Simmie, J., Martin, R.: The economic resilience of regions: towards an evolutionary approach. Camb. J. Reg. Econ. Soc. **3**(1), 27–43 (2010)
Simms, A., Kjell, P., Potts, R.: Clone Town Britain: the survey results on the bland state of the nation. New Economic Foundation, London, pp. 1–38 (2005). http://b.3cdn.net/nefoundation/1733ceec8041a9de5e_ubm6b6t6i.pdf. Accessed 2 July 2017
Slater, T.: The resilience of neoliberal urbanism. openDemocracy (2014). https://www.opendemocracy.net/opensecurity/tom-slater/resilience-of-neoliberal-urbanism. Accessed 1 July 2017
UK Housing Wiki: Sighthill (Glasgow) (2017). http://ukhousing.wikia.com/wiki/Sighthill_(Glasgow). Accessed 20 June 2017
Uyarra, E., Flanagan, K.: Reframing regional innovation systems: evolution, complexity and public policy. In: Cooke, P. (ed.) Reframing Regional Development: Evolution, Innovation & Transition. Routledge, London (2012)

Vale, L.: The politics of resilient cities: whose resilience and whose city? Build. Res. Inf. **42**(2), 191–201 (2013)
Violence Reduction Unit Scotland (VRU): Violence Reduction Unit Scotland (2005). http://www.actiononviolence.org.uk. Accessed 27 June 2017
Wilkinson, C.: Social-ecological resilience: insights and issues for planning theory. Plan. Theory **11**(2), 148–169 (2011)
Wilton, R.: The constitution of difference: space and psyche in landscapes of exclusion. Geoforum **29**(2), 173–185 (1998)
World Bank: World Development Report 2011: Conflict, Security and Development. The World Bank, Washington, D.C. (2011). https://siteresources.worldbank.org/INTWDRS/Resources/WDR2011_Full_Text.pdf. Accessed 16 July 2017
Young Foundation: New communities need services and support, not just buildings|Future Communities (2009). http://www.futurecommunities.net/socialdesign/188/new-communities-need-services-and-support-not-just-buildings. Accessed 9 July 2017

Interview References

Barrett, F.: Interview with F. Barrett on 13th June (2017)
Booker, D.: Interview with D. Booker on 13th June (2017)
Graham, T.: Interview with T. Graham on 13th June (2017)
Harvey, J.: Interview with J. Harvey on 12th June (2017)
Tabbner, C.: Interview with C. Tabbner on 29th June (2017)

Strategic Assessment for the Sustainable Combined Sewer Overflow Problem in Peoria, Illinois, USA

Amir Al-Khafaji[1](✉), Jim Ardis[2], and Scott Reeise[2]

[1] Bradley University, Peoria, IL, USA
amir@fsmail.bradley.edu
[2] City of Peoria, IL, USA

Abstract. Stormwater runoff directly correlates to the amount of impervious surfaces on a property. Impervious surfaces allow the water to fill up pipes and streams faster, increasing the potential for street and/or storm system flooding during heavy rains and increasing maintenance and repair requirements of the storm sewer system. In late 2005, The United States Environmental Protection Agency (USEPA) performed a 'wet-weather inspection' in Peoria, Illinois. They determined that the City needed to make improvements to the combined sewer system. Further, in 2006, USEPA and the Department of Justice (DOJ) found the City of Peoria was in violation of the Clean Water Act (CWA), and offered to enter into a tolling agreement, in lieu of filing suit. The City agreed to negotiate and signed the first tolling agreement. In 2017, the City of Peoria provided the USEPA with its preliminary strategic assessment and proposed a sustainable solution to the CSO problem. This paper highlights various processes, strategies and legal challenges faced by the City of Peoria to meet its obligations to the USEPA. In early 2018, the City of Peoria will complete its strategic assessment and reach a final agreement with the USEPA. This paper will provide insights for other cities and communities to examine and utilize various innovative and sustainable solutions to CSO problems facing similar cities across the globe. Further, the authors will provide meaningful guidance on awareness and strategic communication with the general public to justify expected increases in utility rates. The paper will outline the processes involved in meeting the legal challenges presented by the USEPA and its mandates for clean water in Peoria, Illinois, USA. It is expected that the proposed green solution will require an investment of $200 million to $300 million.

Keywords: USEPA · Peoria · Illinois · CSO · DOJ · CWA
Gray infrastructure · Strategy · Green infrastructure

1 Overview of the Combined Sewer System in Peoria

The City of Peoria has maintained a system of combined sewers and overflows for more than 100 years. Peoria built its first sewers in the late 1800's to carry rainwater and melting snow away from homes, businesses and streets. When indoor plumbing came later, homeowners and business owners hooked their sewage lines to those same

M. Shehata and F. Rodrigues (Eds.): GeoMEast 2018, SUCI, pp. 54–68, 2019.
https://doi.org/10.1007/978-3-030-01905-1_3

sewers, combining stormwater and sewage in one pipe. This was standard practice in many U.S. cities at the time, especially in the Northeast and Midwest. After the Greater Peoria Sanitary District was formed and the combined sewers were connected to the Riverfront Interceptor sewer and treatment plant constructed in 1931, the combined sewers only discharged to the river during rainfall events. Figures 1 and 2 show the condition of no sewage overflow and with sewage overflow.

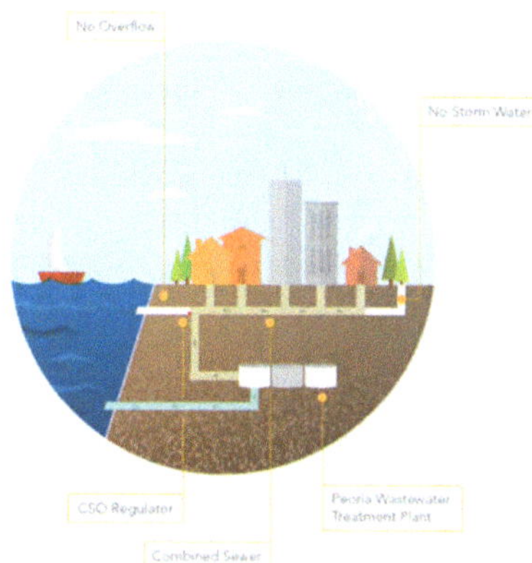

Fig. 1. Graphical illustration of conditions leading to no sewage overflow.

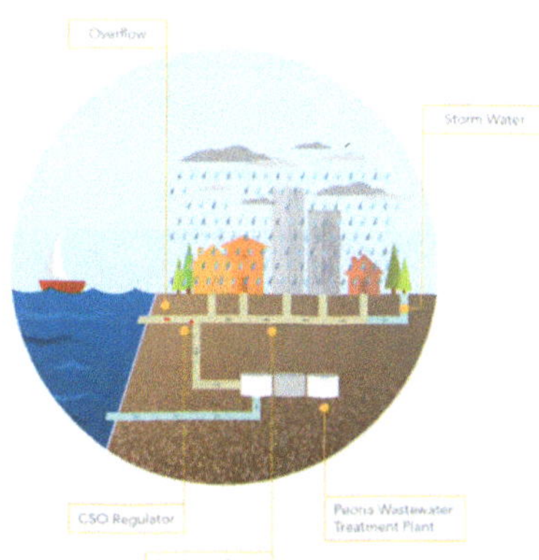

Fig. 2. Graphical illustration of conditions leading to sewage overflow.

Approximately 12.2 square miles of sewer service area currently drains to the Riverfront Interceptor (Department of Public Works 2018a, b). The service area includes 6.8 miles of separate sewers, 3.9 square miles of which flow directly into the Riverfront Interceptor, along with 5.4 square miles of combined sewers. Both the Riverfront Interceptor (along the river) and the Kickapoo Interceptor (to the northwest) drain to the wastewater treatment plant. The combined sewer area is shown in blue in Fig. 3 along with the entire drainage area to the wastewater treatment plant. Figure 3 also outlines how ownership of the sewers is split between the City and the Greater Peoria Sanitary District. The City owns all of the sewers in the combined sewer area and some separate area sewers.

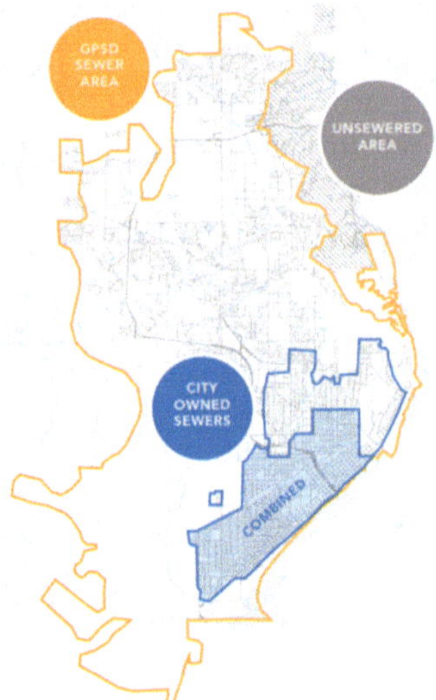

Fig. 3. The Combined Sewer System in Peoria.

Sewage from the combined sewer service area currently enters the Riverfront Interceptor through 19 regulating structures. Excess flow can discharge to the Illinois River through 16 permitted outfalls, although only 8 of the permitted outfalls regularly discharge. The rest are considered emergency outfalls. The cause of sewage overflow is illustrated graphically in Fig. 4.

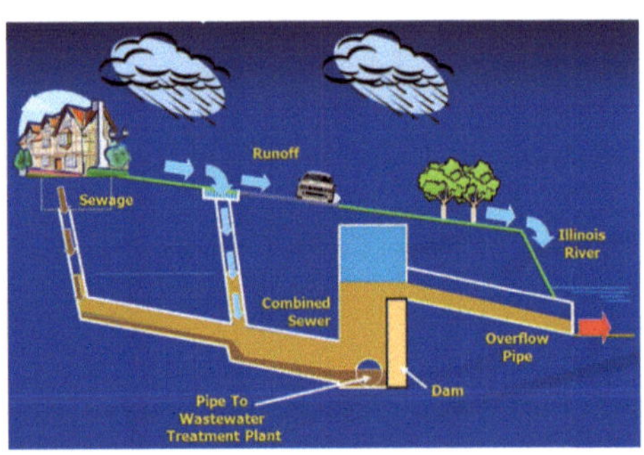

Fig. 4. Graphical illustration of the causes of sewage overflow.

Swirl concentrators are located at the Eaton and Fayette CSO outfalls to treat the CSO flow by removing floatables and heavier solids before discharging to the Illinois River. There is one 5,197-foot-long storage sewer designed to reduce the quantity and frequency of overflows in the area northeast of downtown. The storage sewer has the capacity to store 540,000 gallons of sewage prior to overflowing. Flow from the storage sewer is introduced into the Riverfront Interceptor through a vortex flow regulator. The design capacity of the Riverfront Interceptor is 80 MGD as it flows into the wastewater treatment plant.

2 Record of Major CSO Improvements

In 1985, Peoria had an overflow volume of approximately 830 MG in the rainy season of an average year. In the late 1980s and early 1990s, Peoria implemented about $10 million in CSO improvements. The number of permitted CSOs was reduced from 20 to 16 permitted overflows. Of the 16 overflows only 8 remain very active with the other 8 being emergency overflows. The largest project in the late 1980's was to construct a ¾ mile to store/convey sewage and stormwater. This new sewer is utilized to store large volumes of combined sewer during a stormwater event. Projects in the early 1990's consisted of installing floatables controls downtown and the Riverfront Interceptor sewer was reconfigured to withstand a river elevation increase due to a 25-year flood. Four CSO basins were partially separated. Smart controls and monitoring were installed. Peoria currently has about 243 MG of CSO volume in a typical year. This represents a reduction of about 70% since the CSO improvements were implemented in the 1980s. Figure 5 shows a graph of that reduction in sewer volume starting in 1985 and completing in 1996. Figure 3 also shows the reduction rate proposed to fully reduce the typical year combined sewer overly by 2036.

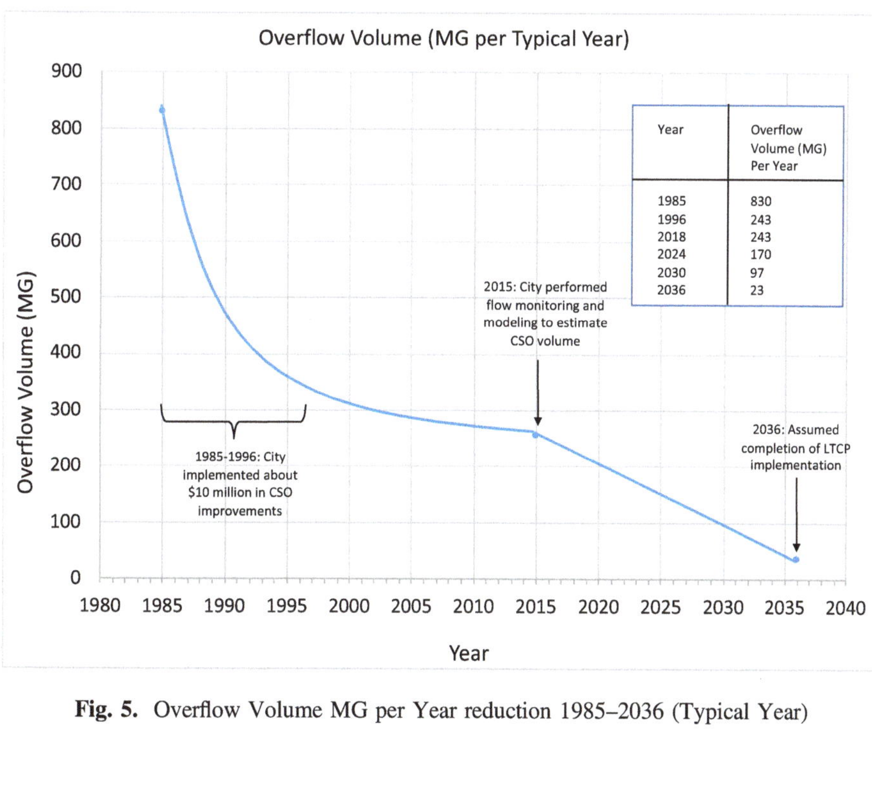

Fig. 5. Overflow Volume MG per Year reduction 1985–2036 (Typical Year)

The volume of combined sewer overflow expected during a typical recreation season in a given year is much less than the volume for an entire year. The associated cost with the number of expected number of overflow events is shown in Table 1.

Table 1. Estimated cost associated with number of overflow events during a typical recreation season and an entire typical year

Typical year recreation season			Entire typical year			Total estimated capital cost ($M)
Number of overflow events	CSO volume (MG)	CSO volume reduced (MG)	Number of overflow events	CSO volume (MG)	CSO volume reduced (MG)	
22	128	0	51	256	0	N/A – existing conditions
4	37	91	7	64	192	$37
1	23	105	2	36	220	$109
0	0	128	1	2	254	$312

Table 1 indicates that allowing for even one overflow event will involve 2 MG of combined sewer overflow. The cost associated with even one overflow event is expected to cost $312 million.

3 City of Peoria CSO Management

The strategy to reduce CSO volumes for the City of Peoria may be divided into three phases. In the first phase (1985–1996), the City of Peoria constructed a ¾ mile storage sewer between Spring Street and Caroline Street that added additional storage volume to the combined sewer system. Swirl concentrators were also constructed at Eaton Street and Fayette Street to collect floatables and also allow additional storage volume. The existing riverfront interceptor was also modified to protect it to a 25-year flood from the Illinois River to prevent surcharging. Additionally 4 of the CSO basins were partially separated as road projects were completed to reduce the overall volume of combine sewer entering the system.

In the second phase (1996–2018), minimal improvements were performed since the City of Peoria was operating under a CSO permit issued from the Illinois EPA. The City would take advantage of road projects to continue working on separating the storm from the sanitary sewers as the opportunities presented themselves.

In the third phase (2018–2036), the City is proposing an adaptive management style for project selection and implementation. The goal is to a meet performance criterion of 33% reduction in CSO volume every 6 years. A public participation process, including input on the type of green infrastructure to be adopted, will be used to identify the projects for implementation and construction. The City has prioritized the construction of projects linked with the Riverfront Interceptor to address regulatory concerns as a top priority. Another priority will be the creation of in-system storage using newly installed weirs and smart controls. Finally, green infrastructure projects will be implemented throughout the City to capture rainfall at the source and to divert it from entering the combined sewer system.

4 Regulatory Drivers and the City's Evolving Response

The City of Peoria's Public Works Staff compiled a summary of the events and legal issues relating to the USEPA and Department of Justice (DOJ) actions regarding the City's CSO permit. These regulatory drivers and actions inform the City of Peoria's evolving response. These actions and the City's responses are chronicled below.

4.1 2006: City Signs the First Tolling Agreement

In late 2005 the USEPA, based on a "wet weather inspection," determined that the City needed to make improvements to the CSO system. In 2006, the USEPA and DOJ found the City of Peoria in violation of the Clean Water Act (CWA) and offered to enter a tolling agreement, in lieu of filing suit. The City signed its first tolling agreement in June 2006. Typically, the Illinois EPA (IEPA) issues CSO permits to communities on a 5-year basis, dictating how the CSO should be operated and how and when the reporting is done. The USEPA was involved in crafting the renewal language for the City's CSO permit when it expired on May 31, 2006. The new language required the City to submit a Long Term Control Plan (LTCP) to further reduce the occurrence and magnitude of CSO events. The LTCP also mandated testing CSO water samples, analyzing the existing sewer system and overflow volume, evaluating alternative solutions, and disclosing community financial ability to fund improvements. The City hired Amec Foster Wheeler to develop the LTCP, and Barnes & Thornburg, LLC as outside legal counsel.

4.2 Spring 2007: Monitoring, Modeling, and Stakeholder Meetings

In spring and summer of 2007, monitoring equipment was placed in the city sewers and at the outfalls. Rain gauges were installed in the CSO area. Samples of river water were obtained and tested for contaminants upstream of CSOs and locations along the CSO area. Samples were taken from the west bank, the middle of the river, and the east bank. Samples were taken from Farm Creek and from Kickapoo Creek. Samples were taken in dry weather and during and after CSOs. Samples were taken of the water that overflowed from the sewers, and the volume of overflow was measured. Data showed that the river has a fecal coliform (bacteria) level higher than IEPA standards even in dry weather upstream of the Peoria CSOs. When CSOs occurred, the bacteria count increased further. The CSOs add more pollution to a polluted river. The flow monitoring data was used to build a computer model of the combined sewer system. Rainfall data from about 50 years was fed through the computer model. A statistically 'typical' year was determined.

During this period, the Clean River Committee (CRC) was formed of volunteer stakeholders. The CRC met about every six weeks to provide input on the LTCP development. These are discussed further in a following section.

4.3 Fall 2008: Formal Public Hearings, Recommended Solutions

Two formal public meetings were held in August 2008. The solution proposed was to build "store/treat" tanks to hold the excess water during a heavy storm and to screen out solids and disinfect the water before it could flow out to the river. It is less expensive to build tanks than to build a long tunnel to send all the water to the GPSD wastewater treatment plant (WWTP). The tanks were sized to allow 4 overflows in a typical year, per the USEPA CSO Policy. The following three options were presented with costs in 2008 dollars:

a. 4 tanks (distributed along river) \$105–127 M
b. 3 tanks (no tanks above Bob Michel Bridge) \$186–223 M
c. 1 BIG tank (no overflows upstream of WWTP) \$218–262 M

Other components of all proposed solutions were litter control and green infrastructure. The public feedback was compiled and the results were that the public chose the least expensive option; 'voted' to use green if the use of green lowered the overall cost; and agreed that the first focus was to get the Spring Street overflow eliminated in Peoria Lake, where people ski and swim. A draft LTCP was submitted to USEPA in November 2008 promoting the lowest cost solution.

4.4 2009: USEPA Reviews First Draft of LTCP

In 2009, the USEPA reviewed portions of the first draft LTCP and challenged the City on affordability, noting that more than a \$500 million LTCP could be afforded by its citizens. One measure US EPA uses to estimate the impact of CSO control to a community is the expected residential share of estimated project costs and associated wastewater treatment costs as a percentage of Median Household Income (MHI). If the calculated percentage is 2% or greater, then the burden is considered heavy. MHI and number of households in the City of Peoria are obtained from US Census data. The projected burden for the recommended alternative was 1.49% based on MHI of \$42,260 and an annual residential share of estimated project costs of \$631.07 (\$631.07/ \$42,260 = 1.49%). US EPA noted that the City could undertake a more significant project before reaching the 2% burden.

US EPA suggested that treatment of CSOs similar to full secondary treatment would be needed due to the thought that screening and disinfection would not be sufficient. The City noted that this would be significantly more expensive and would impose a burden far in excess of 2% of MHI for the City's lower income residents. US EPA noted that this was the case for many CSO communities.

The USEPA also informed the City that it needed to consider additional technologies and alternatives along with a higher level of control. The number of alternatives considered was expanded and detailed comparisons of 19 alternatives were prepared. The City recommended a CSO control alternative with an estimated cost of $92.5 M in 2009 that included screening and disinfection of remaining overflows.

4.5 Spring 2010: City Submits Second Draft of LTCP

In March 2010, the City submitted a second Draft LTCP based on analysis of 19 alternate solutions. The range of alternatives included providing different technologies and different levels of protection. Different technologies included storage only tanks, storage tanks with disinfection (store/treat tanks), sewer separation and other technologies. "Level of protection" refers to the statistical likelihood that a CSO would occur. For example, a "typical year" level of protection would mean that there would be no overflows for an average year or any years drier than the average year, but there would be overflows in any year wetter than the typical year. A 10-year level of protection would mean there would be only a 10% chance that an overflow would occur in any year.

Based on cost-benefit curves, the City recommended "Alternative 16," which included four store/treat tanks and two long pipe runs to convey flow to the tanks at a net present value (NPV) cost of $92.5 Million in 2009 dollars. The preferred plan was presented to the City Council in open session in July 2010.

4.6 2010–2011: USEPA Reviews Second Draft of LTCP

The USEPA reviewed portions of the second Draft LTCP and challenged the City repeatedly on the effectiveness of the chosen technology (store/treat tanks) although many permitted store/treat tanks were operated elsewhere in the country. The USEPA repeatedly stated that the City could afford a higher level of protection, and must provide a higher level of protection, because USEPA determined that the Illinois River at Peoria is a sensitive area. City Staff visited five 5 store/treat facilities in Michigan, and determined that these facilities could be made aesthetically pleasing and could be maintained in our climate.

4.7 November 2011: USEPA & DOJ Bring GPSD into Tolling Agreement

In November 2011, USEPA/DOJ invited GPSD to enter into a tolling agreement or be sued. They explained that they viewed the City sewer system and the GPSD WWTP as one, joint issue. They also alleged other GPSD CWA violations unrelated to CSO. GPSD and the City were signatory to the same tolling agreement. GPSD hired the same outside counsel to represent them.

4.8 July 2012: Legislator Advocacy

In July 2012, the City reached out to Congressman Aaron Schock to express the unfairness of the CSO negotiations. The City provided testimony to a Congressional Sub-Committee about the unfairness. Congressman Schock sent an inquiry letter to USEPA.

4.9 May-December 2013: USEPA Brings Hires Its Own Engineering Consultants

In May 2013 USEPA hired their own consultant to evaluate Peoria CSO alternatives. Their subconsultant suggested that a 4-mile long, 11-foot diameter tunnel from Spring Street to the GPSD WWTP could provide an 'affordable' 10-year, level of protection at a NPV cost of $230 million (2009 dollars). They asked that the City evaluate the preliminary design of the tunnel along Adams Street, Washington Street or Water Street.

4.10 December 2013: DOJ Action

In December 2013, DOJ threatened to file suit unless the City agreed in principle to a 10-year storm level of control for the entire CSO area. They reiterated that the $230 million, 4-mile tunnel option was an affordable solution.

4.11 Spring 2014: Summary of 2010–2014 Alternatives

In February and March, 2014 the City prepared a summary of treatment alternatives, based on the various discussions between the regulatory agencies and the City of Peoria and proposed a 2014 concept with a 10-year peak storm protection plan, taking USEPA and DOJ comments into consideration. The NPV cost comparisons were based on the 160 MG of annual overflow volume and indexed to 2014 dollars.

a. The "Floor": Alternative 16 (2010) based on 4 store/treat tanks. Typical year control to have only 4 overflows - $105 M NPV (Rejected by USEPA)
b. The USEPA "Ceiling": 4 mile tunnel, 10 year storm control - $260 M NPV (Rejected by City)
c. The February Counter City Proposal: 2 levels of protection - $163 M NPV (Rejected by USEPA)
d. 2014 concept: 10 year peak storm protection with Green Infrastructure - $175 to $230 M NPV

Besides being cost competitive with gray infrastructure, green infrastructure has other benefits, including better aesthetics and an opportunity to rebuild visible infrastructure such as roads and parking lots, while addressing the CSO mandates at the same time.

The City decided to pursue a green infrastructure solution as a compromise between USEPA's desire for a higher level of control and the City's desire to use green infrastructure to simultaneously address multiple City challenges. The issues and desires of the various stakeholders are best described in Fig. 6.

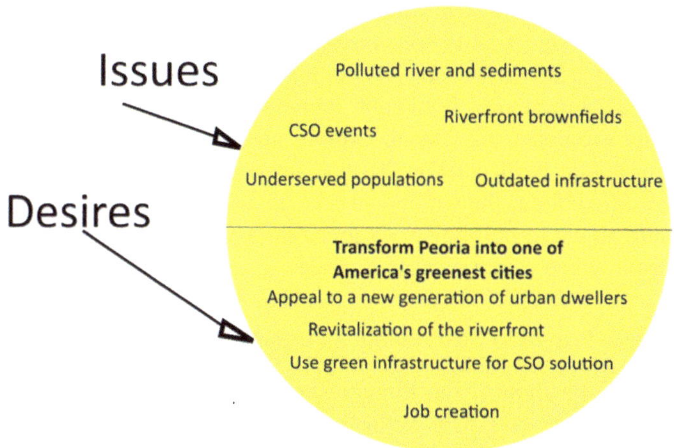

Fig. 6. Issues and Desires of Stakeholders.

Clearly, the City of Peoria saw many positive aspects to its proposed green solution of the combined sewer overflow problem. These include job creation, economic development on its riverfront and appealing to younger professionals to move to the downtown area. The pushback relates to funding of the proposed green solution.

5 The Peoria Preliminary Strategic Assessment

The City was confident that the alternative selected for the revised LTCP submitted in 2010 would meet water quality standards and was cost-effective. The selected alternative limited overflows to four in a typical year and was based on the City's belief that there was no compelling evidence of locations that would be considered sensitive within the CSO study area. Table 2 show the Illinois water quality standards for the Illinois river.

Table 2. Illinois water quality standards for the illinois river

.	.

The US EPA's determination that the entire CSO study area is sensitive meant a higher level of control would be needed than that provided by the City's recommended alternative. The City's initial thought was that the level of CSO control needed would be to eliminate CSOs in the typical year to meet USEPA's sensitive area determination. USEPA insisted that the City would need to provide a level of control for a 10-year storm for the entire CSO study area.

The City proposed different levels of control for upstream and downstream regions of the CSO study area based on potential locations of recreational use. The USEPA insisted the entire CSO study area would need to be protected up to the 10-year level of control and strongly suggested Peoria construct a full-length tunnel. The 10-year level of control provides CSO control for a storm that has a 10% occurrence probability in any given year. The cost of a full-length tunnel and US EPA's estimate of Peoria's burden capacity prompted the City to explore green infrastructure as an alternative (Department of Public Works 2017c, 2018b). The cost per gallon of green versus gray favored green. In addition, the City could take advantage of the naturally sandy soils below the bluff, suited to infiltration via green infrastructure, and reap the co-benefits that a sustainable green solution could provide.

5.1 The Peoria Proposed Sustainable Solution to the CSO Problem

With USEPA's designation of the Illinois River CSO area as "sensitive"—requiring higher levels of control—and their estimation of the City's affordability limit, the City refocused CSO control to a cost-effective sustainable solution using 100% green infrastructure. It is well known that permeable pavements have been successfully used in controlling stormwater runoff and infiltration (Brattebo and Booth 2003; EPA 2000, 2009, 2010, 2017; Hansen 2008). Rather than constructing more capital-intensive "gray" infrastructure (like pipes, tanks or tunnels), the City would install pervious pavement and rain gardens to prevent stormwater from entering combined sewers in the first place. It is expected that such solutions will allow the seamless integration of vegetation into the urban environment (Morgenroth and Visser 2011).

The 100% green solution would not only meet CSO policy requirements but also provide additional related benefits to the City. Green infrastructure co-benefits include:

a. Beautification of public rights-of-way
b. Enhancement public health and wellness by:
    Decreasing pollutants, which can lessen illness
    Improving air quality
c. Reduction in potential crime and increase surveillance
d. Mitigation of localized flooding
e. Connections between people and places
f. Safe and accessible transportation infrastructure
g. Ecological health and productivity
h. Local jobs and workforce training
i. Neighborhood-scale economic development

5.2 Challenges Faced by the City – Financial Aspect

The CSO control requirement is a large unfunded mandate that presents a challenge to the City on how to pay for such a massive project that will not only take years for the initial implementation but will also require yearly appropriations for operation and maintenance. Availability of bonds, attainment of grants and low interest government loans, and increased user fees will all need to be investigated as potential funding options. Additional costs passed on to residents may have a greater impact on lower income residents in the CSO area.

The solution to an issue of this scope and duration must be acceptable to City leaders and beneficial to stakeholders. The solution should be cost-effective and provide additional benefits beyond the CSO control required by US EPA. A 100% green solution can achieve these goals. A 100% green solution will allow the City to demonstrate to the Agencies its commitment to protecting the Illinois River, provide opportunities for local businesses, and serve as a driver for community improvement and revitalization.

The City of Peoria has currently adopted a Storm Water Utility to address wet weather issues (Department of Public Works 2017b). However, that is projected to only generate $8 Million per year and also has to cover existing operating and capital costs the City has. The proposed green solution will cost the City an additional $15–$18 Million per year that is not currently in the budget. The green solution is expected to require a total investment of $200 million to $300 million. The current thought is to utilize $5 Million a year from the Storm Water Utility to fund the CSO solution but that still leaves a financing gap of $10–$13 Million per year (Department of Public Works 2017a, c, 2018b). The City will look to have a mixed financing of sewer increases and bond sales to cover these additional costs. The Stormwater sewer fees will be raised over the course of the 18-year implementation period to cover the debt service of the bond issuance. The City will see the average sewer bill for a resident of the City increase 5–7 times to fund this endeavor.

5.3 The Timeline

A draft implementation schedule was prepared for the 100% green solution. The schedule sets a series of performance criteria goals to be met over the 18-year timeline of the project. Green infrastructure components will be designed and installed to manage stormwater to achieve the agreed upon level of CSO control. Below is the timeline of events and meetings between the US EPA, the US DOJ and the City of Peoria that led to the currently proposed plan.

a. May 2006: IEPA issues NPDES permit that includes requirement for development of CSO LTCP
b. January 2007–December 2008: work conducted for preparation of LTCP
c. November 2008: US EPA determines Peoria's CSO area is "sensitive"
d. December 2008: draft LTCP submitted with recommended alternative with a cost of $105 M–$127 M (2008 dollars)
e. 2009 US EPA requests additional CSO control alternatives be considered and notes City's affordability capacity of $500 million or more
f. March 2010: revised LTCP submitted with recommended alternative with a cost of $92.5 M (2009 dollars)
g. 2010–2014: negotiations with US EPA on required level of control and alternatives
h. 2014: City submits option to provide CSO control with 100% green infrastructure
i. 2014: current – negotiations with US EPA on effectiveness and ability of 100% green solution to control Peoria's CSOs
j. 2017: The Peoria City Council approved new rate increases to the stormwater utility fee to cover the cost of the green solution that took effect in June 2018.

The schedule proposed green infrastructure will be installed to eliminate 33% of the required overflow volume by year 6, and 66% volume to be eliminated by the 12th year and 100% of required volume eliminated by year 18. Flow monitoring and modeling of installed green infrastructure will be used to determine if the required performance criteria are being met throughout the implementation period. Flow monitoring and modeling will be utilized to determine if additional green infrastructure components are needed in a particular area as well as reduce the proposed installation of green components if the required volume reduction is already met.

6 Public Awareness and Strategic Communication

As noted earlier, in May 2007 a citizen's advisory committee known as CRC was formed with the mission of providing advice and recommendations to the Peoria Department of Public Works on topics related to the development of a CSO LTCP. The committee included representatives from the Chamber of Commerce, the business community, health department, park district, school districts, environmental advocacy groups, government agencies, citizens, Bradley University and other key stakeholders. The committee was involved in reviewing the City's analysis and solutions to protect local waterways. The committee also considered the financial impacts to the community as a whole and helping the City reach out to the public.

A project website was developed to provide up-to-date information on sewer overflow warnings and the combined sewer overflow problem. During the public comment period, the web site also was used to post information on the alternatives analysis and allow the public to comment on the alternatives.

A 30-day comment period was held from July 21–August 20, 2008. Two public meetings were held on August 6 and 7 to obtain public input on the CSO control alternatives. More than 20 news releases and media advisories were sent to Peoria area news media in 2007 and 2008.

The CRC had served its purpose as an advisory committee for the development of the LTCP and no further meetings of this group were held after the LTCP submittal. After the shift to a 100% green solution it was decided to re-establish a public outreach program. In 2015 the OneWater advisory committee was formed. The OneWater Committee is a group of community leaders focused on comprehensive wet weather management that makes recommendations to shape City policy and research issues, share ideas and serve as educated advocates.

The processes involved in meeting the legal challenges presented by the USEPA and its unfunded mandates for clean water in Peoria continues. The development of a CSO control solution that meets CSO policy while being cost-effective and beneficial on multiple levels for the City is a challenging task. The City's desire to meet CWA requirements, improve public places and minimize financial impacts to its residents is made even more challenging by trying to meet agency requirements that are undefined and frequently changed. Agency changes in required levels of control, recommended control approaches (gray tanks and tunnels instead of green) and noting the City's capacity to spend up to 2% of MHI have hampered the development of an agreed upon CSO solution to date.

7 Activities in Progress

It is expected that once the City of Peoria Staff and Department of Justice agree on terms of a Consent Decree, a the detailed document will be developed and approved by all parties. After approval internally of the language the Consent Decree will be presented to the City Council for approval and then it will go to a Circuit Judge to sit for 6 months for public comment. After 6 months is complete, if the Judge is agreeable to the terms the Consent Decree, will be certified and approved. That is when the clock starts ticking to implement and the City has to meet the performance criteria outlined in the Consent Decree or face penalties.

The Peoria Public Works staff is working on recommendations to the Council on how to fund the remaining portion of the capital cost. It will most likely be a blend of bonding (borrowing) and sewer rate increases. The sewer rates could increase to as much as 7 times the current cost to generate the revenue to cover the debt service on the bonds. In 2017, the Peoria City Council approved new rate increases to the approved Storm Water Utility fee in order to cover the cost of the green solution that took effect in June 2018. Further, the city provides residents and businesses various incentives to reduce the new rate hike introduced to pay for the green solution of the CSO problem.

The City of Peoria's Storm Water Utility fee is meant to be a driver for new development and behavior so that property owners are aware of how what happens on their property effects the entire City. Property owners can reduce their run-off by considering Rain Barrels, Rain Gardens, Detention Basins, and Permeable Pavement. The City of Peoria is looking to incentivize change by offering reductions of up to 50% for installing such items. Credit of up to 50% can be substantial for large commercial properties that typically have the majority of the run-off due to the size of their developments. A typical car dealership in Peoria will have a yearly bill of $30,000 for the Storm Water Utility but if the dealership install and prove with certified engineering and best management practices, then they have the opportunity to reduce the Storm Water fee by $15,000 per year.

8 Conclusions

The management of combined sewer overflows is a critical issue in many cities across the US. It is important for these cities to integrate their stormwater management plans into their strategy to deal with the CSO problem. This paper illustrates the impact of regulatory agencies and their actions on cities like the City of Peoria. Frequently, as in the case of the City of Peoria, an unfunded mandate imposes a financial burden on the City. Therefore, it is essential to develop creative sustainable solutions, which can satisfactorily address the environmental concerns while adequately meeting the financial, social, and communication/public outreach goals at the same time.

The proposed green solution embraced by the City of Peoria will set new standards for other cities across the USA and beyond to adopt and emulate. The proposed solution challenges the status quo and the old methods used by other cities to address their CSO problems. Although no final agreement has been reached with the federal government, it is expected to be signed in the near future. The City of Peoria has shown the courage to try a complete green solution to spur economic activity, achieve social equality through creation of new employment opportunities, encourage workforce training to develop new skills required, and create the proper environment for a healthy community. In recent years, the City took the initiative to implement several pilot programs involving permeable pavements and other green solutions with great success.

Finally, it is important for any comprehensive CSO strategy to discuss issues and desires of various stakeholders, including the public and underserved groups in the community. Public awareness and outreach is an integral component in finding optimal, green, and creative solutions to the CSO problem. The City of Peoria developed a citizen's advisory group called the Clean River Committee (CRC), which was instrumental in providing initial guidance for the project. In 2015, the OneWater advisory committee was formed as part of the public outreach effort to make recommendations that shape City policy, research ideas, and serve as educated advocates. The proposed green solution will require an investment of between \$200 million and \$300 million, which is close to being finalized. The City has already taken the initiative to implement the green solution, which will be innovative and creative.

Acknowledgments. Special thanks to Prof. Krishnanand Maillacheruvu for reviewing the paper and providing help in its preparation and to Prof. David Spelman for reviewing it.

References

Brattebo, B.O., Booth, D.B.: Long-term stormwater quantity and quality performance of permeable pavement systems. Water Res. (2003). https://doi.org/10.1016/s0043-1354(03)00410-x

Department of Public Works: Credit Manual and Grant Program. City of Peoria (2017a)

Department of Public Works: Wet Weather Management Informational Packet. City of Peoria (2017b)

Department of Public Works: Funding a Better Tomorrow. City of Peoria (2017c)

Department of Public Works: CSO History and Summary. City of Peoria, Peoria (2018a)

Department of Public Works: Sustainable Green Solutions to the City of Peoria Combined Sewer Overflow. City of Peoria (2018b)

EPA: Field Evaluation of Permeable Pavements for Stormwater Management, Olympia, Washington. Document No. EPA-841-B-00-005B (2000)

EPA: National Menu of Stormwater Best Management Practices. Pervious Concrete Pavement (2009)

EPA: Stormwater Best Management Practices (BMP) Performance Analysis. Prepared by Tetra Tech, Inc. Fairfax VA for USEPA Region 1, Boston (2010)

EPA: NPDES Compliance Inspection Manual. Chapter 14. EPA Publication No. 305-K-17-001 (2017)

Hansen, K.: IS-131: Porous Asphalt Pavements for Stormwater Management, p. 16. National Asphalt Pavement Association, Lanham (2008)

Morgenroth, J., Visser, L.: Aboveground growth response of Platanus orientalis to porous pavements. Arboricult. Urban For. **37**(1), 1–5 (2011)

Sustainable and Green Solutions to The City of Peoria Combined Sewer Overflow Problems

Amir Al-Khafaji[1(✉)], Jim Ardis[2], Scott Reeise[2], and Patrick Urich[1,2]

[1] Bradley University, Peoria, IL, USA
amir@fsmail.bradley.edu
[2] City of Peoria, Peoria, IL, USA

Abstract. In late 2005, the United States Environmental Protection Agency (USEPA) performed a "wet-weather inspection" in Peoria, Illinois. They determined that the City needed to make improvements to the combined sewer system (CSO). Further, in 2006, USEPA and the Department of Justice (DOJ) found the city of Peoria in violation of the Clean Water Act (CWA), and offered to enter into a tolling agreement, in lieu of filing suit. The City signed the first tolling agreement. The DOJ would file suit unless the City agreed in principle to a 10-year, peak storm level of control for the entire CSO area. The DOJ suggested a $230 million, 4-mile tunnel option as an affordable solution. The solution involved permeable pavement and other green solutions. This paper presents the proposed sustainable and green solution to the CSO problem in Peoria. The City of Peoria pursued an imaginative approach to initiate a new green strategy and sustainable methodologies to address its CSO challenges. Generally, gray infrastructure is cheaper than green infrastructure in the short-term, the green infrastructure becomes cost competitive in the long term. Green infrastructure has other benefits, including better aesthetics and an opportunity to rebuild visible infrastructure such as roads and parking lots, while addressing the CSO mandates from the USEPA and DOJ at the same time. The expected cost for the proposed green solution is $200 million to $300 million. This paper also provides an overview of the proposed green solutions and the revenues needed to pay for the CSO project.

Keywords: USEPA · Peoria · Illinois · CSO · DOJ · Peak storm
Gray infrastructure · Green infrastructure

1 Introduction

A spring storm rumbles into a city with lightning and thunder and the city gets soaked by rainfall. The clouds move through and we all go about the day. However, in streets, backyards, creeks, ravines, pipes, and tanks, the storm is not over yet, not by a long shot, because all that rainfall did not disappear. Mother Nature does not distinguish between neighborhoods, district boundaries, public/private/not-for-profit properties. Dealing with rain and snow within city limits is the sole responsibility of the City of Peoria and its Department of Public Works.

Over the years, stormwater has caused a myriad of problems throughout the city of Peoria. Stormwater flows into the City's combined sewers and discharges directly into

M. Shehata and F. Rodrigues (Eds.): GeoMEast 2018, SUCI, pp. 69–86, 2019.
https://doi.org/10.1007/978-3-030-01905-1_4

the City's most precious asset, the Illinois River. Stormwater runoff has caused excessive soil erosion, which has threatened residential and commercial properties. In addition, street flooding, basements backing up, and sinkholes are among the many other problems created by stormwater runoff. Increasingly stringent water quality regulations have forced the City to address the stormwater runoff problem and combined sewer overflow issue. The federal government has mandated that the City of Peoria take concrete steps to protect the health of waterways and eliminate dumping sewage into local water bodies.

While conventional wisdom suggests that gray infrastructure would be cheaper compared to a green solution, the City does not believe that additional gray infrastructure is the answer in this case. That is why it has proposed a more sensible and cost-effective approach to control combined sewer overflows (CSOs) through green infrastructure. The City of Peoria has taken a proactive approach to solving the stormwater problem including the creation of a fair funding stream that could help offset green infrastructure cost while funding much-needed drainage repairs in neighborhoods north and south, east and west to meet citizens demands.

For years, erosion, flooding and other damage from stormwater runoff have overwhelmed the City of Peoria's infrastructure—even in newer areas of the City not served by combined storm/sanitary sewers. Recognizing the challenges, Peoria is deeply invested in wet weather management (Department of Public Works 2017c). Solving the CSO problems are essential for Peoria's present and future generations. The proposed solution is a community effort, which builds on the local knowhow to past damage and prevent/minimize future problems resulting from wet weather events.

2 Overview of Peoria and Sewage Overflow

The City of Peoria is divided into five districts with representatives elected from each district. These city council members are engaged in decision-making relative to the CSO problem faced by the City of Peoria. The five districts and locations of various storm water projects are highlighted in Fig. 1. Most of the combined sewer area is in District 1, but also covers parts of Districts 2 and 3 (Greeley and Hansen 1985; Department of Public Works 2018b). The City has been implementing green infrastructure pilot projects in these areas. The future green infrastructure projects that are required for CSO control are not shown on Fig. 1.

The City of Peoria has undertaken multiple pilot projects to initiate green solutions that could serve as the basis for reaching a final agreement with the Federal government on possible solution to the CSO problem. Each pilot project includes different features for point source control of the stormwater. These features include permeable pavers, bioswales, rain gardens, and dry-wells that allow the stormwater to be stored in place and infiltrate, reducing and slowing the amount of rain water from entering the system. Figure 2 shows a pilot project that was completed by utilizing a design/build method that was able to capture one million gallons of stormwater run-off from a 10-year storm event.

The City has 16 CSO locations, although only 8 of them actively discharge on a regular basis and are shown as red triangles on Fig. 3. The other locations, not shown on the figure, are considered emergency overflows and do not overflow in a typical

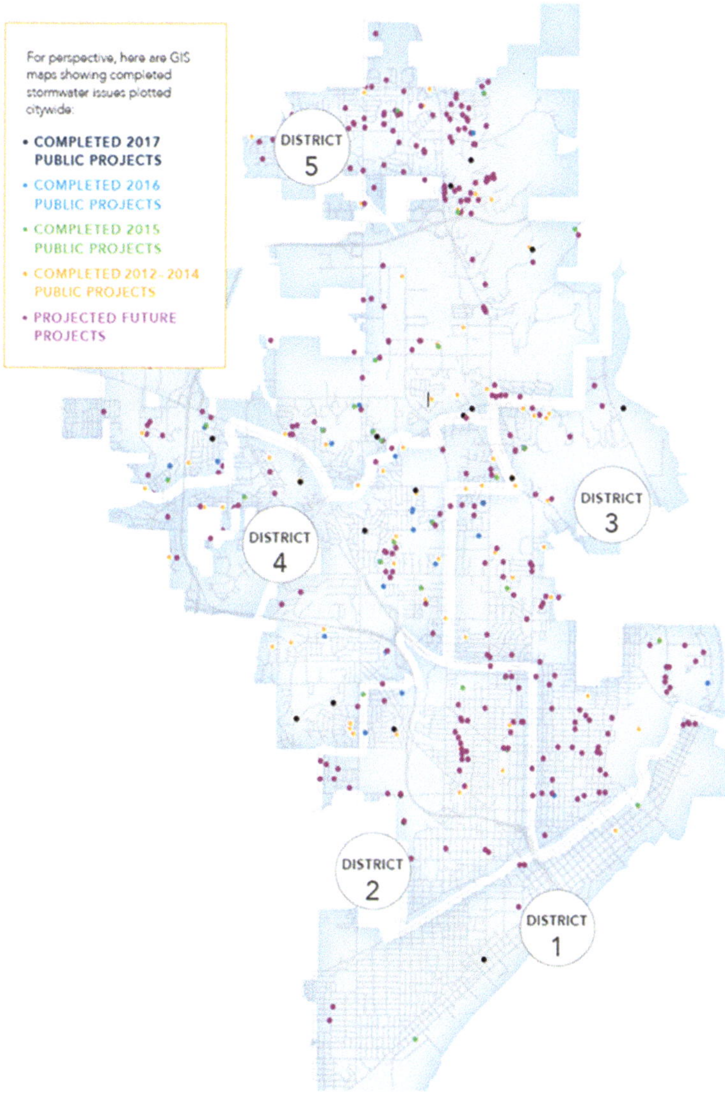

Fig. 1. Map showing the various districts and projects in the City of Peoria. (Source: Department of Public Works 2017c. Wet Weather Management Informational Packet. City of Peoria).

year. Table 1 provides an example of overflow volumes for a typical year. The CSOs contribute to high bacteria levels in the Illinois River during and after storms. The Illinois River is used for boating, swimming, and other recreational activities. Thus, it must remain clean for public usage.

Fig. 2. Adams Street Pilot Project during rain event. (Source: Department of Public Works 2017c. Wet Weather Management Informational Packet. City of Peoria).

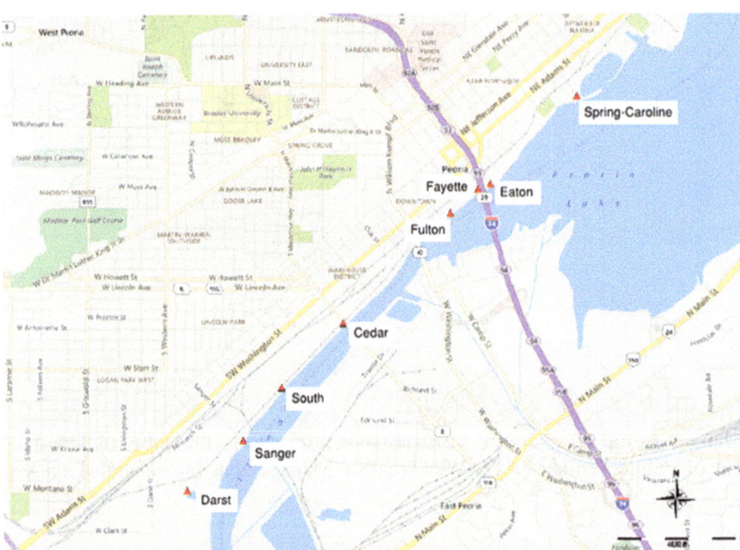

Fig. 3. Map showing the active Peoria Sewage Overflow locations. (Source: AMEC Foster Wheeler (2010). Peoria CSO Long-Term Control Plan).

Table 1. CSO volume in a typical year. (Source: Wood PLC 2016. Wood PLC/symbiont for the City of Peoria: existing conditions model report)

Outfall number	Outfall name	Existing conditions overflow volume for the entire 1949 typical year (MG)
001	Green	0
003	Spring-Caroline	54.7
006A and 006B	Eaton	19.6
07A	Fayette	23.3
008	Hamilton	0
009	Fulton	0.1
010	Liberty	0
011	Harrison	0
013	Walnut	0
014	State	0
016	Cedar	74.6
017	South	16.1
018	Sanger	37.6
019	Darst	16.9
020	Main	0
Total		243

3 City of Peoria Green Initiative

The City has proposed a green infrastructure solution to minimize the effects of CSOs on the Illinois River. At its best, green infrastructure will capture stormwater and infiltrate it into native soil, such that it no longer reaches the sewer system. It does this by having a permeable surface through which water can pass to get to the native soil below (Brattebo and Booth 2003; EPA 2000). The permeable surface is typically pervious pavement or a sandy soil mixture that is capable of growing plants. If the native soil is not capable of absorbing significant amounts of storm water (e.g., clay), then the storm water can be stored in a shallow depression on the surface (e.g., rain garden or bioswale) and/or stored under the permeable surface in the open spaces in a layer of sand or open graded gravel. The storage holds the water and releases it to the sewer system at a lower flow rate so that it can be conveyed to the treatment plant instead of overflowing. Plants and trees can also be used to evapotranspirate the stored water.

Permeable pavements have been shown to have good potential to control stormwater discharge (Hansen 2008; EPA 2009, 2010; Volder et al. 2009). The City of Peoria has already implemented a green solution at Trewyn School to address past flooding issues using new permeable parking lanes. Figure 4 shows a picture of a newly completed green solution in Peoria.

Fig. 4. Picture of a newly completed green project in Peoria.

4 The US Environmental Protection Agency (EPA)

The mission of EPA is to protect human health and the environment. To accomplish this mission, the EPA develops and enforces regulations. When the US Congress writes an environmental law, EPA implements it by writing regulations. Often, EPA sets national standards that states and tribes enforce through their own regulations. The EPA's main task is to ensure that:

a. All Americans are protected from significant risks to human health and the environment where they live, learn and work;
b. National efforts to reduce environmental risk are based on the best available scientific information;
c. Federal laws protecting human health and the environment are enforced fairly and effectively;
d. Environmental protection is an integral consideration in U.S. policies concerning natural resources, human health, economic growth, energy, transportation, agriculture, industry, and international trade, and these factors are similarly considered in establishing environmental policy;
e. All parts of society - communities, individuals, businesses, and state, local and tribal governments - have access to accurate information sufficient to effectively participate in managing human health and environmental risks;
f. Environmental protection contributes to making our communities and ecosystems diverse, sustainable and economically productive

The United States also plays a leadership role in working with other nations to protect the global environment

5 Impact of EPA and Department of Justice on CSO Plan

The National Pollutant Discharge Elimination System (NPDES) permit issued to the City by Illinois EPA in 2006 contained a condition for the City to develop a Combined Sewer Overflow (CSO) Long Term Control Plan (LTCP). The City of Peoria's objectives in completing the LTCP were to remain in full compliance with the terms and conditions of its NPDES Permit and to meet the objectives of USEPA's April 19, 1994 Combined Sewer Overflow Policy (59 *Federal Register* 18688), the Clean Water Act, and the CSO control requirements of the Illinois Administrative Code. The Combined Sewer Overflow Long Term Control Plan is intended to satisfy Special Condition 10 of the City's May 11, 2006 NPDES permit. According to terms of the permit, the LTCP must include, at minimum, the following elements:

a. Characterization, monitoring, and modeling of the Combined Sewer System
b. Consideration of sensitive areas
c. Evaluation of alternatives
d. Cost-performance considerations
e. Revised CSO operational plan
f. Maximizing treatment at the treatment plant
g. Implementation schedule
h. Post-construction compliance monitoring program; and
i. Public participation

During the initial development of the CSO LTCP, the City focused on CSO control solutions designed to allow no more than 4 overflows in a typical year. The typical year was identified based on a statistical analysis of Peoria precipitation data dating back to 1949. A typical year based level of control is typically applied to communities required to address CSOs. In November 2008 EPA determined that the portion of the Illinois River where Peoria's CSO are located is environmentally "sensitive." This mandates a higher level of protection than included in normal CSO control policy. It means CSOs must be eliminated or relocated, to the extent of community affordability. EPA typically considers a CSO to be eliminated if it does not discharge up to and during a 10-year design storm. Figure 5 illustrates a timeline from 2000 until 2015 of the interaction between the United States Department of Justice and the City of Peoria. Since 2015, the City has continued to negotiate for a lower level of control than the 10-year storm dictated by EPA. To this end, the City has completed a cost-performance evaluation, otherwise known as a knee of the curve analysis.

The CSO drainage area encompasses 8 square miles and less than 50% of the roads in this area needs to be permeable pavement to meet the US EPA mandated 10-year design storm. If the City is able to negotiate a lower level of control, then less than 10% of road area will be required to manage CSOs. This provides the City with a safety factor by having additional road area to utilize if they find more is required to meet the design storm.

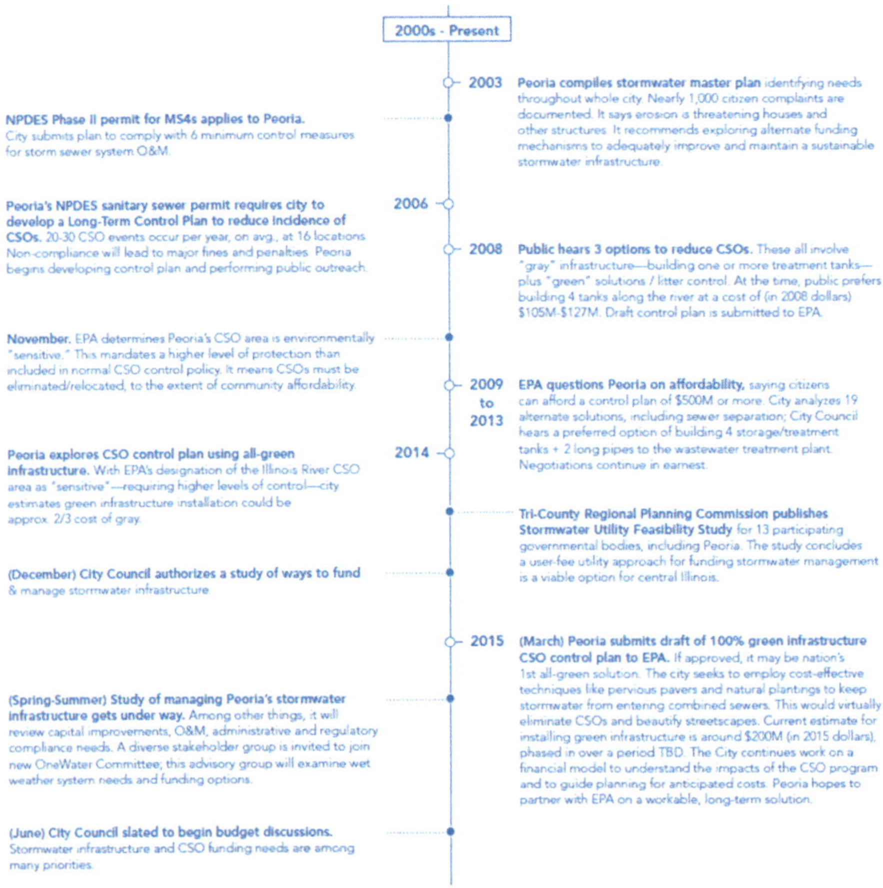

Fig. 5. Regulation-driven evolution of the Peoria green infrastructure plan. (Source: Department of Public Works 2016. Presentation on green infrastructure to USEPA)

6 Cost Performance Evaluation

The CSO Control Policy calls for developing appropriate cost-performance curves to demonstrate the relationship among a comprehensive set of reasonable control alternatives that correspond to a range of controls. The cost-performance evaluation should identify where the increment of pollution reduction achieved in the receiving water diminishes compared to the increased costs. This point is often referred as the knee of the curve. This was among the considerations the City used to help determine the recommended level of service.

Figure 6 provides the knee of the curve analysis for five performance levels ranging from zero to four overflows in the recreational season of the typical year. The curve shows that the knee of the curve is at the one overflow in the typical year level of service. The cost to achieve the performance criteria of one, two, or three overflows in the typical year recreation season is the same for all three criteria. This is due to the similarity between the second, third, and fourth largest storm events during the typical year and that the event controlling GI sizing to prevent overflows varies by sewershed. For the purposes of cost estimating, all green infrastructure was assumed to be pervious pavement. Table 2 shows the estimated capital costs associated with CSO volume reduction in a typical year.

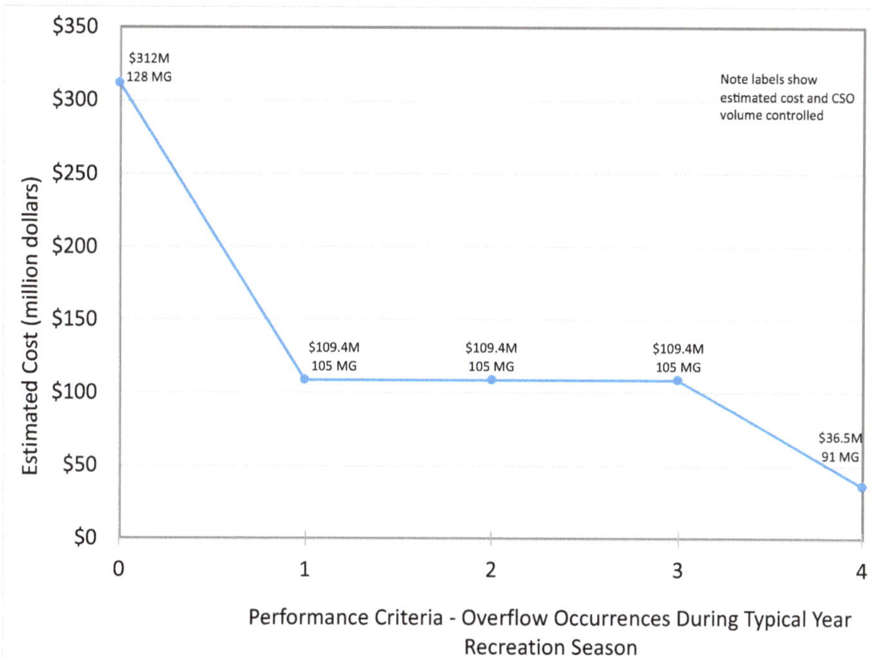

Fig. 6. Cost vs. overflow occurrences performance criteria. (Source: Wood PLC 2018. Wood PLC (2018). Knee of the curve analysis.)

Figure 6 indicates that the cost to achieve one, two, or three overflows in the typical year recreation season is the same due to the similarity between the second, third, and fourth storm events during the typical year recreation season and that event controlling the green infrastructure sizing to prevent overflows varies by sewershed. The water quality standard only applies during the recreation season from May–October since that is the most likely opportunity for bodily contact to the water. All modeling data is based on this assumption (Wood PLC 2014).

Figure 7 provides a direct relationship between the expected cost and CSO volume reduction in millions of gallons.

One measure US EPA uses to estimate the impact, or affordability, of CSO control to a community is the expected residential share of estimated project costs as a percentage of median household income (MHI). If the calculated percentage is 2% or greater, then the burden is considered heavy. The MHI and number of households in

Table 2. Overflow volume reduced w/cost. (Source: Wood PLC 2018. Wood PLC (2018). Knee of the curve analysis.)

Typical year recreation season			Total estimated capital cost ($M)
Number of overflow events	CSO volume (MG)	CSO volume reduced (MG)	
22	128	0	N/A - existing conditions
4	37	91	$37
1	23	105	$109
0	0	128	$312

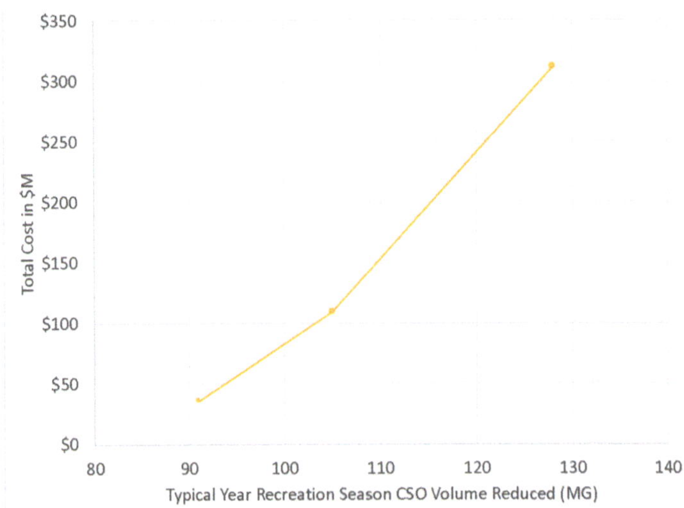

Fig. 7. Expected cost vs CSO volume reduction. (Source: Wood PLC 2018. Wood PLC (2018). Knee of the curve analysis.)

the City of Peoria are obtained from US Census data. The projected burden on residents in the CSO area for the recommended alternative submitted in 2018 was 4.45% based on MHI of $28,569 and an annual residential share of estimated project costs of $1,272.70 ($1,272.70/$28,569 = 4.45%).

7 The Combined Sewer System (CSO) Problem

The City of Peoria has maintained a system of combined sewers and overflows for more than 100 years. Peoria built its first sewers in the late 1800s to carry rainwater and melting snow away from homes, businesses and streets. When indoor plumbing came later, homeowners and business owners hooked their sewage lines to those same sewers, combining stormwater and sewage in one pipe. This was standard practice in many U.S. cities at the time, especially in the Northeast and Midwest.

After the Greater Peoria Sanitary District was formed and the combined sewers were connected to the Riverfront Interceptor sewer and the treatment plant, which was constructed in 1931, the combined sewers only discharged to the river during rainfall events. However, the sewers still retain their ability to overflow when sewage levels get too high. If they didn't have this escape valve, combined sewage would back up into people's basements and streets. In new neighborhoods today, this problem is avoided by building separate sewers for stormwater and sewage.

On average, the City of Peoria currently experiences between 20 and 30 CSO events per year. CSOs can occur throughout the year and can occur with as little as 0.15 inches of rain. CSOs contribute to elevated bacteria levels in the Illinois River. The Illinois River bacteria levels exceed state water quality standards.

The distribution of flow between the overflows to the river and the riverfront interceptor (RI) to the treatment plant occurs at regulator and swirl concentrator structures. Flow from a regulator or swirl concentrator to the interceptor (via a throttle pipe) is limited only by the physical configuration of the regulator or swirl concentrator and throttle pipe, and related hydraulics. Once a certain minimum flow rate is reached, the flow depth in the regulator or swirl concentrator starts to increase due to the hydraulic head required to get the flow into and through the throttle pipe. Once the flow depth in the regulator or swirl concentrator reaches a certain minimum level, flow starts to be diverted into the overflow pipe, starting an overflow.

The GPSD owned RI and associated combined sewer regulators, swirl concentrators (swirls), and throttle pipes in Peoria are not currently maximizing flow to the WWTP. The CSOs start to occur when the flow through the regulators and swirls is about 48 MGD, even though the nominal RI capacity is 80 MGD.

8 Peoria Area Ecosystems and the Clean Water Act

The Illinois River and Peoria Lake within the CSO study area supports a range of activities including boating, jet skiing, fishing, and commercial transportation. One of the required components of the LTCP is a review of the CSO study area for the consideration of sensitive areas (AMEC Foster Wheeler 2010).

According to the US EPA CSO control policy, sensitive areas include:

a. Outstanding National Resource Waters
b. National Marine Sanctuaries
c. Waters with threatened or endangered species or their designated critical habitat
d. Primary contact recreation waters, such as bathing beaches
e. Public drinking water intakes or their designated protection areas
f. Shellfish beds

Literature searches determined that the CSO study area does not contain Outstanding National Resource Waters, National Marine Sanctuaries, waters with threatened or endangered species or their designated critical habitat, or public drinking water intakes or their designated protection areas. A mussel survey was performed and no shellfish beds were found within the CSO study area. No bathing beaches are located within the CSO study area.

Despite this, the US EPA has determined that the entire CSO study area is "sensitive" based on their assessment that primary contact recreation waters are present. However, the majority of recreational activities take place north of the CSO area where conditions are more favorable.

Based on the definition of sensitive areas provided in the USEPA CSO Control Policy and guidance provided by US EPA ("Combined Sewer Overflows, Guidance for Long-Term Control Plan" (EPA 832-B-95-002) September 1995, Section 3.3.3.6), the City believes there are no locations designated as sensitive areas for development of control alternatives and plan implementation.

9 Peoria Sustainability Challenges

Peoria's selection of a 100% green solution presents an opportunity for the City to implement a sustainable way to address the CSO issue. A tank or tunnel option would not provide a long term environmentally or economically sustainable solution. These methods collect and convey impacted water that would still need treatment before discharge to the river. The green solution provides CSO control by keeping storm water out of the system and eliminating the need for treatment. The green solution is a chance to reinvest in the community, protect the environment, and achieve the co-benefits discussed previously.

A hydrologic and hydraulic model was constructed and used to develop the green solution. Flow meter data was collected for both dry weather and wet weather sewer flows in all sewersheds in the CSO area. Smaller parcels or subcatchments were delineated within the sewersheds and modeled to estimate the volume of storm water needed to be managed in each subcatchment to provide the required level of CSO control. Considering the CSO problem on a subcatchment level allows the flexibility to select green infrastructure components that are best suited for a particular area from residential neighborhoods to schools to areas that are more business centered.

10 The 4-Mile Tunnel Option

US EPA determined Peoria's CSO area is environmentally "sensitive." This determination mandated a higher level of protection than included in normal CSO control policy. It means CSOs must be eliminated/relocated, to the extent of community affordability. EPA typically considers a CSO to be eliminated if it does not discharge up to and during a 10-year design storm.

The City argued that such a high level of control is not necessary to meet water quality standards and is beyond what the City can afford. In response, the US EPA proposed a tunnel to convey all flow from the upstream reaches of the CSO area to the GPSD wastewater treatment plant. US EPA used a consultant to prepare a project description and cost estimate, noting that a four mile tunnel was feasible and well within the City's MHI burden capacity noted by US EPA. The estimated cost of the tunnel project was $260 million.

11 Innovation and Sustainable Growth

Faced with escalating CSO management costs due to EPA's insistence on a 10-year design storm, the City began exploring a more cost-effective green infrastructure solution while continuing to negotiate for a lower level of control. The implementation of a green solution would provide the flexibility to incorporate an innovative approach to address wet weather flows on a neighborhood-by-neighborhood basis. Solutions could be incorporated to be complementary with other infrastructure improvements proposed for specific areas. A green solution could also be incorporated into the City's Complete, Green Streets concept.

A green solution would encourage the use of local contractors for initial construction and future operation and maintenance where a gray solution would most likely require outside contractors due to the specialized nature and size of these types of construction projects. The implementation of a green solution would take almost 20 years to complete with the desire to use local labor to complete this work.

12 Solutions Considered, Companies and Other Cities

Various technologies have been reviewed and visits made to multiple communities by the City staff and their consultants to gather knowledge on green ideas and see how they could potentially fit into Peoria's sustainable CSO solution. Visits included a tour of an Eco-Machine greenhouse installation in Vermont and stops in Indianapolis, Philadelphia, Chicago, Seattle, Portland, Denver, and Evansville, IN. First-hand knowledge was gained by observing how installed green infrastructure was incorporated into the surrounding environment, how it was functioning, how it was being used by residents, and lessons learned from local staff that were happy to share their green knowledge.

12.1 Permeable Pavement and Other Green Solutions

The tools available to provide CSO control through green infrastructure are numerous. The wide range of options allow the City design flexibility and encourages stakeholder input. Green infrastructure components include, but are not limited to:

a. Pervious pavers
b. Native landscaping
c. Green, permeable alleyways
d. Community gardens and rain gardens
e. Boulevards with natural plantings
f. Urban planters and vegetative planter boxes
g. Porous asphalt
h. Green roofs

12.2 Gray Versus Green Infrastructure

Construction of gray infrastructure includes components such as tunnels, tanks, piping and treatment equipment. Gray infrastructure typically requires specialized labor to install and operate and has little secondary value. Green infrastructure could be installed and maintained utilizing local labor. In addition to employment opportunities, green infrastructure co-benefits include:

a. Beautification of public rights-of-way
b. Enhancement public health and wellness by:
   - Decreasing pollutants, which can lessen illness from contact or polluted drinking water.
   - Improving air quality.
c. Reduction in potential crime and increase surveillance
d. Mitigation of localized flooding
e. Connections between people and places
f. Safe and accessible transportation infrastructure
g. Ecological health and productivity
h. Neighborhood-scale economic development

12.3 Economic Feasibility

Green infrastructure is a demonstrably more cost-effective solution for Peoria. Green infrastructure can extend the useful life of existing infrastructure. By keeping more water out of the system and soaking it into the ground, not as much demand is put on existing drains and pipes.

Green infrastructure would enable Peoria to involve local businesses—including diverse-owned engineers and contractors—in planning, design, construction, hauling and labor. Local business involvement would also be needed for maintenance and upkeep once green components are installed.

12.4 Timeline

A tentative schedule shows an 18-year time frame for complete implementation of green infrastructure to control for Peoria's CSOs. Proposed progress goals from project start date (to be determined) are 33% complete after six years, 66% complete after 12 years and 100% complete after 18 years.

12.5 The Next Step

The US EPA has recently verbally committed to further discussion on the level of control. The City has recently submitted a memo detailing proposed level of service cost and a knee of the curve analysis that we believe will clearly show performance criteria of one overflow in the typical year provides the maximum benefit for the cost. These discussions are ongoing.

Once the City of Peoria Staff and Department of Justice (DOJ) agree on terms of a Consent Decree the detailed document will have to be written and approved by all parties. This is a legal document that has a number of conditions and stipulations in it. After approval internally, the Consent Decree will be presented to the City Council for approval and then it will go to a Circuit Judge to sit for six months for public comment. After six months is complete and if a Judge is agreeable to the terms in the consent decree it will be certified and approved. That is when the clock starts ticking to implement and the City has to meet the performance criteria or face penalties as outlined in the Consent Decree.

The City Staff will make a recommendation to the Council on how to fund the capital costs. Some of the costs will be covered by the storm water utility fees that are described below. The remaining costs will likely be covered by a blend of bonding (borrowing) and sewer rate increases. The sewer rates could increase to as much as 7 times the current cost to generate the revenue to cover the debt service on the bonds.

While the remaining details of the Consent Decree are being negotiated, the City is continuing to incorporate green infrastructure for CSO control into its road improvement projects and is looking at partnering with private property owners to design and install green infrastructure on underutilized lots. The City is also developing a geographic information system for dynamic green infrastructure planning, tracking, and reporting.

Once the Consent Decree is signed, the City must build enough green infrastructure every year to meet interim overflow reduction milestones and the agreed to level of control at the end of the implementation period, which is expected to be about 18-years from the signing of the Consent Decree.

13 Understanding the Stormwater Utility Fee

The initial capital cost of Permeable Pavers is typically 30% higher than asphalt or concrete pavements and is covered by the City through capital budget. The City operating budget is $24 million and $20–$25 million for capital projects. The CSO will generate between $15–$18 million a year in new infrastructure need that is not covered in the either current budget. In December 2017, the Peoria City Council created a Stormwater Utility to help the City adequately handle wet weather and create innovative solutions for the future. The money raised through this process can only be used to deal with stormwater issues, including the CSO problem. The utility pays for the management, education, planning, inspection, enforcement, protection, control, regulation, use, construction, maintenance and enhancement of the stormwater system. The Stormwater Utility that will cover up to $5 million a year of the new CSO need and the remaining funds will come from either increasing sewer fees or raising the property taxes.

Over 1,500 stormwater utilities are known to exist in the United States, including 22 within the State of Illinois. These numbers have been steadily increasing since the first stormwater utility was created in 1974. Why are stormwater utilities popular? A stormwater utility is considered a fair and equitable approach to generate funds, as it is based on the amount of stormwater runoff each property contributes to the stormwater system. Since the amount of runoff is directly related to the area of impervious surface, each property owner is billed based the square feet of impervious area. The initial Stormwater fee in Peoria 2018 is $3 per billing unit. Residents are encouraged to install one of the so-called best management practices (BPMs)—which are essentially different types of green infrastructure. Such BMPs for stormwater runoff control have been in use for many years (EPA 2014). They include:

13.1 Rain Barrel

These are ideal for both residential properties and smaller businesses. Rain barrels collect water from a roof and store it for later use on gardens or lawns. It is a good way to conserve water and reduce stormwater runoff. Rain barrels can be purchased at local home improvement stores.

13.2 Rain Garden

This is ideal for both residential and commercial properties. Rain gardens are shallow beds planted with perennial plants that collect rainwater. Rain gardens can reduce flooding, absorb pollutants and sustain wildlife. Any size rain garden, even a small one, will make a difference.

13.3 Detention Basins

These are ideal for subdivisions and large developments. Detention ponds are designed to capture, retain and slow the release of stormwater runoff from impervious areas. Detention basins are generally used for flood control and are dry between storms.

13.4 Permeable Pavement

This is ideal for both residential and commercial properties. Permeable pavement allows stormwater to seep into the pavement. This BMP can either help with detention (holds water for a short period of time) or retention (holds water for longer), depending on how the pavement is designed.

The City of Peoria's Stormwater Utility is meant to be a driver to change development behavior so that property owners are aware of how what happens on their property affects the entire City. All of the green solutions listed above (e.g., Rain Barrels, Rain Gardens, Detention Basins, Permeable Pavement) are methods that property owners can use to reduce run-off (Department of Public Works 2018a, c). The City of Peoria is looking to incentivize change by offering reductions of up to 50% for installing these solutions (Department of Public Works 2017a, b). A 50% credit or incentive can be substantial for large commercial properties that typically have the majority of the run-off due to the size of their developments. For example, a typical car dealer in Peoria will have a yearly bill of $30,000 for the Stormwater Utility but if the dealership installs and proves with certified engineering any stormwater BMP the bill can be reduced by $15,000 per year.

14 Conclusions

The City of Peoria has embarked on an ambitious program to solve its CSO problems by opting to go with green solutions rather than gray water initiatives. The City's imaginative green solutions will set new standards for the rest of the USA to follow and emulate. The proposed solution has already been initiated with pilot projects that show great promise for the future. Although a final agreement has not yet been reached with the US Department of Justice and the US Environmental Protection Agency, it is anticipated one will be reached in the next few months. The proposed green solution will save the City money over time because it reduces water treatment cost, defers capital investment, improves infrastructure lifecycle and performance, reduces regulatory violations, reduces energy and reduces waste.

The discussions between the City of Peoria, the Department of Justice, and the USEPA are critical in arriving at a mutually agreeable decision regarding the level of control. Currently, less than 50% of the roads in the CSO drainage area need to be permeable pavement to meet the US EPA mandated 10-year design storm. If the City is able to negotiate a lower level of control, then less than 10% of road area will be required to address the CSO problem. This provides the City to position itself to meet potentially additional stringent regulations in the future.

The City has already developed a master plan to solve its CSO problems and introduced new fees to help generate the funds required for implementing its proposed green solution. The new fees went into effect in June 2018. The level of fee incurred by resident and businesses can be reduced using incentives to help reduce overflow by reducing impervious surfaces associated with properties owned by residents.

Acknowledgments. Special thanks to Prof. Krishnanand Maillacheruvu for reviewing the paper and providing help in its preparation and to Prof. David Spelman for reviewing it.

References

AMEC Foster Wheeler: Peoria CSO Long-Term Control Plan. AMEC Foster Wheeler, London (2010)

Brattebo, B.O., Booth, D.B.: Long-Term Stormwater Quantity and Quality Performance of Permeable Pavement Systems. Water Res. **37**, 4369–4376 (2003). https://doi.org/10.1016/s0043-1354(03)00410-x

Department of Public Works: Funding a Better Tomorrow. City of Peoria, Peoria (2017a)

Department of Public Works: Stormwater Utility Credit Manual and Grant Program. City of Peoria, Peoria (2017b)

Department of Public Works: Wet Weather Management Informational Packet. City of Peoria, Peoria (2017c)

Department of Public Works: Co-benefits of Green Infrastructure. City of Peoria, Peoria (2018a)

Department of Public Works: CSO History and Summary. City of Peoria, Peoria (2018b)

Department of Public Works: Sustainable Green Solutions to the City of Peoria Combined Sewer Overflow. City of Peoria, Peoria (2018c)

EPA: Field Evaluation of Permeable Pavements for Stormwater Management, Olympia, Washington. Document No. EPA-841-B-00-005B (2000)

EPA: National Menu of Stormwater Best Management Practices. Pervious Concrete Pavement. EPA, Washington, D.C. (2009)

EPA: Permeable Pavers. Stormwater Management Best Practices. EPA, Washington, D.C. (2010)

EPA: Greening CSO Plans: Planning and Modeling Green Infrastructure for Combined Sewer Overflow (CSO) Control. United States Environmental Protection Agency. Publication No. 832-R-14-001 (2014)

Greeley and Hansen: Greeley and Hansen/Rand & Associates Inc. Report on Combined Sewer Overflow Control Plan (1985)

Hansen, K.: IS-131: Porous Asphalt Pavements for Stormwater Management, p. 16. National Asphalt Pavement Association, Lanham (2008)

Volder, A., Watson, T., Viswanathan, B.: Potential use of pervious concrete for maintaining existing mature trees during and after urban development. Urban For. Urban Green. **8**(4), 249–256 (2009). https://doi.org/10.1016/j.ufug.2009.08.006

Wood PLC: Wood PLC/Symbiont for the City of Peoria: Existing Conditions Model Report (2016)

Wood PLC: Knee of the Curve Analysis. Report to City of Peoria (2018)

Long-Term Strategies for Sustainable Funding of Infrastructure in the USA

Amir Al-Khafaji[1(✉)], Dan Meckes[2], and Dan Gallagher[3]

[1] Bradley University, Peoria, IL, USA
amir@fsmail.bradley.edu
[2] Crawford, Murphy & Tilly, Kansas City, MO, USA
[3] National Asphalt Pavement Association, Springfield, IL, USA

Abstract. The majority of the money spent on infrastructure in the USA go directly to fund transportation systems and projects. The Departments Of transportation (DOT) in each of the fifty States set performance targets for each program within each State. The US Department of Transportation's 30-year framework for the future, identified trends and choices for our transportation system. Some of the most pressing issues include population growth and pervasive congestion across all modes of transportation. As policy makers and industry leaders, the challenge is to identify a sustainable, equitable, and flexible funding solution that enables the right capital investments for our desired future. Transportation funding plans incorporating Motor Fuel Tax (MFT) increases were passed in eight states. Many of these plans contain a combination of transportation related fee increases, including increased vehicle registration fees, taxes on alternative fuel or electric vehicles, and indexing fuel taxes to inflation. However, MFT is a temporary solution because of electric cars, fuel efficiencies mandated by the US government on car manufacturers and demographics. This paper offers long-term funding options for a reliable transportation system in the USA. Both traditional and innovative funding options are proposed for the short-term while long-term solutions are developed. Emerging trends across the nation, as well as internationally, are focused on effective funding solutions for transportation infrastructure. The authors content that any solution must be sustainable, equitable, flexible and reflect future driving trends. States are increasingly exploring user fees, dynamic priced managed lane systems, toll roads, public–private-partnerships, and Mileage-based User Fee (MbUF) systems as solutions to the infrastructure funding inadequacies. It is the right time now to raise the motor fuel tax and index it to inflation, while utilizing other options to supplement and plan for long-term sustainable funding of infrastructure; so that the US will continue to compete economically at the global level.

Keywords: USA · Transportation · Motor fuel tax · Mileage-based user fee Trends · Infrastructure

M. Shehata and F. Rodrigues (Eds.): GeoMEast 2018, SUCI, pp. 87–105, 2019.
https://doi.org/10.1007/978-3-030-01905-1_5

1 Overview

At the earliest stages of United States history, the founding fathers understood the importance of the federal government's role in the development of a strong transportation system. Article 1, Section 8 of the U.S. Constitution states that it is the responsibility of the federal government to "establish post offices and post roads" (U.S. Const. art. 1, § 8) as shown in Fig. 1.

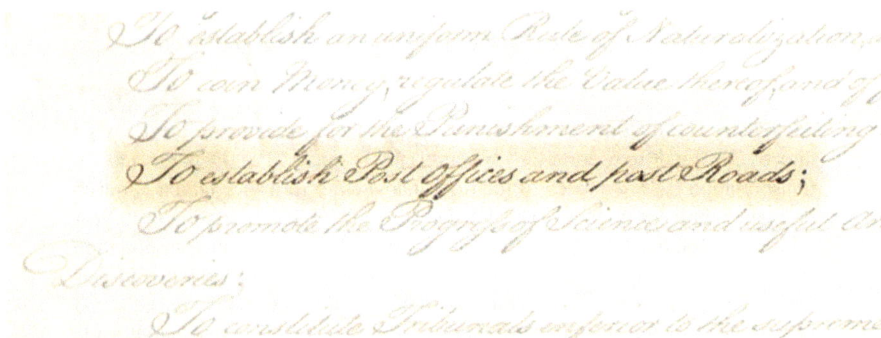

Fig. 1. The U.S. Constitution establishes post offices and subsequent transportation systems as a right of citizens.

It is truly remarkable that the founding fathers of the United States understood early that the economic vitality and growth of their newly formed nation is dependent on the quality of its transportation systems and infrastructure.

Americans recently celebrated the sixtieth anniversary of the Interstate Highway System, which is one of the proudest achievements of twentieth century America. The Interstate Highway System is named after President Dwight D. Eisenhower, who championed its formation. However, the US transportation system has begun to deteriorate across all modes. This deterioration has diminished the ability of the transportation system to support America's growth. Prior to the 2008 recession, bottlenecks in all transportation modes had begun to stall and compromise both the quality of people's lives and America's global competitiveness.

2 History of U.S. Transportation Infrastructure

In order to appreciate America's transportation infrastructure, it is important to understand its history. The following Pfeiffer (2006) research provide summarizes for the U.S. National Archives. In 1953, Dwight D. Eisenhower was elected as the 34th President of the United States. One of his top priorities was to secure legislation for an interstate highway system. Its construction was authorized by the Federal Aid Highway Act of 1956. The network has since been extended and, as of 2013, it had a total length of 47,856 miles (77,017 km). Several events in President Eisenhower's life before becoming elected motivated his policy and direction regarding America's interstate highway system. In 1919, as a Lieutenant Colonel in the army, Eisenhower was involved in a transcontinental convoy across the United States. Based on this arduous journey, he believed that extended trips by trucks through the middle and western parts of the country were impractical until roadways were improved. Later, during World War II, Eisenhower, saw the benefit of a good roadway system, as the autobahn was used to move allied troops quickly and efficiently through Germany to aid in the defeat of the Nazi army.

Based on personal experiences, President Eisenhower led efforts to fund infrastructure improvements, and only a year after his election, Congress passed the Federal Highway Act of 1954, which authorized $175 million for an interstate highway system to be distributed on a 60-40 federal- state basis. President Eisenhower did not believe this was enough and continued to lobby for more federal involvement, including through his state of the union speech in 1956. This time, the Bureau of Public Roads released a book, "General Location of National System of Interstate Highways", which showed where interstates would be located in and around the nation's largest metropolitan areas.

Opponents to the funding mechanism in 1955 agreed and supported a gas tax in 1956. Congress went back to work and produced legislation that called for 90% federal funding, with revenue being generated from the federal gasoline tax. The final version also reflected a compromise on how the funds would be apportioned among the states, and it contained provisions on uniform design standards, inclusion of existing toll roads, and the wage rate to be paid on federal construction projects (Pfeiffer 2006).

According to the Batic Institute, transportation infrastructure has traditionally been financed primarily through a combination of state and local taxes and fees; and—for major projects—federal grants funded by national motor fuels taxes. In their Transportation Funding and Financing report, they noted that (Batic Institute):

i. These resources are typically combined to fund projects on a "pay-as-you-go" basis, meaning that projects have often been built in phases or increments as funds become available over a period of years. Project funding has been tied closely to federal and state cash management policies, with nearly exclusive responsibility for the process vested in state and local public transportation agencies.
ii. Motor fuel and vehicle taxes are deposited into the Federal Highway Trust Fund (HTF) from which Federal-aid grants are provided, typically on an 80-20 federal-to-state matching ratio. However, state and local funding provides the majority of revenue available to highway projects through state motor fuels and vehicle taxes, tolls, local property taxes, sales taxes and other special assessments, general fund appropriations and bond-issue proceeds. Tax-exempt municipal bond issues have a long history of helping to finance public infrastructure, including local transportation improvements.
iii. Federal aid for transit projects, through formula funds and discretionary grants, is also provided by the HTF, as well as from general fund appropriations. Again, most of the funding is non-federal, with a majority derived from local sources including various taxes, public funds, and fare revenues (Batic Institute n.d.).

Over the last several decades, the federal surface transportation programs and funding have been legislated in multi-year transportation bills. These bills provide the direction and programs for highways and transit on the federal system, as well as the federal share of the funding for the transportation projects. Transportation bills have historically been passed for 2–6 years; however, as one bill expires, it is not uncommon for those programs and funding levels to be extended, often numerous times, until the next long-term bill can be passed.

The primary revenue source for the HTF comes from the Federal Gas Tax. It was increased to 18.4 cents per gallon in 1993 and remains there today. In recent years, the solvency of the HTF has become an issue. The lack of an increase in the federal gas tax, in tandem with an increase in average vehicle fuel efficiency, will eventually lead to the HTF's inability to meet the funding demands of our transportation infrastructure. In recent years, transfers from the General Fund have helped to maintain the solvency of the HTF; but a long-term solution is needed and must be a topic of discussion for every transportation bill.

3 The Sustainable Funding of Infrastructure Group ($Fig)

The Sustainable Funding of Infrastructure Group ($FIG) was established to promote investment in infrastructure and explore creative funding options. The $FIG has held numerous meetings and discussions with members of the US Congress from both the Republican and Democratic parties to discuss possible funding and implementation options for infrastructure challenges in the United States. At the 2014 Executive Summit held in Peoria, Illinois in conjunction with the Innovations in Construction, Asphalt and Transportation (ICAT) conference, Congresswoman Cheri Bustos (D) and Congressman Rodney Davis (R), from USA House Committee on Transportation and Infrastructure, attended and participated in the discussion. At this meeting, many funding options were examined. These discussions included raising and indexing the gas tax, vehicles miles travelled, tolling, Public Private Partnerships, vehicle registration fees, sales tax and other potential options. At the meeting, Congressman Davis voiced his reluctance for a VMT tax, saying that it would adversely impact his constituents because of his rural district.

In April 2015, Congressman Darin LaHood (R) attended the Executive Summit and provided input on the discussions being held in Congress regarding FAA reauthorization and the FAST Act. Congressman LaHood also participated in the discussion regarding the need for a long-term strategy to fund the nations surface transportation needs.

Dr. Amir Al Khafaji, founder of ICAT and $FIG, has had numerous meetings with Congressman LaHood and had Congressman LaHood speak at the fall planning meeting in the fall of 2017 to discuss infrastructure funding, gas tax and tax reform.

Members of $FIG continue to meet with USA House and Senate representatives to discuss the need to fix the Highway Trust Fund (HTF) by providing new and sustainable revenue sources for transportation programs. On April 17, 2018, $FIG chair, Dan Meckes, and other members of $FIG and the American Council of Engineering Companies (ACEC), met with USA Senator Dick Durbin, and staff members from the offices of Senator Roy Blunt, Senator Clare McCaskill, and Congressman Sam Graves, who serves as Chairman of the House Subcommittee on Highways and Transit, to discuss the status of legislation to fix the HTF. The consensus of both Republicans and Democrats of both the House and the Senate was that they did not foresee any legislation or authorization of additional surface transportation funding until after the mid-term elections or possibly not until reauthorization of the FAST Act, which would be in late 2019 or early 2020. Members of Congress certainly understand the need for additional revenue and are seeking creative methods to expand financing options to leverage additional state, local and private investment, but there does not seem to be the political will at this time to increase federal taxes to support the need for additional revenue.

4 Transportation Funding Bills in the U.S.A.

Over the last two decades, the US congress passed several significant highway bills to fund transportation and infrastructure across the USA. These included

i. Transportation Equity Act for the 21st Century (TEA-21), 1998
ii. Safe, Accountable, Flexible, Efficient Transportation Equity Act: A Legacy for Users (SAFETEA-LU), 2005
iii. Moving Ahead for Progress in the 21st Century Act (MAP-21), 2012
iv. Fixing America's Surface Transportation Act (FACT Act), 2015

A detailed discussion of the various US bills passed by congress is beyond the scope of this paper. However, a summary of these bills are outlined below.

4.1 The Transportation Equity Act for the 21st Century (TEA-21)

This bill was signed into legislation by President Clinton on June 9, 1998. It authorized the federal surface transportation programs for highways, highway safety, and transit for a 6-year period from 1998 to 2003. The priorities of TEA-21 included improving safety, protecting public health and the environment, and creating opportunities for all Americans.

a. Record levels of investment for highways and transit (a guaranteed $198 billion)
b. Investment in highways, transit, intermodal projects and technologies—such as Intelligent Transportation System—with flexibility for states and locals in how the funds were used.
c. Incentives to increase seat belt usage and state adoption of the 0.08 blood alcohol content standards to reduce drunk driving
d. Improved safety through programs for pipeline safety, railroad crossings improvements, and truck safety programs
e. Supported environmental improvements including:
f. Expanded the Congestion Mitigation and Air Quality Improvement program and the Transportation Enhancements program and continued programs for National Scenic Byways and bike, pedestrian and recreational paths

g. Contained provisions to develop clean, fuel efficient trucks
h. Increased tax-free transit benefits to encourage ridership
i. Provided opportunities to move individuals from welfare to work
j. Continued the Disadvantage Business Program and provided transportation workers with labor protections

The key components of the TEA-21 bill included guaranteed a level of federal funding for the 6-year period, through 2003, totaling $198 billion. It also set the annual floor for highway funding, was keyed to the receipts in the highway trust fund—which was the highest source of funding for the bill and is itself highly funded by the federal motor fuel tax—and fixed transit funding. Finally, it extended all highway user taxes at the same rates as when the bill was enacted

4.2 The Safe Accountable, Flexible, Efficient Transportation Equity Act

This bill was signed into law by President George W. Bush on August 10, 2005. This $244.1 billion, 4-year transportation bill was set to expire on September 30, 2009. However, the funding formulas were extended ten times until its replacement by MAP-21 in 2012. SAFETEA-LU sought to improve safety, reduce traffic congestion, improve the efficiency of freight movement, increase intermodal connectivity and protect the environment, all while growing the transportation infrastructure. Key Components of SAFETEA-LU included

a. Established the Highway Safety Improvement Program with an emphasis on reducing fatalities
b. Created an Equity Bonus Program that, among other things, set a minimum on a states' return on investment to the Highway Trust Fund
c. Promoted private-sector participation in financing projects using private activity bonds and additional flexibility for the use of tolling and broader use of TIFIA and SIB loans
d. Allowed flexibility for the states in use of congestion pricing and promoted real-time management to address congestion issues
e. Instituted three tolling pilot programs to help manage congestion and finance improvements
f. Contained several environmental initiatives
g. Extended the users fees in place at the time it was enacted, keeping the 18.4 cent federal gas tax from 1993 and maintaining the Highway Trust Fund as the primary funding source for surface transportation.

4.3 The Moving Ahead for Progress in the 21st Century Act (MAP-21)

This bill was signed by President Obama on July 6, 2012. This was the first long-term transportation bill enacted since 2005. The legislation provided funding and authorization for federal surface transportation spending, including highways and transit. The bill passed both the House and the Senate by wide margins, 373-52 and 74-19, respectively. This 2-year bill provided for a $105 billion investment in transportation, which did not significantly change the funding levels from TEA-21. But, MAP-21 did provide much-needed funding and include several major reforms to existing policy and programs. Key components of MAP-21 included the following:

a. Introduced performance management to highway programs, with a focus on national transportation goals
b. Reduced several programs by two-thirds by consolidating existing programs
c. Worked to shorten delivery time for transportation projects
d. Streamlined the funding for bikes and pedestrian trails through consolidation into Transportation Alternatives—one, broader program—which provided half of its funding to states and half to Metropolitan Planning Organizations (MPO)
e. Provided that a National Freight Policy be developed
f. Expanded the National Highway System by including principle arterials not previously included
g. Contained provisions to reform tolling on new highways and to expand and repair existing highways
h. Provided financial assistance on certain projects and increased private sector investment through the Transportation Infrastructure Financing and Innovations Act (TIFIA)
i. Extended the users fees once again, keeping the 18.4 cent federal gas tax from 1993 and maintaining the Highway Trust Fund as the primary funding source for surface transportation

4.4 The Fixing America's Surface Transportation Act (FAST Act)

This bill was signed into law by President Obama on December 4, 2015. It was easily passed in the House, by a vote of 359-65, and in the Senate, by a vote of 83-16. The FAST Act was a 5-year bill providing $305 billion for funding surface transportation from 2016 through 2020. The FAST Act was intended to build on the streamlining, performance-based and multimodal aspects of MAP-21. This bill, again, did not increase the user fees, and maintained the Highway Trust Fund as its primary source of funds, keeping the federal gas tax at the 1993 level of 18.4 cents per mile. Key components of the FAST Act are

a. Continues the performance goals, measures and targets, along with the long-range planning efforts in MAP-21
b. Reinforces the freight performance targets and establishes actions to meet targets
c. Works to accelerate Project Delivery by adding flexibility to increase efficiency in environmental approvals and refining existing authorities.
d. Builds on the FHWA's "Everyday Counts" program
e. Establishes a National Highway Freight Program to improve the movement of freight on the National Highway Freight Network and requires states to have a State Freight Plan to compete for these funds
f. Establishes FASTLANES grants, a $4.5 billion competitive grant program to provide funding for projects of national or regional significance
g. Places an emphasis on multimodal freight policy and planning

h. Extends the users fees once again, keeping the 18.4 cent federal gas tax from 1993 and maintaining the Highway Trust Fund as the primary funding source for surface transportation

An infrastructure system launched with a bold and historic vision in the 1950s, followed by continuous transportation legislation and foresight, has provided the United States with a world-class transportation system. However, the future maintenance and expansion of the system will continue to be a challenge, and funding the system is of utmost concern. Traditional federal funding sources, such as the Highway Trust Fund (HTF), which pays for our road and transit system upkeep, are stagnant at best, and the prospects for future funding are uncertain, given gasoline consumption fluctuation. A reduction in vehicle miles traveled, an increase in more fuel-efficient vehicles, and a lack of a national gas tax increase ongoing for over 20 years (meaning decreased purchasing power due to inflation), has led to the decline of the amount of revenues available for transportation improvements. According to the U.S. Energy Information Administration, combined federal and state gas taxes as a percentage of the cost of a gallon of gas dropped from 28% in 2000 to 12% in 2013 (Auxier 2014). The Gas Tax per Gallon for various countries demonstrates that the Gas Tax in the USA is extremely low when compared with other industrialized nations as shown in Fig. 2.

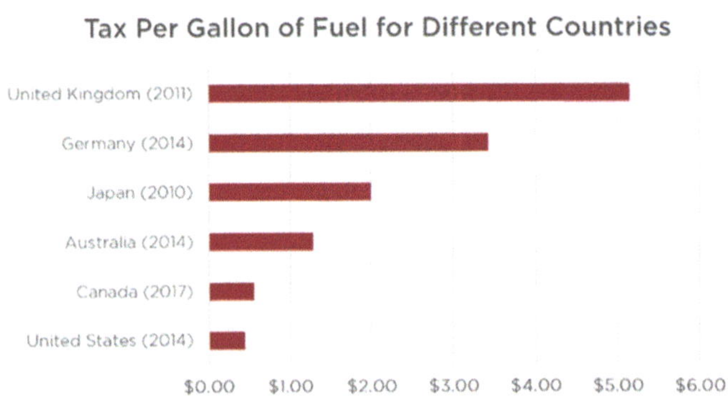

Fig. 2. Investment in transportation by other nation's fuel tax; Sources (ENO Center for Transportation and the Federal Highway Administration, Ontario Ministry of Finance).

The recent decline in gasoline prices has resulted in an increase in gasoline consumption, with a corresponding increase in gas tax revenue. Nationwide motor fuel consumption grew by 3% from January through August 2015, compared to the same period in 2014, according the Federal Highway Administration's data (Holan 2016). In addition, according to Holan (2016) of ARTBA, "Total fuel volume…was 94.8 billion gallons for the first eight months of 2015, also a record for the last 11 years…Volume dropped to a low of 60 billion gallons from January to August 2007." According to Dr. Alison Premo Black, American Road and Transportation Builder's (ARTBA's) chief economist, "This is in line with the increases that we've seen in vehicle miles traveled data. As the economy recovers and more people are working again, they are also driving more, and gasoline consumption has increased." (Holan 2016).

The impact of infrastructure deterioration in the USA will impact its ability to compete at the global level with other industrialized nations. The Global Competitiveness Index lists the USA at number 10 in the world below Portugal and Austria as shown in Fig. 3. This trend cannot continue if the USA is to maintain its economic growth and competitive edge at the global level.

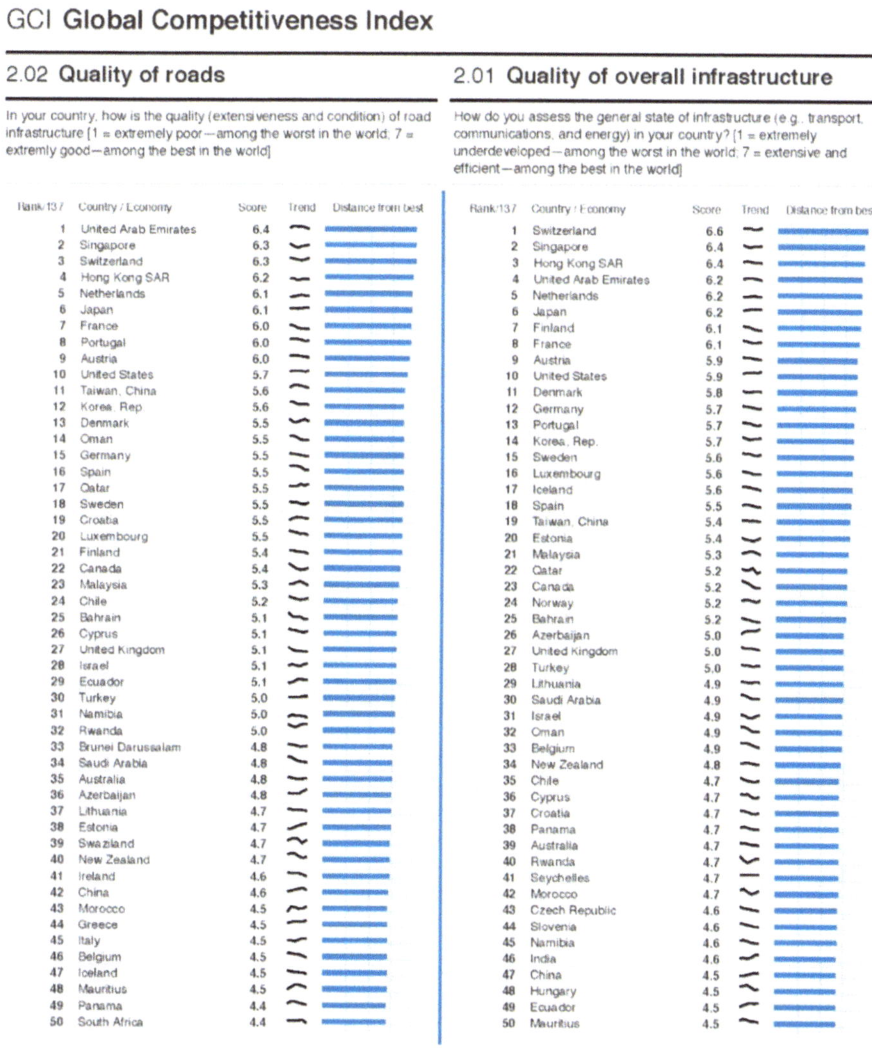

GCI **Global Competitiveness Index**

2.02 **Quality of roads**

In your country, how is the quality (extensiveness and condition) of road infrastructure [1 = extremely poor—among the worst in the world; 7 = extremly good—among the best in the world]

Rank/137	Country / Economy	Score
1	United Arab Emirates	6.4
2	Singapore	6.3
3	Switzerland	6.3
4	Hong Kong SAR	6.2
5	Netherlands	6.1
6	Japan	6.1
7	France	6.0
8	Portugal	6.0
9	Austria	6.0
10	United States	5.7
11	Taiwan, China	5.6
12	Korea, Rep.	5.6
13	Denmark	5.5
14	Oman	5.5
15	Germany	5.5
16	Spain	5.5
17	Qatar	5.5
18	Sweden	5.5
19	Croatia	5.5
20	Luxembourg	5.5
21	Finland	5.4
22	Canada	5.4
23	Malaysia	5.3
24	Chile	5.2
25	Bahrain	5.1
26	Cyprus	5.1
27	United Kingdom	5.1
28	Israel	5.1
29	Ecuador	5.1
30	Turkey	5.0
31	Namibia	5.0
32	Rwanda	5.0
33	Brunei Darussalam	4.8
34	Saudi Arabia	4.8
35	Australia	4.8
36	Azerbaijan	4.8
37	Lithuania	4.7
38	Estonia	4.7
39	Swaziland	4.7
40	New Zealand	4.7
41	Ireland	4.6
42	China	4.6
43	Morocco	4.5
44	Greece	4.5
45	Italy	4.5
46	Belgium	4.5
47	Iceland	4.5
48	Mauritius	4.5
49	Panama	4.4
50	South Africa	4.4

2.01 **Quality of overall infrastructure**

How do you assess the general state of infrastructure (e.g., transport, communications, and energy) in your country? [1 = extremely underdeveloped—among the worst in the world; 7 = extensive and efficient—among the best in the world]

Rank/137	Country / Economy	Score
1	Switzerland	6.6
2	Singapore	6.4
3	Hong Kong SAR	6.4
4	United Arab Emirates	6.2
5	Netherlands	6.2
6	Japan	6.2
7	Finland	6.1
8	France	6.1
9	Austria	5.9
10	United States	5.9
11	Denmark	5.8
12	Germany	5.7
13	Portugal	5.7
14	Korea, Rep.	5.7
15	Sweden	5.6
16	Luxembourg	5.6
17	Iceland	5.6
18	Spain	5.5
19	Taiwan, China	5.4
20	Estonia	5.4
21	Malaysia	5.3
22	Qatar	5.2
23	Canada	5.2
24	Norway	5.2
25	Bahrain	5.2
26	Azerbaijan	5.0
27	United Kingdom	5.0
28	Turkey	5.0
29	Lithuania	4.9
30	Saudi Arabia	4.9
31	Israel	4.9
32	Oman	4.9
33	Belgium	4.9
34	New Zealand	4.8
35	Chile	4.7
36	Cyprus	4.7
37	Croatia	4.7
38	Panama	4.7
39	Australia	4.7
40	Rwanda	4.7
41	Seychelles	4.7
42	Morocco	4.7
43	Czech Republic	4.6
44	Slovenia	4.6
45	Namibia	4.6
46	India	4.6
47	China	4.5
48	Hungary	4.5
49	Ecuador	4.5
50	Mauritius	4.5

Source World Economic Forum, Executive Opinion Survey. For more details, refer to Appendix C of this Report

Fig. 3. Global competitiveness index of U.S. infrastructure (World Economic Forum).

5 Current Discussions of Infrastructure Improvement

As shown below, the U.S. is rapidly falling behind other nations on investment in transportation (Craighead and Manzo 2016), which puts the U.S. at a competitive disadvantage in the global economy. The World Economic Forum's latest Global Competitive Index ranked the U.S. 10th for quality of overall transportation infrastructure and quality of roads, well below other advanced economies, as seen in World Economic Forum (2018) as shown in Fig. 3.

The federal government plays an important role in oversight and funding of the nation's transportation infrastructure. The nation needs a strong federal partner that provides predictable funding to allow state transportation agencies to effectively plan, develop, construct, and maintain transportation projects. At the same time, the need for increased state investment is clear. States across the nation are stepping up to increase funding for critical infrastructure improvements. An updated report by the ARTBA Transportation Investment Advocacy Center™ shows that numerous states have passed legislation to support transportation investment since 2017. Figure 4 shows a map of motor fuel tax rates for each of the states in the USA.

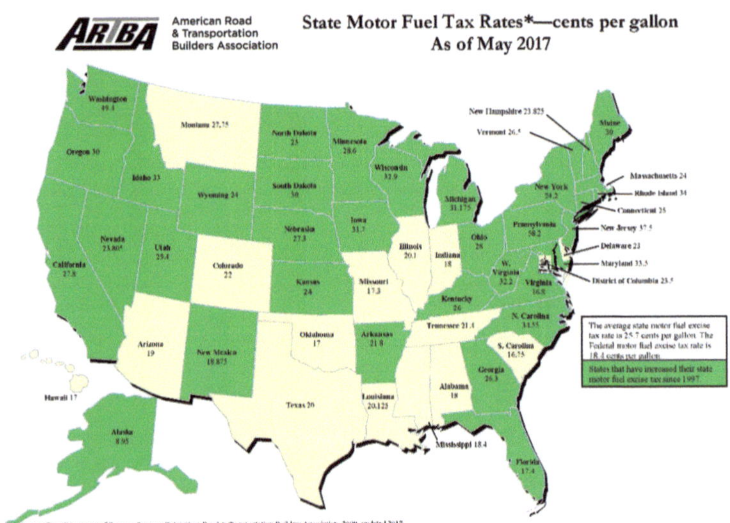

Fig. 4. Fuel tax rates by state (ARTBA).

Clearly, the United States has a history of building and funding programs resulting in a world-class transportation system. However, as the United States looks to the future of transportation, funding continues to be the greatest challenge, with a user-based system reliant on a gas tax that has not been increased since 1993 or indexed to keep pace with inflation. A long-term sustainable source of revenue is needed to maintain and improve a system that is well beyond its design life. Legislative action is now in need of fundamental overhaul. Without a transformed funding mechanism and an accompanying increase in investment, America's infrastructure will continue to decay.

6 The ASCE Infrastructure Report Card

Current conditions, as reported by the American Society of Civil Engineers (ASCE), reflect the lack of investment over the last several decades. The 2017 Infrastructure Report Card reveals that we have made some incremental progress toward restoring our nation's infrastructure, but it has not been enough. As in 2013, America's cumulative GPA is a D + , indicating poor or at-risk infrastructure across the board. The 2017 grades range from a B—indicating "adequate for now" condition—for rail to a D—indicating at risk and nearly critical condition—for transit, underscoring the clear impact of investment, or lack thereof, on infrastructure condition grades. ASCE's (2017) full report card can be seen below:

AVIATION: D
U.S. airports serve more than two million passengers every day. Airports struggle to keep up with investment needs, creating an anticipated $42 billion funding gap between 2016 and 2025.

ENERGY: D+
Most electric transmission and distribution lines were constructed in the 1950s and 1960s with a 50-year life expectancy. Additionally, the 640,000-plus miles of high-voltage transmission lines in the lower 48 states' power grids are at full capacity. Energy infrastructure is undergoing increased investment to ensure long-term capacity and sustainability.

INLAND WATERWAYS: D
The United States' 25,000 miles of inland waterways and 239 locks form the freight network's "water highway." This intricate system, operated and maintained by the U.S. Army Corps of Engineers, supports more than half a million jobs and delivers more than 600 million tons of cargo each year, about 14% of all domestic freight.

PORTS: C+
The United States' 926 ports are essential to the nation's competitiveness, serving as the gateway through which 99% of overseas trade passes. Ports are responsible for $4.6 trillion in economic activity, roughly 26% of the U.S. economy. As ships get bigger, congestion at landside connections to other components of the freight network increasingly hinders ports' productivity.

RAIL: B
For more than 150 years, the rail network has been a critical component of the U.S. transportation system and economy. Today, it carries approximately one-third of U.S. exports and delivers five million tons of freight and approximately 85,000 passengers each day.

BRIDGES: C+

The U.S. has 614,387 bridges, almost four in 10 of which are 50 years or older. In 2016, 9.1%—or 56,007—of the nation's bridges were structurally deficient, and on average, there were 188 million trips across a structurally deficient bridge each day. While the number of bridges considered structurally deficient is decreasing, the average age of America's bridges is increasing, and many of the nation's bridges are approaching the end of their design life. The most recent estimate puts the nation's backlog of bridge rehabilitation needs at $123 billion.

DRINKING WATER: D

Many of these pipes were laid in the early to mid-20th century, with a lifespan of 75–100 years. While water consumption is down, there are still an estimated 240,000 water main breaks per year in the United States, wasting over two trillion gallons of treated drinking water. According to the American Water Works Association, an estimated $1 trillion is necessary to maintain and expand service to meet demands over the next 25 years.

ROADS: D

More than two of every five miles of America's urban interstates are congested, and traffic delays cost the country $160 billion in wasted time and fuel in 2014. After years of decline, America's roads have also become more dangerous. Traffic fatalities increased by 7% from 2014 to 2015, with 35,092 individual deaths.

TRANSIT: D−

Transit in America continues to grow, carrying 10.5 billion trips in 2015 and adding new lines and systems every year. Despite increasing demand, the nation's transit systems have been chronically underfunded, resulting in aging infrastructure and a $90-billion rehabilitation backlog.

WASTEWATER: D+

The nation's 14,748 wastewater treatment plants are the most basic and critical infrastructure systems for protecting public health and the environment. Years of treatment plant upgrades and more-stringent federal and state regulations have significantly reduced untreated releases and improved water quality nationwide. It is expected that more than 56 million new users will be connected to centralized treatment systems over the next two decades, requiring at least $271 billion to meet current and future demands.

Clearly, the scores outlined above for US infrastructure earned the U.S. an overall D+ condition rating in 2013 and 2016. Without significant changes in funding and approaches to caring for our infrastructure, these conditions will continue.

7 Current Discussions of Infrastructure Improvement

The current administration has discussed the need to improve the infrastructure in the United States. According to Andrew Garbutt, KPMG's U.S. Lead Infrastructure Principal (2018), President Donald Trump's "broad infrastructure financing and policy blueprint includes a $200-billion program that aims to incentivize a larger, $1.5-trillion investment." The following covers a summary of the key elements of this program, as highlighted by Garbutt:

Considering just how realistically individual states and municipalities can respond —knowing that the federal government is contributing just $200 billion to this 10-year program—will be key to this program's successful implementation and sustainability. To achieve the goal of $1.5 trillion in infrastructure investments over the next decade, the bulk of funding to rebuild roads, bridges, waterways, energy projects, rural infrastructure, public lands and other projects needs to come from state and local governments or the private sector, based on the proposed outline for Rebuilding Infrastructure in America.

This new game plan would significantly alter, if not reverse, the long-standing federal public policy of Washington contributing most of the funding needed for national infrastructure priorities. Instead, it will require states, municipalities and private investment to up the ante as never before. Some highlights of the infrastructure proposal are as follows:

Incentive Program

$100 billion is dedicated to encouraging increased state, local, and private investments by awarding incentives to project sponsors who demonstrate innovative funding strategies. The administration hopes this will assure long-term performance while modernizing infrastructure project technology, delivery, efficiency, and accountability.

Rural Infrastructure Program – $50 billion is earmarked for rural infrastructure needs and challenges, primarily offering states a block grant formula to spend as they see fit. States can petition for additional funds based on performance criteria.

Transformative Projects Program – $20 billion is dedicated to funding ambitious, exploratory, or groundbreaking projects proposed among key areas such as transportation, drinking water, energy, commercial space, and broadband sectors.

Financing Programs

$20 billion will go toward expanding existing financing programs such as the Transportation Infrastructure Finance and Innovation Act (TIFIA), Railroad Rehabilitation & Improvement Financing (RRIF), Water Infrastructure Finance and Innovation Act (WIFIA), and the United States Department of Agriculture (USDA) Rural Utilities Service. The goal is to increase the capacity of federal credit programs to fund investments and broaden eligibility for the use of Private Activity Bonds. TIFIA, for example, would be expanded to allow eligibility for airports, ports, and non-federal waterway projects.

Federal Capital Revolving Fund

$10 billion is set aside to establish a revolving fund to finance purchases, construction, or renovation of federally owned property. The goal is to make it easier for federal agencies to buy real estate needed for federal purposes, rather than relying on the more-expensive option of leasing.

Public Lands and Federal Assets

Proposed statutory changes will direct revenue generated from public lands toward capital costs and maintenance needs related to infrastructure on public lands. It also allows for the sale or lease of federally owned assets.

Improving Infrastructure Permitting
This proposal will expedite the federal permitting process for projects by creating a "One Agency, One Decision" approval process, reducing the permitting process time to two years or less by amending statutory provisions and eliminating redundancies between state and federal authorities. The proposal will also broaden existing authorities to delegate environmental review and permitting duties to the states. This proposal will be considered in 2018.

In addition to these proposal elements, President Trump has indicated that he would consider a 25-cent federal gas tax increase. Many industries—including the U.S. Chamber, labor organizations, trucking associations, and engineering and construction advocates—support a gas tax increase, although many believe that this is a short-term solution.

8 The Cost of Doing Nothing

Without an immediate infusion of funds to the Highway Trust Fund (HTF), the cost to repair grows exponentially as shown in Fig. 5. Pavement deteriorates exponentially at a certain point. For example, a 14-year old pavement where needed maintenance is delayed for another 4 years will diminish very poor condition. Consequently, it will cost 4–5 times more the cost if the maintenance is performed in time.

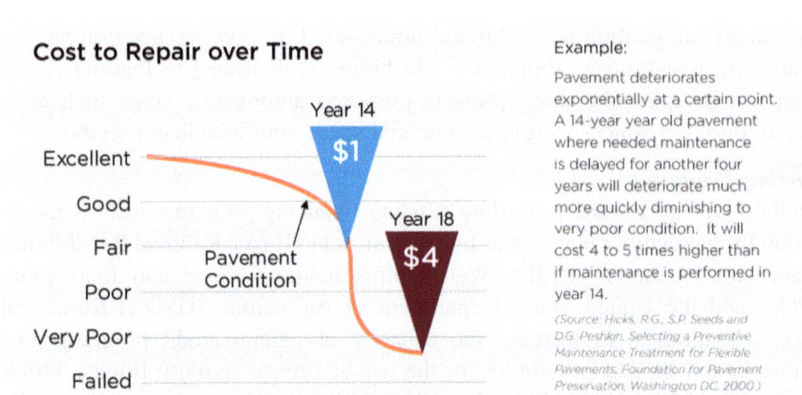

Fig. 5. Cost to repair pavements increases over time.

Furthermore, Investing in Infrastructure Returns $3 to $4 in Economic Activity for Each Dollar Invested in Infrastructure as Shown in Fig. 6.

The various costs associated with delays in investment and maintenance of infrastructure are described below

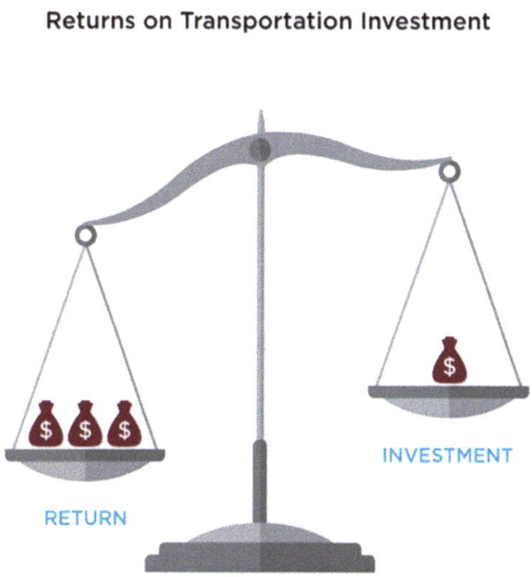

Fig. 6. Transportation investment returns $3 to $4 for every $1 invested.

8.1 Congestion Costs

In urban areas people spend many hours in congestion, which means gallons of fuel wasted, an adverse effect on air quality, a loss in productivity, hours lost away from family or planned activities and an overall adverse effect on people's quality of life.

8.2 Increased Risk of Failure

Delaying maintenance makes transportation infrastructure more susceptible to life-threatening failure, even without catastrophic conditions.

8.3 Increased Medical and Health Insurance Costs

Accidents and fatalities will continue to continue to grow without needed safety improvement to our transportation

8.4 Lost Potential Economic Development

Studies consistently conclude that transportation investments return $3 to $4 for every $1 invested

9 Options and Trends in Funding Solutions

Emerging trends across both the U.S. and the world are focused on effective funding solutions for transportation infrastructure. Any solutions must be sustainable, equitable, and flexible, as well as reflect future driving trends. Beyond what has already been discussed above, other options to provide this effective, sustainable funding that have been considered by legislative and industry leaders include:

9.1 Increasing and Indexing the Gas Tax

The mechanism with the highest potential for impact is a change in our fuel tax policy. First and foremost is the need to increase the gas tax to a level more commensurate with the need and in line with what other states are doing. Second would be the need to index the rate to inflation.

This has been accomplished at the state level. In fact, a 2016 analysis of over 2,500 state legislators from 16 states found that 91% of those who voted to increase state gas taxes for transportation funding were returned to office in the next general election. This included 89% of Democratic legislators and 95% of Republican lawmakers (American Road and Transportation and Transportation Investment Advocacy Center).

It must be noted that an increase in the gas tax will not provide a long-term solution to the funding issue alone, due to the rise of electric and high-efficiency vehicles and the need for fewer gallons of gas per mile travelled.

9.2 The BOLD Act

This is a federal transportation revenue and tax reform infrastructure investment plan for the new congress and the Trump administration. The BOLD act states: (i) maintain the current gas tax; (ii) eliminate heavy truck sales, usage and tire excises; (iii) initiate 6.25% highway transportation services tax (air cargo tax parity); and (iv) initiate $1,200 electric motor battery highway use parity sales tax (10-year gas tax equivalent).

9.3 Road User Fee and Vehicle Miles Traveled (VMT) Tax

A VMT tax or a Mileage-based User Fee (MbUF) is the process of charging motorists a fee based on how many miles they have travelled. Emerging trends across the U.S. and internationally are focused on effective funding solutions for transportation infrastructure. As this option has been explored around the country, privacy concerns have been expressed that tying a fee to a GPS would provide real-time information on where drivers are and have been. There are options to ameliorate these concerns through fee-structure options and substantial public awareness.

Internationally, similar user-fee based systems are already in operation to fund the transportation system, including the Salik Toll System in Dubai, United Arab Emirates; the congestion pricing system in the commercial hub of London, England; and the Road User Charge (RUC) system for drivers of light diesel vehicles and diesel-powered heavy vehicles in New Zealand.

9.4 Public–Private Partnerships (P3)

The P3s represent agreements that allow private companies to take on traditionally public roles in infrastructure projects, while keeping the public sector ultimately accountable for a project and the overall service to the public. P3s are a way to finance a project may not necessarily provide additional funding for a project. The risk can be transferred from a public entity to a private company or at a minimum shared between the two entities. A total of 33 US States have enacted P3 Statutes for transportation infrastructure development as shown in Fig. 7.

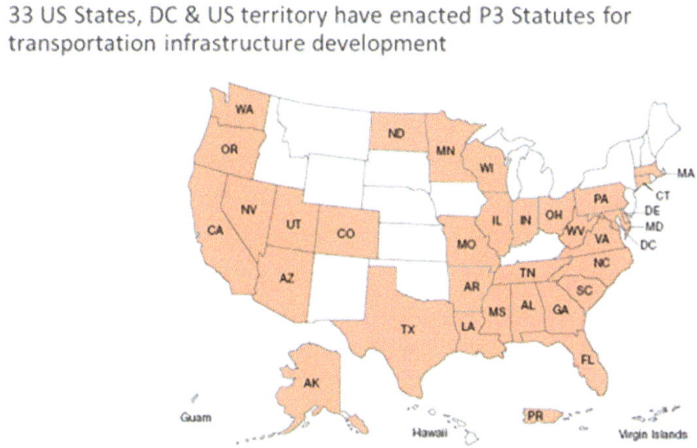

Fig. 7. States that have enacted public–private partnership statues (Federal Highway Administration).

9.5 Managed Lanes

A managed lane is a type of highway lane that is operated with a management scheme, such as land-use restrictions or variable tolling, to optimize traffic flow, vehicle throughput, or both. This concept is being used in several states and is gaining popularity across the nation to provide the necessary funding to add capacity to existing facilities.

9.6 Tolling Interstates

The utilization of tolls on selected roadways would supply a much-needed infusion of funding to state departments of transportation (DOTs). Existing federal rules allow for tolling additional capacity and the initial construction of bridges on the interstate system. There are also some existing FHWA pilot programs that allow tolling under certain situations. Tolling is an option that could be utilized to help fund these improvements. Possible interstates could include I-80, I-55 and I-290.

10 Conclusions

The purchasing power of the USA federal gas tax of 18.4 cents per gallon and 24.4 cents per gallon for diesel fuel has diminished since it was last raised in 1993. Our parents and grandparents paid more for transportation as a percentage of their wages than we pay today. Given the deteriorating condition of the USA transportation system, the cost of doing nothing will cause the USA to fall behind developed countries. The exceptional value drivers receive for the price currently being paid for transportation other types of public infrastructure is not sustainable. The time has come to significantly increase investment in infrastructure at a national level. The USA needs to, once again, to invest in its infrastructure system at a level of investment equal to what generations before us have paid. It is proposed that for the short-term that the motor fuel tax is raised and indexed to inflation. We need to utilize other creative funding options to supplement and plan for long-term sustainable funding of infrastructure. States are increasingly exploring user fees, dynamic priced managed lane systems, toll roads, public–private-partnerships, and Mileage-based User Fee (MbUF) systems as solutions to the infrastructure funding inadequacies. Our children and grandchildren's quality of life, safety and economic well-being depends on it. More importantly, the United States ability to compete economically at the global level will continue to diminish, if we fail to act soon.

Acknowledgements. The authors wish to thank the President of the Illinois Senate, the Honorable John J. Cullerton for his imaginative leadership and willingness to explore creative funding options.

References

American Jobs Act, S. 1549, 112th Congress (2011)

American Society of Civil Engineers: Infrastructure Report Card (2017). https://www.infrastructurereportcard.org/wp-content/uploads/2016/10/2017-Infrastructure-Report-Card.pdf

American Road & Transportation Builders Association: State Motor Fuel Tax Rates (2017). https://www.artba.org/economics/research/gas-tax-history/

Auxier, R.C.: Reforming state gas taxes: how states are (and are not) addressing an eroding tax base. The state and local finance initiative brief. The Urban Institute, Washington D.C. (2014). https://www.urban.org/sites/default/files/publication/49811/413286-Reforming-State-Gas-Taxes.PDF

Batic Institute: Transportation Funding and Financing. http://www.financingtransportation.org/funding_financing/financing/ (n.d.)

Craighead, M., Manzo, J.: Gas taxes in the United States and globally: failing to address transportation infrastructure. Midwest Economic Policy Institute (2016). https://illinoisepi.org/site/wp-content/themes/hollow/docs/infrastructure-investment/gas-tax-final.pdf

Federal Highway Administration: State P3 Legislation. https://www.fhwa.dot.gov/ipd/p3/legislation/ (n.d.)

Garbutt, A.: Trump's infrastructure plan—broad but facing challenges on funding and implementation, 62nd edn. Foresight (2018). https://assets.kpmg.com/content/dam/kpmg/xx/pdf/2018/02/foresight-trump-s-infrastructure-plan.pdf

Holan, M.: U.S. Motor Fuel Consumption Grew Record 3 Percent Jan.–Aug. 2015. Washington Newsline (2016). http://newsline.artba.org/2016/01/08/u-s-motor-fuel-consumption-grew-record-3-percent-jan-aug-2015/

Ontario Ministry of Finance: Gasoline Tax. https://www.fin.gov.on.ca/en/tax/gt/index.html (n.d.)

Pfeiffer, D.A.: Ike's interstates at 50: anniversary of the highway system recalls Eisenhower's role as catalyst. Prologue Magazine **38**(2) (2006). https://www.archives.gov/publications/prologue/2006/summer/interstates.html

Transportation Investment Advocacy Center: Analysis Shows Voting for a State Gas Tax Increase does not Hurt Reelection Chances (2016). http://transportationinvestment.org/wp-content/uploads/2017/04/ARTBA-TIAC-2013_2016-Legislator-Reelection-Analysis.pdf

The U.S. Constitution, Art. 1, § 8

World Economic Forum: Global competitiveness index: rankings. Global Competitiveness Index 2017–2018 (2018). http://reports.weforum.org/global-competitiveness-index-2017-2018/competitiveness-rankings/

Strategies for Sustainable Funding of Infrastructure in Illinois, USA

Amir Al-Khafaji[1(✉)], Dan Meckes[2], and Dan Gallagher[3]

[1] Bradley University, Peoria, IL, USA
amir@fsmail.bradley.edu
[2] Crawford, Murphy & Tilly, Kansas City, MO, USA
[3] National Asphalt Pavement Association, Springfield, IL, USA

Abstract. As a crossroad to the nation, the Illinois transportation system has been instrumental to the historic growth and stature of Illinois as a hub of commerce. The Illinois transportation network includes interstates, highways, local roads and streets, bridges, airports, transit, passenger and freight railroads and ports and waterways. Together, these facilities move Illinois travelers, businesses and freight and drive economic growth. The Illinois Department of Transportation (IDOT) has primary responsibility for sustaining, strengthening, expanding, and maintaining the Illinois transportation system. Additionally, the Illinois Tollway plays a vital role in the Chicagoland area and invest substantially in infrastructure. The task is carried out with local, federal and other state stakeholders to ensure the transportation network fosters and supports economic growth.

This paper examines short-term and long-term alternative funding options to support future investments in Illinois infrastructure and the political dimensions involved. The authors lead the Illinois based Sustainable Funding of Infrastructure Group ($FIG) that has become a leading voice for funding of infrastructure. It recently embarked on an imaginative process to enlist and educate Illinois legislators and government official on the need for funding mechanism of infrastructure in Illinois. The authors contend that continued underinvestment in Illinois' system will cripple the state's long-term growth and investment needs. Further, a national transportation research group finds that driving on roads in need of repair, costs Illinois motorists $3.7 billion a year in extra vehicle repairs and operating costs, or $441 per motorist annually. This level will grow with deferred maintenance on the order of $4 in 10 years for every $1 invested today, instead.

Keywords: Illinois · Transportation · IDOT · $FIG · Highways
Infrastructure · Tollway · Economic growth · Repairs

1 Introduction

The citizens of Illinois own a tremendous asset. In many ways, the State of Illinois transportation system is one of the most important assets the state has to ensure a vibrant economy and general well-being of its citizens. The infrastructure asset in Illinois including roads, bridges, rail, airports, and waterways provide reliable and efficient transportation which is vital to the state's economy and quality of life. This

M. Shehata and F. Rodrigues (Eds.): GeoMEast 2018, SUCI, pp. 106–124, 2019.
https://doi.org/10.1007/978-3-030-01905-1_6

asset though is beginning to diminish due to the ever-increasing gap between basic infrastructure needs and infrastructure revenue streams. This gap presents serious concerns for mobility, safety and continued job growth. Table 1 shows that Illinois is one of the largest transportation hubs in the USA.

Table 1. The state of Illinois ranking within the USA.

Mode	Rank within the USA
Public Roads IDOT maintains 15,967 miles including 2,100 miles of interstates	3rd largest
Bridges IDOT maintains 7,847 bridges	3rd largest
Transit system The Chicago transit authority operates this public transportation system	2nd largest
Railroad system	2nd largest
Airports for cargo and passenger flights	3rd largest
Navigable waterways for large cargo	8th largest
Freight tonnage	3rd largest

The Illinois Department of Transportation (IDOT) maintains the majority of the transportation infrastructure. Additionally, the Illinois Tollway invests approximately $1.4 billion annually on maintenance and new transportation infrastructure.

With almost 400,000 businesses and an employment base of more than 5.7 million, Illinois leads the Midwest in volume of both companies and jobs. (Illinois Innovation Network (IIN)). The Chicago region exported goods worth $53.9 billion in 2010, supporting over 375,000 jobs in the region. Both in the value of exports and the number of exporting jobs, the Chicago region ranked only behind New York City and Los Angeles as the largest exporting center in the country.

For Illinois to retain its position as a premier business destination, and to provide a foundation for our continued economic prosperity, there must be an immediate and sustainable funding solution to support the transportation Investments needed to keep Illinois competitive in a global economy. The solution needs to allow Illinois to maintain the system in a state of good repair.

2 The Sustainable Funding of Infrastructure Group ($FIG)

The Sustainable Funding of Infrastructure Group ($FIG) with help from the Illinois Senate President John Cullerton and other government officials in April of 2014. The purpose of $FIG continues to be the exploration of innovative and long-term sustainable funding solutions for the state of Illinois. The group has positioned itself as a trusted and credible source for making recommendations to lawmakers and government officials. $FIG is non-partisan, does not have a sunset and places a special emphasis on educating the public about the need for additional revenue to help fund transportation challenges across the state.

Since the formation of the group, there have been four Executive Summits focused on sustainable funding of infrastructure in Illinois. The Executive Summit involves elected officials (local, state and federal) agency officials including local, state and federal including leaders from around the globe, Illinois Department of Transportation, Illinois Tollway, O'Hare Airport and the Federal Highway Administration, Illinois business leaders, and other stakeholders. Approximately 50 to 65 guests attended Summit each year. The Executive Summits have been chaired by Illinois Senate President John Cullerton. In 2015 and following the April Executive Summit, President Cullerton released a complex senate bill that would gradually replace the gas tax with a Vehicle Miles Travelled tax (VMT) tax. The public caught wind of the bill, which was reported that the VMT tax would be in addition to the gas tax. The bill was quickly removed from further consideration due to the negative publicity that it received primarily because of privacy concerns and also because the bill had been misrepresented. In addition to the Executive Summits, the $FIG sponsored the following major events:

a. Meeting with Illinois Senate President Cullerton in his senate office in June of 2015 to discuss sustainability funding as well as the state budget impasse, which was completely holding up in legislative discussion regarding potential sources for new infrastructure revenue.
b. Senator Cullerton agreed to another meeting with $FIG to discuss potential solutions to the infrastructure funding as well as potential solutions to the budget impasse. A meeting was set for mid-October in Chicago. Over 300 business leaders representing vertical construction, horizontal construction, transit, environmental groups, airports and waterways including labor and business were considered. 43 people representing these groups attended the summit on October 14, 2015 in downtown Chicago.
c. $FIG founder, Dr. Amir Al Khafaji and $FIG vice-chair, Dan Gallagher met with the Governor's office in March 2016 to discuss potential funding sources and the budget impasse.
d. Republican House Minority Leader Jim Durkin and six other republican house members met with 50 $FIG members on December 11, 2017 in Chicago to discuss sustainable funding of infrastructure for Illinois. In addition to the House members there were individuals from the largest engineering firms in the world, contractors, soy bean association executive, Midwest Trucking Association, Illinois Auto Dealers, the national chair of the National Asphalt Paving Association, a Caterpillar and Harley dealer and other Illinois business leaders.
e. Meetings are planned in July 2018 with Illinois Senate President John Cullerton and other democratic leaders to continue discussions for potential solutions for new revenue for infrastructure as well as the timing of potential future legislation. A similar meeting is also planned with Governor Rauner to discuss the same topics in August 2018.

The purpose of the meetings with Minority Leader Durkin, Illinois Senate President John Cullerton and Governor Rauner are to attempt to find common ground between the executive office, the House and the Senate. The $FIG is determined to pave the way for legislators to find common grounds to invest in Illinois infrastructure.

3 The Current Situation

IDOT sets performance targets for both the Illinois Roads and Bridges programs. For Illinois Roads, IDOT's target is to maintain 90% of state-maintained highway system mileage in acceptable condition. Today, only 83% meets acceptable condition. Further, according to IDOT's 2018–2023 Multi-Year Program, in FY 2023, if the current rate of investment and deterioration remain the same, the state-maintained mileage will drop from 83 to 65% acceptable condition.

Not only Illinois roads are suffering. According to the American Road and Transportation Builders Association, in 2016, Illinois ranked as the 6th highest state in the number of structurally deficient bridges.

In addition, because these statistics were determined before IDOT lost $300 million from the FY 2018 Program and every year thereafter (Illinois Road and Transportation Builders Association), following the state budget approval in July 2018, the percentage of roads and bridges in acceptable condition will continue to decline. The life cycle of Illinois' roads is greatly affected by the state and local governments' ability to perform timely maintenance and upgrades to ensure that road and highway surfaces last as long as possible.

The pavement data in this paper, which is for all arterial and collector roads and highways, is provided by the Federal Highway Administration (FHWA), based on data submitted annually by the Illinois Department of Transportation on the condition of major state and locally maintained roads and highways. Pavement data for Interstate highways and other principal arterials is collected for all system mileage, whereas pavement data for minor arterial and all collector roads and highways is based on sampling portions of roadways as prescribed by FHWA to ensure that the data collected is adequate to provide an accurate assessment of pavement conditions on these roads and highways.

Statewide, nearly one-third of Illinois' major roads are in poor or mediocre condition. Seventeen percent of Illinois' major locally and state-maintained roads are in poor condition and 14% are in mediocre condition. Thirteen percent are in fair condition and the remaining 56% are in good condition.

As mentioned earlier, the condition of state-maintained roads and highways in Illinois is expected to deteriorate further, based on current levels of state funding. A 2017 IDOT report found that the share of state-maintained roads and highways that have deteriorated to the point where an improvement is needed now is anticipated to increase from 23% in 2018 to 35% in 2023 (from 3,596 miles to 5,588 miles of the 15,968 miles). Table 2 below details pavement conditions on major urban roads in the State of Illinois largest urban areas.

Driving on Illinois roads that are deteriorated, congested and that lack some desirable safety features costs Illinois drivers a total of $16.4 billion each year as shown in Table 3. TRIP has calculated the cost to the average motorist in the state's largest urban areas in the form of additional vehicle operating costs (VOC) as a result of

Table 2. Pavement conditions on major roads in Illinois' largest urban areas. (TRIP analysis of Federal Highway Administration data).

Location	Poor	Mediocre	Fair	Good
Chicago	28%	35%	14%	22%
Champaign–Urbana	24%	31%	13%	32%
Metro east	9%	33%	22%	36%
Peoria–Bloomington	31%	24%	10%	34%
Rockford	32%	29%	15%	25%
Springfield	20%	31%	14%	35%
Illinois statewide	27%	9%	10%	54%

driving on rough roads, the cost of lost time and wasted fuel due to congestion and the financial cost of traffic crashes.

Table 3. Pavement conditions and its impact on drivers.

Location	VOC	Safety	Congestion	Total
Chicago	$627	$374	$1,484	$2,485
Champaign–Urbana	$544	$629	$299	$1,472
Metro east	$374	$879	$1,048	$2,301
Peoria–Bloomington	$597	$600	$363	$1,560
Rockford	$639	$723	$573	$1,935
Springfield	$493	$651	$295	$1,439
Illinois statewide	$3.5 billion	$4.7 billion	$8.2 billion	$16.4 billion

Nine percent of Illinois' bridges are structurally deficient, meaning there is significant deterioration of the bridge deck, supports or other major components as shown in Table 4. The condition of state-maintained bridges in Illinois is anticipated to decline through 2023 based on current funding.

Congested roads choke commuting and commerce and cost Illinois drivers $8.2 billion each year in the form of lost time and wasted fuel. Drivers in the state's largest urban areas lose thousands of dollars and as much as two-and-a-half days each year in congestion (Table 5).

Nearly 5,000 people were killed in traffic crashes in Illinois in the last 5 years. Traffic crashes in which roadway features were likely a contributing factor imposed $4. 7 billion in economic costs in 2016 (Table 6).

The overall condition of transportation and other infrastructure in Illinois is shown in Fig. 1. The score card is not limited to roads and bridges but also other infrastructure assets as reported by the American Society of Civil Engineers.

Table 4. Illinois bridges conditions

Location	Number structurally deficient	Share structurally deficient	Total bridges
Chicago	323	10%	3,271
Champaign–Urbana	10	10%	102
Metro east	54	8%	698
Peoria–Bloomington	67	15%	458
Rockford	29	8%	377
Springfield	12	7%	181
Illinois statewide	2,290	9%	26,770

Table 5. Illinois roads congestion

Location	Hours lost to congestion	Annual cost per driver
Chicago	63	$1,484
Champaign–Urbana	13	$299
Metro east	44	$1,048
Peoria–Bloomington	16	$363
Rockford	24	$573
Springfield	13	$295

Table 6. Illinois traffic safety and fatalities

Location	Average fatalities 2014–2016	Safety cost
Chicago	424	$374
Champaign–Urbana	20	$629
Metro east	66	$879
Peoria–Bloomington	45	$600
Rockford	35	$723
Springfield	19	$651
Illinois statewide	1,000	$4.7 Billion

3.1 Aviation

Illinois has 11 commercial service airports, serving over 42 million passengers. The number of passengers at large htib airports-including O'Hare and Midway International Airports-are predicted to grow at an annual rate of 2.3% through 2040. O'Hare is forecasted to have the highest number of aircraft operations in the U.S. by 2040, and a forecasted 72% increase in operations at Midway International Airport will make it the 23rd busiest airport in the nation by 2040, up from its current position at 28th.

2014 Report Card, Illinois Infrastructure

Based on 8 criteria (capacity, condition, funding, future need, operation and maintenance, public safety, resilience and innovation).

A = Exceptional: Fit for the Future
B = Good: Adequate for Now
C = Mediocre: Requires Attention
D = Poor: At Risk
F = Failing: Critical: Unfit for Purpose

Aviation	C+
Bridges	C+
Dams	C
Drinking Water	C-
Navigable Water	D-
Rail	C
Roads	D+
Transit	D+
Wastewater	D+
Illinois 2014 Grade	**C-**

Fig. 1. The ASCE report card for the state of Illinois

3.2 Bridge

Illinois is the cross roads for the movement of goods for a large part of the country, and Illinois has the 3rd largest bridge inventory in the nation. Regrettably, 16% of Illinois' bridges are classified as structurally deficient or functionally obsolete. While it is encouraging that the percentages of structurally deficient and functionally obsolete bridges in Illinois are lower than the national average, inevitable increases in the number of trucks and axle loads will degrade roads and bridges more rapidly.

3.3 Dams

Dams in Illinois supply communities with water, create recreational lakes, and provide much needed flood control. Illinois has 1,759 State-regulated dams. The overall condition of dams in Illinois is well above average and the State Dam Safety program is strong. However, there are a number of older dams that were not issued permits. This is a concern because the condition and level of risk posed from these dams are unknown.

3.4 Drinking Water

In 2011, Illinois estimated that the drinking water systems needed an investment of $19 billion over the next 20 years to replace aging facilities and comply with safe drinking water regulations. However, over that same time period the water quality compliance rate has increased from 93.9% to 96.5%. Through programs such as the Clean Water Initiative, Public Water Supply Loan Program, and the Green Infrastructure for Clean Water Act, Illinois has begun to make steps to improve its drinking water infrastructure. The City of Chicago Department of Water Management's recently adopted 10-year plan for replacing aged water lines is a good step forward.

3.5 Navigable Waterways

The Mississippi River, Illinois River, Ohio River, and Lake Michigan locks near Chicago provide vital commercial shipping links for goods to travel throughout Illinois as well as the entire country. About 116 million tons of commodities are shipped annually on the 1,118 miles of Illinois' inland waterways, totaling more than $23 billion. Much of the system was built in the 1930s with a 50-year design life and is now in desperate need of upgrades, rehabilitation, and repair.

3.6 Rail

The Illinois rail network is the 2nd largest in the country, and Chicago is the single largest rail hub in the nation. Ridership in Illinois has increased to 85% between 2006 and 2011. Every day 500 freight trains with 37,500 cars and 700 passenger and commuter lines pass through Chicago. In 2013, Amtrak recorded its highest year of ridership ever with 6 million Illinois riders. Both freight rail and passenger rail have been investing heavily in their tracks, bridges, and tunnels as well as adding new capacity for freight and passengers bringing. a grade improvement. However, rail freight volume is expected to double by 2025, causing congestion and delays unless continued investments are made. Congestion on the State's rail system costs millions of dollars in shipping delays and causes substantial noise and air pollution as trains idle for hours waiting for track clearance. Future rail investment needs are estimated at $20.5 billion.

3.7 Roads

Between 1990 and 2012, vehicle travel on Interstate highways in Illinois increased 25%, while the population grew by 13% and lane miles added grew by just 11%. Severe traffic congestion costs Illinois' economy billions of dollars in lost productivity each year. Congestion is estimated to cost $4 billion annually for the Chicago area alone, and 42% of Illinois' major roads are in poor or mediocre condition. Driving on these roads costs Illinois motorists $3.7 billion a year in extra vehicle repairs and operating costs. Significant investments have been made by the Illinois Tollway and IDOT since the 2010 Report Card, but these funding sources must be supplemented with a long term sustainable funding source at the State and federal level.

3.8 Transit

The current capital needs for transit systems in Northeastern Illinois are over $2 billion annually, and transit ridership in the region is 5% higher than in 2010. The dramatic growth in intra-suburban commuting is a major transit challenge; however, the cost of gasoline, growing environmental concerns, and worsening roadway congestion make efficient and upgraded transit systems valuable. Much of the infrastructure for a world-class transit system, particularly in Northeastern Illinois, is in place, but it has suffered as a result of age and lack of funding. The Chicago region under-spends on transit operations and capital compared to both national and international cities. The estimated need to bring the Chicago region's transit to a state of good repair over the next 10 years has risen to $30.9 billion in 2012 from $26.1 billion in 2010.

3.9 Wastewater

Many wastewater management systems in Illinois are more than 100 years old and continue to serve Illinois families and businesses. Currently, the EPA estimates that Illinois must invest $17.5 billion over the next 20 years to replace existing systems and build new ones to meet increasing demands. Federal assistance alone cannot be expected to meet Illinois' needs. Financial shortfalls will need to be funded by the State of Illinois and its ratepayers if Illinois hopes to maintain the necessary infrastructure to meet its growing population. The Illinois Clean Water Initiative has recently been expanded and time will tell if it can lead to progress toward meeting this vast need.

4 Why are We Not Keeping Pace

The purchasing power of the gas tax and other conventional funding sources has been declining for years.

4.1 Vehicle Efficiencies

The primary state revenue source, the 19-cents-per-gallon fuel tax, has not changed since 1991. Average vehicle fuel efficiency has since increased substantially (24%), diminishing revenues as less fuel is purchased. Government efficiency standards, alternative-fuel vehicle trends, and rising fuel costs will only continue to push efficiencies upward.

As illustrated in Fig. 2, the value of each dollar raised through the fuel tax is diminished over time due simply to the continued increase in vehicle fuel efficiency. It is also important to note that the fuel tax is "per gallon" and does not change with the price of gas. The state collects 19 cents per gallon, whether the price at the pump is $2.00 or $4.00 a gallon.

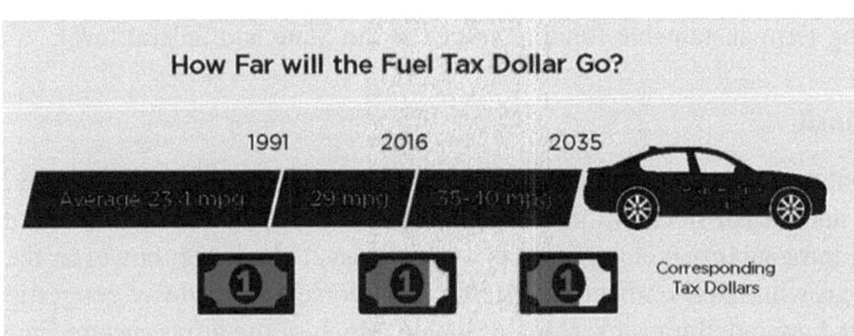

Fig. 2. Vehicle fuel efficiency impact on revenue available for infrastructure.

4.2 Inflation Effects

With inflation, the purchasing power of fuel tax dollars simply cannot keep up. The price tag of infrastructure maintenance—including construction labor, equipment, and materials costs—continues to grow. All the while, according to IDOT, "The rate of deterioration far exceeds the available funding to improve the system to acceptable conditions." As deterioration increases with delayed maintenance, the ultimate price tag only increases further. The purchasing power, since the last gas tax increase in 1991, has declined over 40%. The Motor Fuel Tax, as currently set up, is not sufficient to provide sustainable revenue for the state's transportation system.

4.3 Budgetary Competition

Even with steps forward, like the $2.2 billion 2018-2023 program passed in July of 2017, the General Assembly passed legislation to transfer out $303 Million originally obligated to the Road Fund from the General Revenue – a 13% reduction. In addition, the state recently imposed a 2% surcharge on sales taxes used to fund the Chicago Regional Transportation Authority (RTA). This further cripples the authority's ability to catch up with significant needs for what is the 2nd largest transit system in the US. In addition to the region's $16.6 billion 10-year program, there is a separate $19.5 billion of deferred investment known as the "capital backlog."

Illinois has been investing less and less in transportation—less than 8% of our budget in 2014, compared to 13% in 1991—resulting in higher-than-ever backlogs. Meanwhile, the need and cost of preserving our transportation system continues to grow exponentially. The "Backlog" is a category IDOT uses to describe miles of roadway or number of bridges that have deteriorated to the point where an improvement is needed now. The IDOT chart below compares historical backlog needs for IDOT-maintained roads, beginning in 2000, with estimated backlog needs through the FY 2017–2022 program timeframe. The backlog continues to grow at an exponential rate, while funding is in a trajectory of decline as show in Fig. 3.

As shown in Fig. 3, Illinois' backlog of infrastructure investment needs is extensive, across all sectors: highways, bridges, transit, schools, state buildings, and locks and dams, just to name a few. The state's capital program must consider Illinois' infrastructure holistically to be effective. While highways and bridges are the backbone of our transportation system, we must develop a smart sustainable capital program that addresses the needs of our diverse infrastructure system to remain competitive. This includes infrastructure needs in the following areas:

a. Health care
b. State facilities
c. Environment
d. Energy
e. Technology
f. Education

The Capital Program should be informed by the goals we have established for Illinois to have a prominent position in the global economic marketplace and

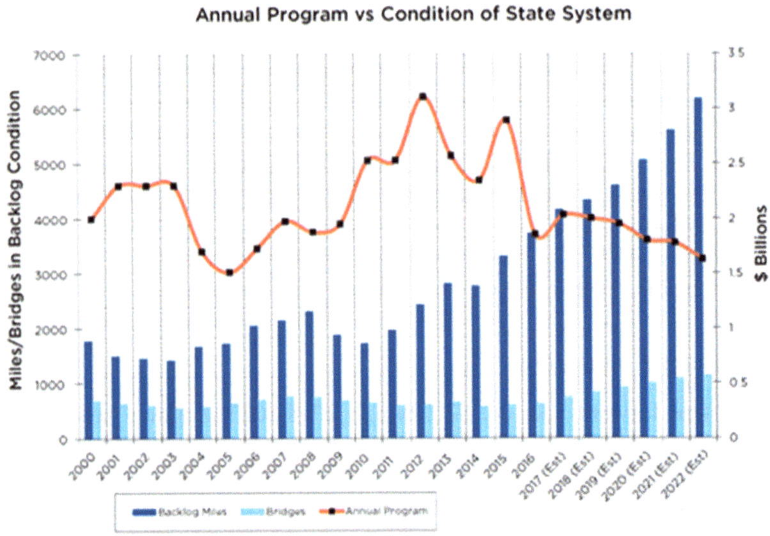

Fig. 3. Annual program funding and condition of transportation system

underscored by the necessary funding to get us to our desired future. The overarching goals for Illinois' future are

a. Enable safer travel
b. Reduce freight choke-points
c. Reduce barriers to new technology
d. Retain and expand our highly qualified, skilled, and diverse work force
e. Create more resilient infrastructure that is adaptive to our changing climate
f. Provide smart investments that align infrastructure decisions with funding

It is interesting to note that the State of Illinois had a capital program of $30.8 Billion that has not been renewed since 2009. A breakdown of the investment made in infrastructure is outlined in Fig. 4.

The sustainable Funding of Infrastructure Group ($FIG) has taken a leadership role in bringing together elected officials and business leaders to address future investments in Infrastructure. To that end Executive Summits were held with the president of the Illinois senate and the Republican leader in the Illinois House to help bridge the divide between the various leaders. Reports were also produced on sustainable funding option to address Illinois' infrastructure system needs. These reports were developed after discussions and collaboration between representatives of the Federal and Illinois State legislature, leadership of the Illinois Department of Transportation, infrastructure industry leaders, and executives from the business community.

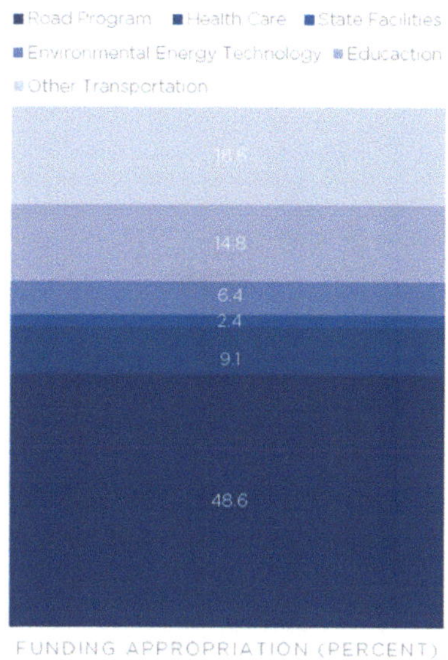

Fig. 4. The Illinois capital bill of 2009.

5 The Cost of no Action and Delay

Increased Vehicle Costs—TRIP, a national transportation research group finds that driving on roads in need of repair, costs Illinois motorists $3.7 billion a year in extra vehicle repairs and operating costs, or $441 per motorist. The would be $37 dollars per month. This level will grow with deferred maintenance.

"The number of cars coming in with damage from disintegrated roads has grown steadily over the years," says the owner of Vanderlaan Automotive Repair in Decatur. "Cars lose ball joints, struts, sway bar links—this stuff breaks all of the time, we're just continually fixing them." (Clay Jackson, Herald and Review 2016)

Congestion Costs—ASCE estimates that congestion cost $4.2 billion annually for the Chicago area alone (ASCE Illinois 2014 Report Card for Illinois Infrastructure)

Increased Risk of Failure—Delaying maintenance makes transportation infrastructure more susceptible to life-threatening failure, even without catastrophic conditions.

Increased Construction Costs—Delaying maintenance can require agencies to perform more extensive and expensive maintenance actions earlier than they otherwise would have. The extensive work can increase an asset's life cycle cost. Any cost saved up front from delaying maintenance is ultimately canceled out or exceeded by pricier repair costs and increased risk later. What costs $1 now will cost $4 if we wait as shown in Fig. 5.

Lost Potential Economic Development—Studies consistently conclude that transportation investments return $1.5 to $3 for every $1 invested as shown in Fig. 6. It stands to reason that, even if Illinois is deferring this investment, other

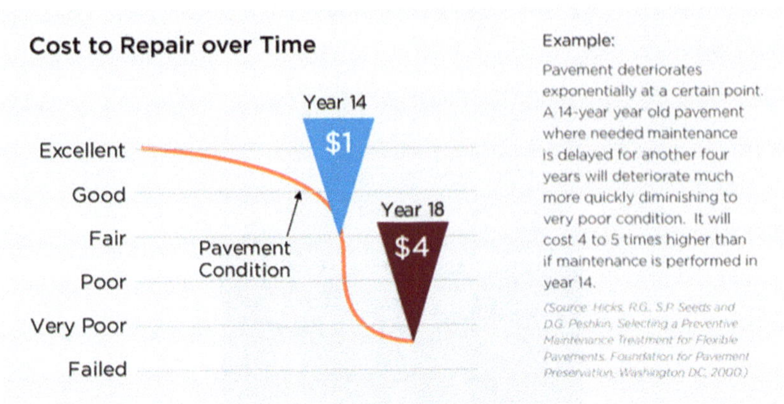

Fig. 5. The cost of inaction on infrastructure.

states are not. According to the American Jobs Act, every $1 billion in highway and transit investment supports 13,000 jobs for a year. The transportation design, construction and maintenance workers in Illinois earn $6.5 billion annually, and many are now targeting their expansions outside of Illinois.

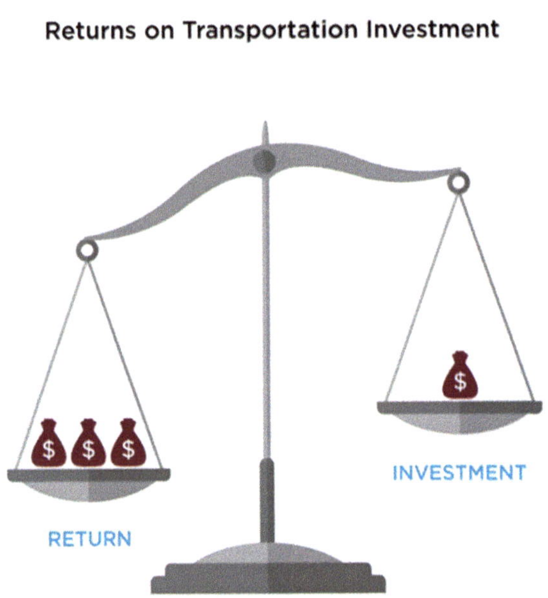

Fig. 6. Economic impact of investment in infrastructure.

Many industries in Illinois are also indirectly reliant on the state's transportation network. 2,632,272 full-time jobs in Illinois, in key industries like tourism, retail sales, agriculture and manufacturing, currently earn $111.0 billion in wages and contribute an estimated $20.3 billion in state and local income, corporate and unemployment insurance taxes, and the federal payroll tax. If these jobs move out of state, so will their tax revenue, crippling the potential for future economic development in Illinois.

Additionally, a substantial federal infrastructure rebuilding program is being planned in Washington, D.C. Though its details are not yet known, it will undoubtedly result in a substantial increase in the flow of federal matching dollars to Illinois. Most federal highway programs offer an 80 federal/20 state match. Such additional federal funds will be lost unless Illinois has sufficient state matching funds.

> **Budgetary Competition**—Even with steps forward, like the $2.2 billion 2018-2023 program passed in July of 2017, the General Assembly passed legislation to transfer out $303 Million originally obligated to the Road Fund from the General Revenue – a 13% reduction. In addition, the state recently imposed a 2% surcharge on sales taxes used to fund the Chicago Regional Transportation Authority (RTA). This further cripples the authority's ability to catch up with significant needs for what is the 2nd largest transit system in the US. In addition to the region's $16.6 billion 10-year program, there is a separate $19.5 billion of deferred investment known as the "capital backlog."

Illinois has been investing less and less in transportation—less than 8% of our budget in 2014, compared to 13% in 1991—resulting in higher-than-ever backlogs. Meanwhile, the need and cost of preserving our transportation system continues to grow exponentially. The "Backlog" is a category IDOT uses to describe miles of roadway or number of bridges that have deteriorated to the point where an improvement is needed now. The IDOT chart below compares historical backlog needs for IDOT-maintained roads, beginning in 2000, with estimated backlog needs through the FY 2017–2022 program timeframe. The backlog continues to grow at an exponential rate, while funding is in a trajectory of decline.

6 The Path to Sustainability of Future Funding

6.1 Recognize that Illinois Voters Care About Transportation Investment

In the fall of 2016, voters in Illinois approved a constitutional amendment that essentially requires all transportation funding to be placed in a "lockbox," where revenues cannot be deferred from transportation investments. This amendment protects existing revenue streams, but it does not provide any new revenues. However, the success of the "lockbox" amendment shows the public's belief that transportation revenues should be used to support transportation investment and that supporting transportation is a major concern. Any funding proposal should meet the requirement of the "lockbox."

6.2 Address Fuel-Related Revenues

Increase and Index—The mechanism with the highest potential for impact is a change in our fuel tax policy. First and foremost is to increase the tax on fuel to a level more commensurate with the need and in line with what other states are doing. Second would be to index the rate to inflation. Below is one possible scenario based on 6.1 billion gallons of gas consumed in Illinois annually as shown in Fig. 7.

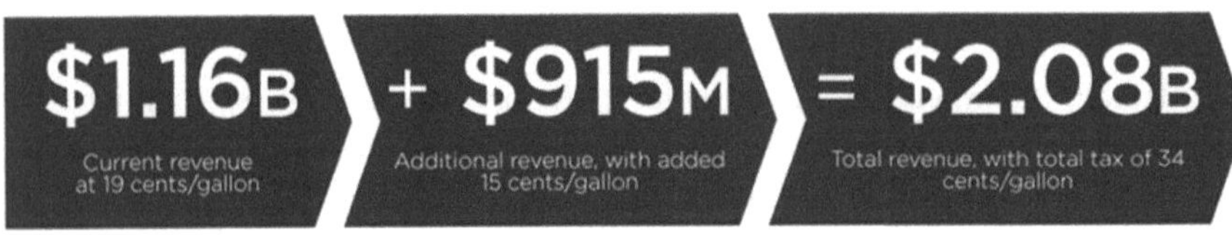

Fig. 7. Economic impact of investment in infrastructure.

However, an increase in the gas tax will not provide a long-term solution to the funding issue alone, due to the rise of electric and high-efficiency vehicles and the need for fewer gallons of gas per mile travelled. A 2016 analysis of over 2,500 state legislators from 16 states found that 91% of those who voted to increase state gas taxes for transportation funding were returned to office in the next general election. This included 89% of Democratic legislators and 95% of Republican lawmakers (American Road and Transportation and Transportation Investment Advocacy Center). The number of States that have increased gas tax stands at 26.

6.3 The BOLD Act

This is a Federal transportation revenue and tax reform infrastructure investment plan for the new congress and the Trump administration. The BOLD act states: (i) maintain the current gas tax; (ii) eliminate heavy truck sales, usage and tire excises; (iii) initiate 6.25% highway transportation services tax (air cargo tax parity); and (iv) initiate $1,200 electric motor battery highway use parity sales tax (10-year gas tax equivalent).

6.4 Road User Fee and VMT

A vehicle miles travelled tax, referred to as a VMT tax or road user fee, is the process of charging motorists a fee based on how many miles they have travelled. As this option has been explored around the country, privacy concerns have been expressed that tying a fee to a GPS would provide real-time information on where drivers are and have been. There are options to ameliorate these concerns through fee-structure options and substantial public awareness campaigns.

6.5 Managed Lanes

Managed Lanes are a concept used in several states and are gaining popularity across the nation to provide the necessary funding to add capacity to existing facilities.

A managed lane is a type of highway lane that is operated with a management scheme, such as land-use restrictions or variable tolling, to optimize traffic flow, vehicle throughput or both.

6.6 Tolling Interstates

The utilization of tolls on selected roadways on the IDOT system would supply a much-needed infusion of funding to IDOT. Existing federal rules allow for tolling additional capacity and the initial construction of bridges on the interstate system. There are also some existing FHWA pilot programs that allow tolling under certain situations. Tolling is an option that could be utilized to help fund these improvements. Possible interstates could include I-80, I-55 and I-290.

6.7 Consider What Other States Have Been Doing

How does the State of Illinois compare to other states in the Midwest in terms of Gas Tax and Sale Taxes are shown in Fig. 8.

Clearly, the State of Illinois is among the lowest in the Midwest in terms of gas taxes collected to pay for infrastructure. This observation continues to be the focus of the Illinois legislature as a possible short-term solution to investing in Illinois infrastructure.

6.8 Educate People on the Value they Receive from Their Transportation System

The people of Illinois receive exceptional value for the price being paid for their transportation infrastructure and would continue to do so with a modest increase in motor fuel taxes. Figure 9 illustrates the cost individuals incur on monthly basis.

It is interesting that individuals pay less that 50% of what they normally spend on coffee and less that 30% of what they spend on telephone bills

6.9 Prioritize Long-Term Thinking

The Vehicle Miles Traveled Tax (VMT) tax, a concept that would charge motorists based on how many miles they have traveled, needs to be explored further in Illinois. Several states are currently researching the topic or have pilot programs in place, including 19 states that currently have an electric vehicle fee. While there are currently many concerns over the use of VMT, these may well be resolved though pilot programs, technology and further research. Illinois should investigate a pilot program and work with other states on possible implementation.

Full acceptance and utilization of VMT tax is still years off, which is why other options need to be investigated and implemented now. However, with the increase in the CAFÉ standards and prevalence of electric vehicles and hybrids, the gas tax cannot be a long-term solution.

The short-term solution to Illinois infrastructure investment must include among other options a gas tax. This short-term solution will enable the state and legislators to

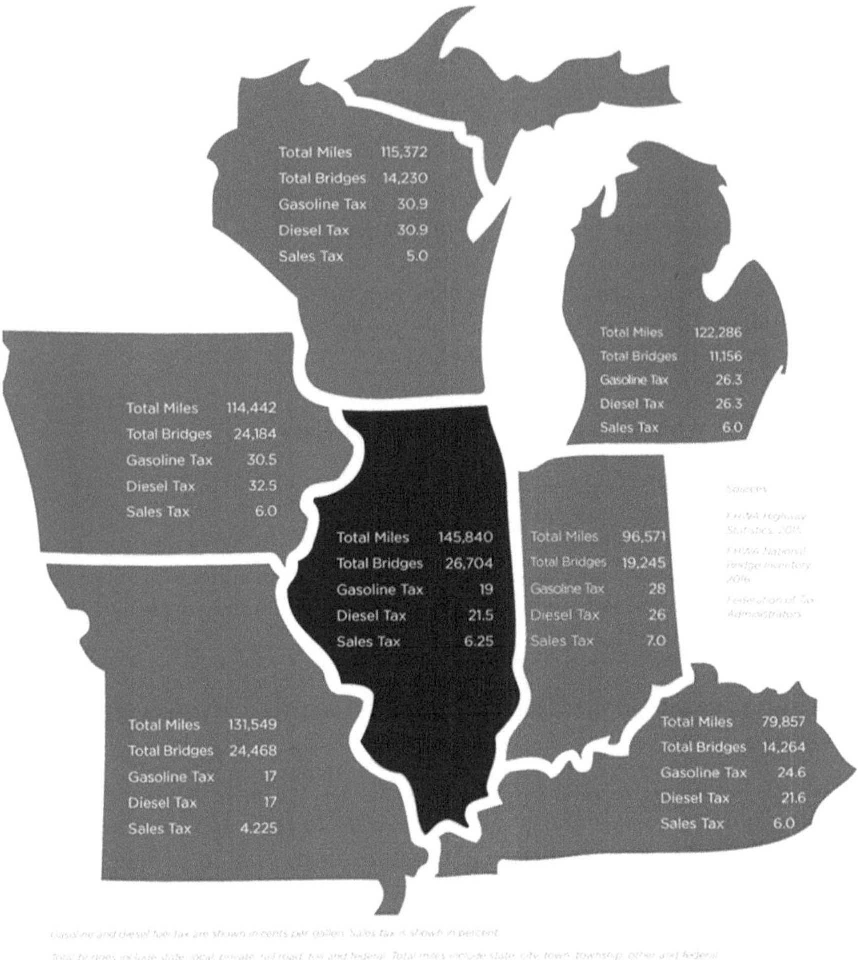

Fig. 8. Gas and sales taxes in the midwest states.

develop long-term sustainable solutions to invest in Illinois Infrastructure. Figure 10 illustrates that the people of Illinois pay less today for gasoline than they did 5 years ago. In fact, the pay almost 50% of what they once did in 2013. The $FIG has called on modest increase in the gas tax that would amount of less than $0.25. It further embraced the concepts of introducing taxes on electric vehicles.

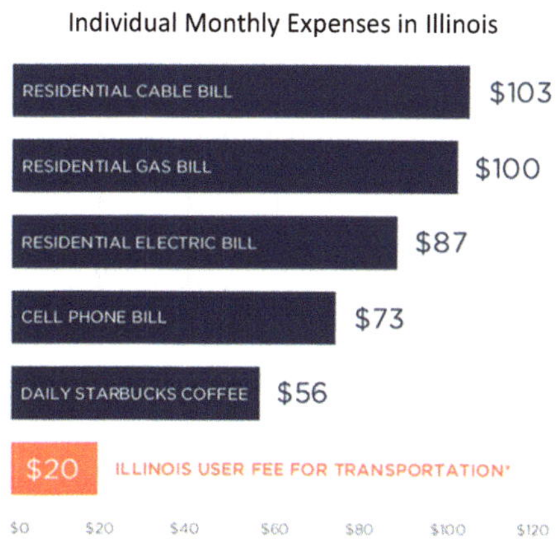

Fig. 9. The average individual expenses for resident in Illinois.

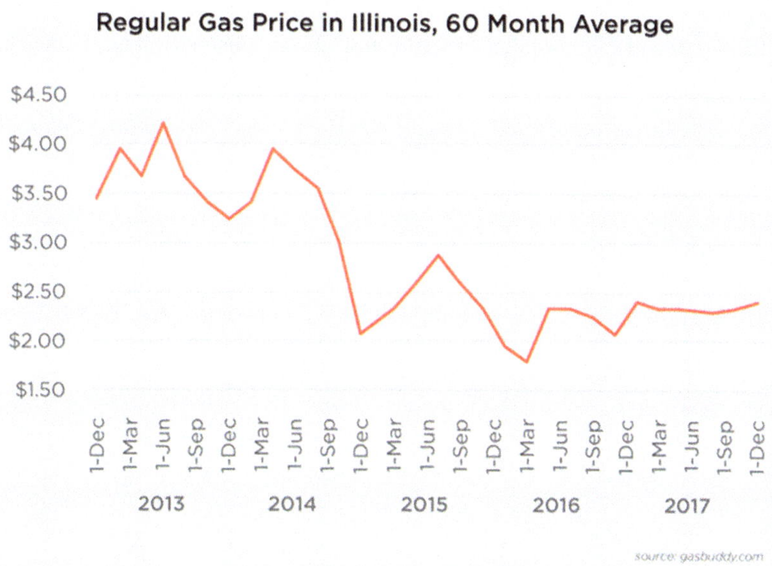

Fig. 10. The average price of gasoline in Illinois since 2013.

7 Conclusions

Given the current lower price of fuel and its upward trajectory in the near future, the time is now to take important action for a short-term solution that puts Illinois on the path to long-term sustainability. The short-term solution to Illinois infrastructure investment must include among other options a gas tax. This short-term solution will enable the state and legislators to develop long-term sustainable solutions to invest in Illinois Infrastructure. Such solutions may include the Vehicle Mile Traveled concept. The VMT is still years off, which is why other options need to be investigated and implemented now. The main objections relate to opposition from people living in rural areas and privacy concerns. However, with the increase in the CAFÉ standards and prevalence of electric vehicles and hybrids, the gas tax cannot be a long-term solution.

The people of Illinois pay less today for gasoline than they did 5 years ago. In fact, they pay almost 50% of what they once did in 2013. The $FIG has called on modest increase in the gas tax that would amount to less than $0.25. The $FIG further embraced the concepts of introducing taxes on electric vehicles.

Acknowledgements. The authors wish to thank the President of the Illinois Senate, the Honorable John J. Cullerton for his willingness to meet with business leaders and other elected officials to address infrastructure investment in Illinois. His genuine desire to invest in Illinois infrastructure is truly appreciated.

References

ASCE Illinois: Report card for illinois infrastructure (2014). http://www.isasce.org/report-card

Risks Affecting the Delivery of Construction Projects in Egypt: Identifying, Assessing and Response

Ahmed Mohammed Abdelalim(✉)

Faculty of Engineering at Mataria, Helwan University,
P.O. Box 11718, Cairo, Egypt
dr.ahmed.abdelalim@gmail.com

Abstract. The construction industry is often considered a risky business due to its complexity and strategic nature. It involves numerous project stakeholders and internal and external factors, which lead to enormous risks. Due to the massive risk factors and the widespread changes in the Egyptian construction sector over the last decade, risk in construction has been the object of great attention. The main objective of this paper is to identify and assess the significant risks in the Egypt construction industry based on their risk rating (impact and probability). The paper also compares effective risk management techniques to cope with risks associated with construction activities and to implement the projects in accordance with project objectives. Using a carefully-selected set of 77 attributes, this research first identified the key factors impacting delay in Egyptian construction industry and then established the relationship between the critical attributes for assessing the impacts of these factors. A questionnaire was carried out then extensive personal interviews were conducted to form the basis of this research. The factor analysis technique is used to examine the significance of the risk factors in addition to the AHP and simulation techniques. The three techniques used to assess risks were compared and evaluated using three case studies. From the factor analysis, most critical factors of construction risks were identified as: (1) lack of experience; (2) lack of owners' commitment; (3) lack of clarity in project scope; (4) Egyptian economic crises; (5) lack of contractor's commitment; and (6) Inefficient site management. The paper also suggests the risk response strategies for each type of identified risk.

Keywords: Risk management · Construction projects · Factor analysis AHP · Simualtion

1 Introduction

Risk analysis and management are important parts of the decision making process in a construction company. The construction industry and its clients are widely associated with a high degree of risk due to the nature of construction business activities, processes, environment and organization. Risk in construction has been highly considered because of time delays and cost overruns associated with construction projects.

M. Shehata and F. Rodrigues (Eds.): GeoMEast 2018, SUCI, pp. 125–154, 2019.
https://doi.org/10.1007/978-3-030-01905-1_7

The main objective of this research is to identify and assess the significant risks in Egypt construction industry based on their risk rating (impact and probability). A comprehensive literature survey was conducted to build up general background knowledge of risk factors in construction projects and particular attention was paid to summarise the previous research findings. A construction industry survey was conducted through a questionnaire amongst contractors, consultants and owners. A total of 73 responses were received. Descriptive Statistical analysis was executed to analyse the responses and present the findings from the survey. The research compares effective risk assessment techniques through case studies to cope with risks associated with construction activities and to implement the projects in accordance with project objectives.

Different risk assessment techniques are evaluated by running case studies from building construction projects comparing Analytic Hierarchy Process (AHP) analysis and Crystal Ball software with the descriptive statistics. The comparison was conducted through three case studies from construction projects. The analysis of Case Study (1) revealed that there is a variation in cost as actual cost increased from the budgeted cost with 9.26% and the predicted cost increased from the budgeted cost by 6.76%–10.52%. Similarily Case Study (2) results revealed that there is a variation in cost as actual cost increased from the budgeted cost with 147.67% and the predicted cost increased by 143.84% to 154.59%. For Case Study (3); there is a variation in cost as actual cost increased from the budgeted cost with 10% and the predicted cost increased by 9.89%–11.70%. Finally, the comparison results found that the prediction of actual cost using different techniques was different.

2 Risk Definition and Causes of Risks

2.1 Risk Definition

Risk management is essential to construction activities in minimizing losses and enhancing profitability. In general, Risk management is a systematic process to define, analyze and respond to system's risk. It's a decision making process enables the organization to exploit the opportunities by increasing the probability and frequency period of desired occurrences and decreasing the negative consequences of undesired occurrences. The PMBOK-6th 2017, summarizes the management of risks into five sequential levels respectively are; Risk Management Planning, Risk Identification, Risk Analysis, Risk Response Planning and Auditing the Risk Management Performance, these processes encounter procedural steps to be accomplished (Smith 1999). The process of RM can be broken down into three essential components, they are risk identification, risk analysis, and risk responses. The construction industry is widely associated with a high degree of risk due to the nature of construction business activities, processes, environment and organization.

There were many different attempts to define risk, among which was that risk is "the potential for unwanted or negative consequences of an event or activity" (Rowe 1977).

Risk can also be generally recognized among those within the construction industry as the phenomenon of continually facing a variety of situations involving many unknown, unexpected, frequently undesirable and often unpredictable factors (Fong 1987). But the latter definition of risk tends to ignore its double- edged nature, which was recognized in defining risk as "the chance of something happening that will have an impact on objectives; may have a positive or negative impact" (AS/NZS 4360 2004).

This paper aims at identifying the top major risks regardless of their nature of impact, through considering the probability of their occurrence and their impact or magnitude of the consequences.

2.2 Egyptian Studies

Several articles have discussed the causes of risk and delays in construction projects; some studies identified the main causes of risk and ranked them, while other studies discussed the analysis methods and the proposed ways to mitigate them. Studies in Egypt were incorporated in this study to compile a list of risk factors. (Amer 1994), studied the major delay causes for construction projects which they are: poor contract management, unrealistic scheduling, lack of owner's financing/payment for completing work, design modifications during construction, and shortages in materials such as cement and steel. Abd El- Razek 2008, considered several delay causes in construction projects in Egypt as; financing by the contractor during construction, delays in contractor's payment by the owner, design changes by owner or his agent during construction, partial payments during construction, and non-utilization of professional construction/contract management. Marzouk 2014, stated that Finance and payments of completed work by owner, variation orders of scope by the owner during construction, effects of subsurface conditions, Low productivity level of labors and Ineffective planning and scheduling of the project were the most five delay causes of construction projects in Egypt. Aziz 2013 ranked factors perceived to affect delays factors and according to their importance level on delay, especially in the last decade. The data were analyzed using Relative Importance Index (RII) and the most important factors were: Delay in progress payments (Funding problems), Different tactical patterns for bribes, Shortage of equipment, Ineffective project planning and scheduling, poor site management and supervision. Khodir 2015, Identified the latest top major risk probabilities in construction projects in Egypt, according to political and economic variables between the time period Jan 2011 and Jan 2013 and then suggested a group of risk response strategies that suit each of the identified key risks. Currency price changes, new tax rates, Lack of fuel, unsecured roads, Official changes, Workers' strikes and Fire risk were the most important risk factors. Marzouk 2014, studied delays that relate to engineering factors which arise due to design development, workshop drawings, and change then he developed a knowledge based expert system for assessing the engineering related delay claims. Further studies have been conducted in the MENA region and internationaly to investigate the most common risks in construction industry, they have been studied and summarized during the questionnaire design, [11–20]. Table 1 summarizes the ranking of the most important risk factors affecting construction projects in Egypt due to recent researching works.

Table 1. Risk ranking according to previous studies in Egypt

Author	Abd El-Razek et al. (2008)	Aziz (2012)	Aziz (2013)	Marzouk et al. (2012)	Khodeir et al. (2015)
1	Financing by contractor during construction	Lowest bidding procurement method owner originated	Delay in progress payments (funding problems)	payments of completed work by owner	Currency fluctuation
2	Delays in contractor's payment by owner	Additional work owner originated	Different tactics patterns for bribes	Variation orders owner during construction	Change in taxation/new tax rates
3	Design changes by owner or his agent during construction	Bureaucracy in bidding/tendering method owner originated	Shortage of equipment	Effects of subsurface conditions (e.g., soil..)	Change energy cost/lack of fuel
4	Partial payments during construction	Wrong method of cost estimation	Ineffective project planning and scheduling	Low productivity level of labors	Safety/unsecure roads
5	Non-utilization of professional construction/contractual management	Funding problems owner originated	Poor site management and supervision	Ineffective planning and scheduling of project	Official changes
6	Slow delivery of materials	Inaccurate cost estimation Designer originated	Poor financial control on site	Difficulties in financing project by contractor	Workers' strikes
7	Miss-Coordination between various parties (contractor, subcontractor, owner, consultant) working on the project	Mode of financing and payment for completed work by owner	Rework due to errors	Type of project bidding and award (negotiation, lowest bidder)	Fire risk
8	Slowness of the owner decision making process	Unexpected ground conditions miscellaneous	Selecting inappropriate contractors	Shortage of construction materials in market	Bad communications between stakeholders
9	The relationship between different subcontractors' schedules	Inflation miscellaneous	Sudden failures actions	Late approval of design documents by owner	Poor documentations
10	Preparation of shop drawings and material samples	Fluctuation in prices of raw materials	Inadequate planning	Unqualified workforce	Poor project planning and control
11	Lack of database in estimating activity duration and resources	Inadequate planning owner originated	Incompetent project team		Owner hesitation about design
12	Shortage in construction materials	Poor contract management owner originated	Inadequate contractor experience		Lack of decision making
13	Poor organization of the contractor or consultant	Unstable cost of manufactured materials miscellaneous	Frequent equipment breakdowns		Poor material management and planning

(continued)

Table 1. (*continued*)

Author	Abd El-Razek et al. (2008)	Aziz (2012)	Aziz (2013)	Marzouk et al. (2012)	Khodeir et al. (2015)
14	Controlling subcontractors by main contractor in the execution of work	Scope changes/ inadequate pre-contract study designer originated	Global financial crisis		Poor equipment management & planning
15	Changes in materials types and specifications during construction	Inadequate site investigations contractor originated	Complexity of project (project type, project scale, etc.)		Poor labor planning
16	Obtaining permits from municipality	Inappropriate government policies miscellaneous	Legal disputes between project participants		Replacement of consultant
17	Waiting for approval of shop drawings and material samples	Inappropriate preconstruction study designer originated	Change orders		Increased material waste
18	Poor labor productivity	Inappropriate contractual procedure Owner originated	Inappropriate construction methods		Force majeure
19	Errors committed due to lack of experience	Inappropriate contractors owner originated	Unqualified/inadequate experienced labor		Geo-technical risks
20	Design errors/incomplete made by designers	Shortening in project period by owner	Conflicts between joint-ownership		New governmental acts or legislations

3 Research Objectives

The main objective of this research is to identify and assess the latest top major risk factors that affected construction projects in Egypt. This was fulfilled through obtaining feedback from different practitioners on the different aspects of risk management that aimed at:

- Construct a general risk register that includes the most common risks facing the construction contractors using heuristic data gathering.
- Based on the probability of occurrence and impact of each risk, ranking is conducted using qualitative risk assessment techniques. The purpose of the ranking is to highlight the risky areas and obtain the priority list of project risks.
- Set up a risk profile; quantitative assessment/response actions.
- Analyze the relationship of these factors and thereby enhance understanding of construction risk factors.

4 Research Methodology

To achieve the study objectives, the following procedure was carried out

- Review literature research to examine previous research and identify the gaps in current knowledge.
- Explore initial list of risk variables that deem important and affecting construction projects through literature review and experts' interviews.
- Conduct a questionnaire survey to assess the probability and the impact of the identified variables on construction projects in Egypt.
- Rank these risk factors using first procedure descriptive statistics based upon the feedback of the questionnaire survey.
- Suggest a strategy to manage risk for each type of identified risk depending on the questionnaire results.
- Collect historical data records for previous projects and prepare these data. Conduct Exploratory Factor Analysis (EFA) by using principal component analysis with VARIMAX rotation through Statistical Package for the Social Sciences (IBM-SPSS- V.23) software to isolate statistical variables that influence the risk factors and are named group (B); and define latent variables.

The sequence of work is illustrated in Fig. 1.

The first step was to identify risks in construction projects. This was done primarily through literature review. A comprehensive list of 77 risks was developed based on previous studies; (Marzouk 2008, Al-Khalil and Al-Ghafly 1999, Assaf and Al-Hejji 1995, 2006, Doloi 2012, Aziz and Remon 2013, Faridi et al. 2006, Abuwarda and Zinab 2016, Gündüz et al. 2013 and Sambasivan 2006). A questionnaire was then structured to get the perceptions of construction experts in Egypt. The questionnaire consisted of two sections. The first section was intended to gather information about the respondents' profile. The second section was intended to get the perception on the rating of each risk and the suggestion of the suitable risk response strategy. Each risk had three questions;

- The first question relates to the probability of the risk event occurring on construction projects. The respondents were asked to choose between rare, low, moderate, high and very high.
- The second question refers to the consequence on project objectives once the risk event occurs.
- The third question relates to the practical actions for managing these risks.

The survey presents seventy-seven (77) factors. These factors were classified into four (4) major categories based on the previous studies and the Egyptian code of project management. Every category extracted to minor categories as explained in Table 2:

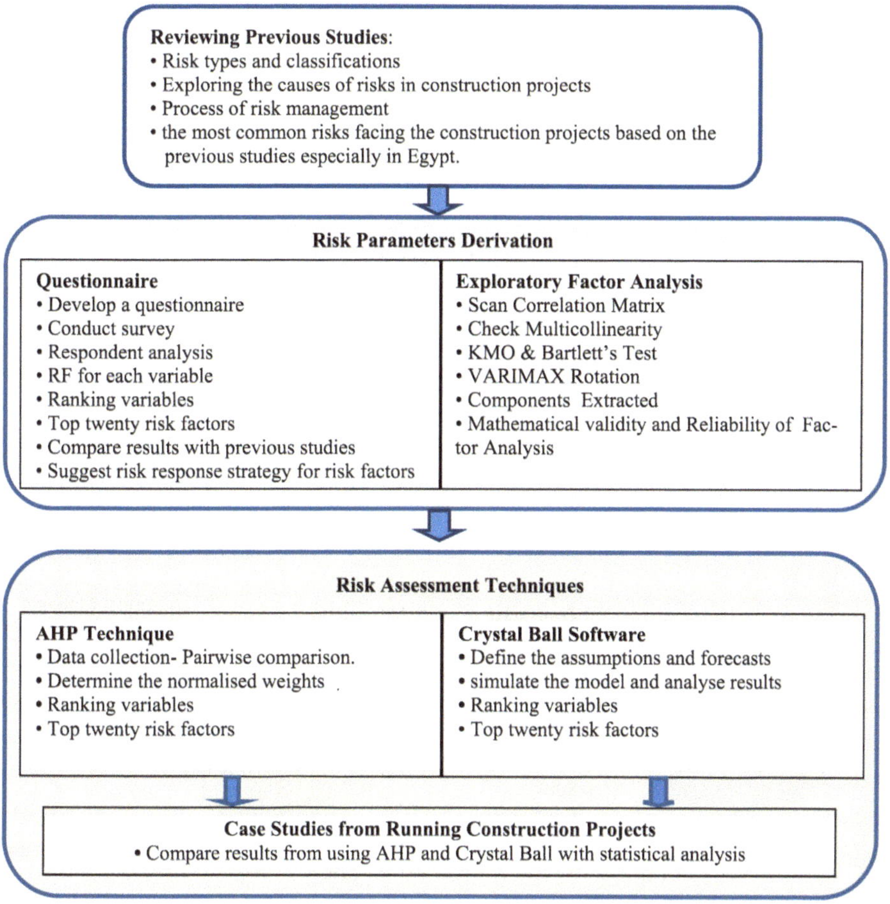

Fig. 1. Research methodology

4.1 Respondent's Profile

The questionnaire survey forms were distributed to construction professionals associated with Egyptian construction industry. The completed responses were collected either personally, or received through regular postal mails, e-mails, and faxes. Out of 100 distributed questionnaires, 73 have returned complete and used in the analysis. The following charts summarize the respondents' profile as illustrated in Fig. 2.

Table 2. Categorized factors that cause delay in construction projects

No.	Major category	Category	ID	No.
1	Project life-cycle category	Feasibility and strategy stage	01:07	7
		Tendering and contract stage	08:12	5
		Construction stage	13:18	6
		Testing, handover and guarantee stage	19:20	2
2	Project parties	Owner related factors	21:28	8
		Contractor related factors	29:32	4
		Consultant related factors	33:35	3
		Project management related factors	36:42	7
3	Resources	Project site related factors	43:48	6
		Equipment related factors	49:58	10
		Labor related factors	59:63	5
4	External	Environmental related factors	64:67	4
		Financial related factors	68:72	5
		Regulation related factors	73:77	5

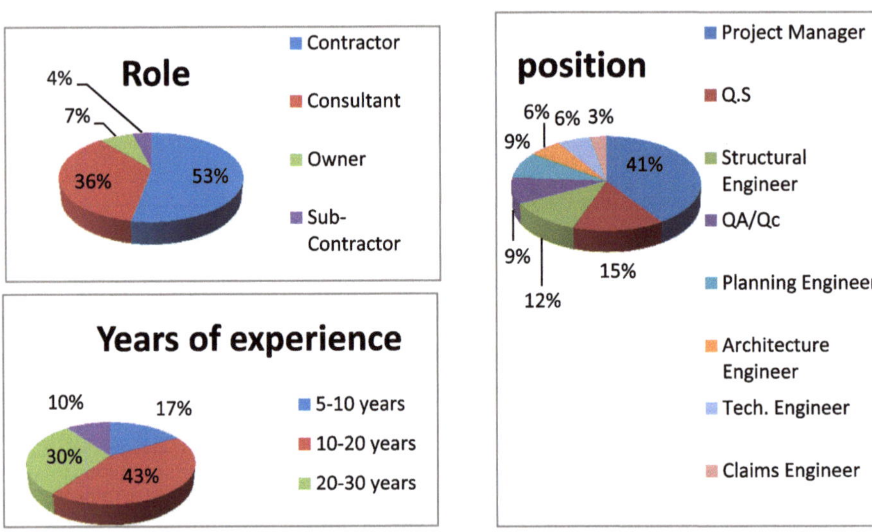

Fig. 2. Respondents' profile

4.2 Data Analysis Approach

The experts were requested to judge two attributes of each risk: the probability of occurrence, denoted by (P), and the degree of consequence, denoted by (C). The risk factor or index, denoted by (RF), is the function of these two attributes, Eqs. (1) and (2):

$$Risk\,Factor;\ RF = P * C \tag{1}$$

$$Risk\,Factor;\ RF = (P + C) - (P * C) \tag{2}$$

The first form is common in some forms of risk analysis, but the second is better because it identifies items with high likelihoods or high consequences or both, so the chance of high consequence but low likelihood items being ignored is reduced greatly (Cooper 2005; Khodier et al. 2015). In this research, the two equations were used to rank the variables to compare the results and choose the most realistic one.

The three variables (RF), (P) and (C) are all measured numerically. The respondents judged the Probability of occurrence using the five-level judgment scale of: very high, high, medium, low and rare. The same scale was also applied to the degree of consequence of the risk.

To apply the model, the opinion judgment scale was converted into numerical scales. The "very high" took a value of 0.9, and the "high," "medium," "low" and "rare" took values of 0.7, 0.5, 0.3 and 0.1 respectively, while the consequence numerical scale was the "very high" took a value of 0.8, and the "high," "medium," "low" and "rare" took values of 0.4, 0.2, 0.1 and 0.05 respectively as the per the risk impact scale involved in the (PMBOK-6th 2017).

4.3 Findings of the Survey

The attributes ranked using the two equations; attribute with highest RF or rank 1 indicates that it has the maximum risk impact while the attribute with lowest rank indicates that it has the least impact on construction projects (Table 3). The Risk responses were associated with the questionnaire to get expert judgement.

Table 3. Overall risk significant factors, with highly significant risks highlighted.

ID	Description	P × C		(P + C) − (P * C)		Response
		RF	Rank	RF	Rank	
R1	Improper project feasibility study	0.16	26	0.64	36	Avoid
R2	Scope changes arising from redesign and extensive variations	0.12	54	0.58	57	Mitigate
R3	Inappropriate overall organizational structure linking to the project	0.14	39	0.62	42	Mitigate
R4	Insufficient data collection and survey before design	0.13	47	0.60	51	Avoid
R5	Lack of design team experience in construction projects	0.18	18	0.67	20	Mitigate
R6	Mistakes and delays in producing design documents	0.13	46	0.65	27	Mitigate
R7	The technical specifications of the project is not achieved	0.14	38	0.62	41	Mitigate

(continued)

Table 3. (*continued*)

ID	Description	P × C		(P + C) − (P * C)		Response
		RF	Rank	RF	Rank	
R8	Improper type of project bidding and award (negotiation, lowest bidder)	0.13	50	0.59	55	Avoid
R9	Inadequate quality/ambiguity of contract documents	0.13	52	0.60	50	Avoid
R10	Inadequate definition of substantial completion and responsibilities	0.13	49	0.60	49	Avoid
R11	Improper risk allocation over parties in contract	0.13	43	0.64	38	Mitigate
R12	Ineffective delay penalties	0.11	62	0.56	63	Avoid
R13	Delay in the project scheduling	0.26	2	0.79	2	Mitigate
R14	Lack of the contractor's commitment to safety rules and regulations	0.22	7	0.75	4	Mitigate
R15	Work suspensions owing to conflicts	0.11	61	0.56	66	accept
R16	Delay in performing inspection and testing	0.10	66	0.57	60	Mitigate
R17	Lack of the contractor's commitment to maintenance and repair of defects that appear during the contractual warranty period	0.10	70	0.56	65	transfer
R18	Unrealistic inspection and testing methods proposed in contract	0.14	41	0.61	46	Mitigate
R19	Non-payment of all dues contractors and reported the final insurance or what is left of it to him	0.09	72	0.54	72	Avoid
R20	Delay in performing final inspection and certification by a third party	0.11	60	0.57	59	Avoid
R21	Owner Type	0.16	23	0.67	22	accept
R22	Owner's financial problems	0.26	3	0.76	3	Avoid
R23	Delays in contractor's progress payment (of completed work) by owner	0.22	6	0.72	8	Mitigate
R24	Delay to furnish and deliver the site to the contractor	0.11	59	0.58	58	Mitigate
R25	Delay in settlement of contractor's claim by the owner	0.18	19	0.70	13	Mitigate
R26	Suspension of work by owner	0.15	33	0.63	39	Avoid
R27	Slow decision making by owner	0.19	11	0.69	15	Mitigate
R28	Variation orders/changes of scope by the owner during construction	0.19	12	0.72	9	Mitigate
R29	Inadequate contractor experience	0.13	44	0.61	47	Avoid
R30	Un-use of advanced engineering design software and modern equipments	0.13	51	0.61	44	Mitigate
R31	Shortage of sub-contractors	0.13	48	0.60	53	Mitigate
R32	Conflicts between contractors and sub-contractors	0.18	20	0.70	14	Mitigate
R33	Inadequate experience of consultant	0.19	14	0.69	17	Avoid

(*continued*)

Table 3. (*continued*)

ID	Description	P × C		(P + C) − (P * C)		Response
		RF	Rank	RF	Rank	
R34	Delay in approving shop drawings and sample materials	0.15	30	0.66	25	Mitigate
R35	Conflicts between consultant and contractor	0.16	27	0.64	32	Avoid
R36	Poor site management and supervision	0.22	8	0.72	7	Mitigate
R37	Lack of communication between the parties	0.16	28	0.65	26	Mitigate
R38	Change in key staffing throughout the project	0.15	32	0.64	37	Avoid
R39	Ineffective planning of project	0.28	1	0.80	1	Mitigate
R40	Shortcoming of the measure and value process	0.17	22	0.67	23	Mitigate
R41	Lack of human resources planning	0.15	34	0.65	28	Mitigate
R42	Poor quality assurance/control	0.16	25	0.67	21	Mitigate
R43	Restricted access at the site	0.08	77	0.53	75	Accept
R44	Lack of protection on a construction site	0.14	42	0.61	45	Mitigate
R45	Unavailability of utilities in the site or Delay in providing services from utilities such as (water, etc.)	0.14	40	0.62	43	Mitigate
R46	Effects of subsurface conditions (e.g., soil, high water table, etc.)	0.17	21	0.66	24	Avoid
R47	Traffic control and restriction at job site	0.11	58	0.56	64	Mitigate
R48	Effect of social and cultural condition	0.09	73	0.54	71	Accept
R49	Shortage of construction materials, equipment and labors in market	0.16	29	0.64	35	Mitigate
R50	Improper storage of materials leading to damage	0.11	63	0.57	62	Mitigate
R51	Thefts done on site	0.10	69	0.54	73	Avoid
R52	Damage of sorted materials	0.10	67	0.55	70	Mitigate
R53	Delay in material delivery	0.22	5	0.74	5	Mitigate
R54	Changes in material types and specifications during construction	0.16	24	0.65	31	Avoid
R55	Change in material prices or price escalation due to monopoly	0.18	15	0.68	18	Accept
R56	Delay in material delivery	0.15	31	0.65	30	Mitigate
R57	Low efficiency of equipment	0.11	57	0.59	54	Mitigate
R58	Delay in equipment delivery	0.12	53	0.60	52	Transfer
R59	Unqualified workforce	0.18	17	0.69	16	Mitigate
R60	Shortage of labors	0.13	45	0.61	48	Mitigate
R61	Low productivity level of labors	0.15	36	0.64	34	Mitigate
R62	Labor injuries on site	0.09	75	0.53	74	Mitigate
R63	Labor strikes due to revolutions	0.11	64	0.55	68	Avoid
R64	Force Majeure as war, revolution, riot, strike, and earthquake, etc.	0.10	68	0.64	33	Accept

(*continued*)

Table 3. (*continued*)

ID	Description	P × C		(P + C) − (P * C)		Response
		RF	Rank	RF	Rank	
R65	Weather effect (hot, rain, etc.)	0.11	65	0.55	67	Accept
R66	pollution due to work	0.10	71	0.53	76	Avoid
R67	Archaeological area	0.08	76	0.55	69	Avoid
R68	Fluctuations in cost/currency	0.22	4	0.73	6	Share
R69	Delay in approval of completed work by client	0.20	10	0.71	12	Mitigate
R70	Financing by contractor during construction	0.21	9	0.71	11	Mitigate
R71	Delays in the owner advanced payment	0.09	74	0.52	77	Avoid
R72	Insufficient data collection and survey before design	0.18	16	0.68	19	Avoid
R73	Slow permit by government/municipality	0.12	55	0.58	56	Mitigate
R74	Delay in performing final inspection and certification by a third party	0.15	35	0.62	40	Avoid
R75	Incessant variation order	0.15	37	0.65	29	Mitigate
R76	Changes in government regulations and laws	0.12	56	0.57	61	Accept
R77	Wars and Revolutions	0.19	13	0.71	10	Accept

The top 20 risk factors affecting construction project in Egypt are bolded in Table 2, it seems to be the same results from the different analysis with different ranking; Conflicts between. Ineffective planning of project, Delay in the project scheduling and Owner's financial problems are the first, second and the third factors with high probability and high consequence on projects.

4.4 Ranking Top Major Risks

By analyzing the top risks from Table 2 using the two equations mentioned before, it is obvious that risk factors related to project management have the highest impact in the construction projects in Egypt followed by factors related to the owner, financial, consultant and construction stage. The other categories ranked as shown in Fig. 3 starting from the risk with highest risk impact.

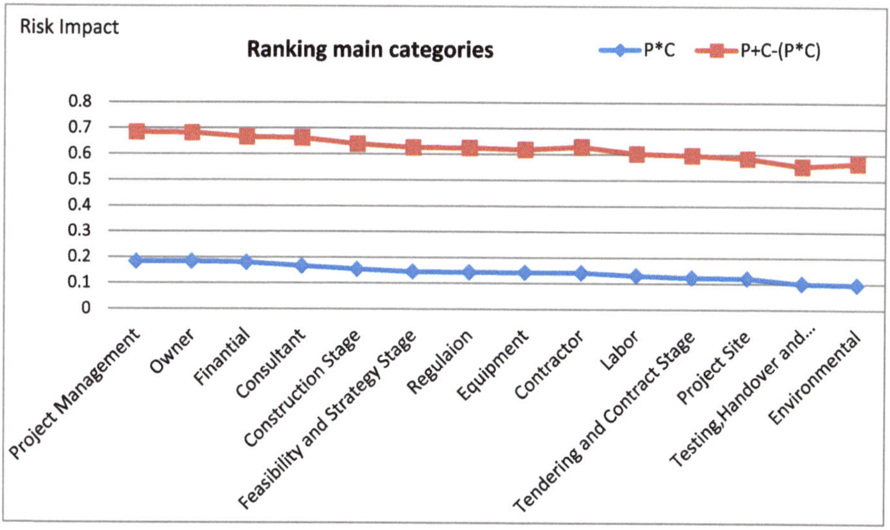

Fig. 3. Ranking main categories

5 Comparative Study

It was essential to correlate similar studies for Egypt and synchronize them to do the comparative study so that each significant cause reflects the most representative cause of delays and risk for construction projects in Egypt. This was necessary because each researcher had identified the attributes based on the local conditions and the prevailing problems. The comparative study depends on the following researches' work; (Abd El-Razek et al. 2008, Aziz 2013, Marzouk et al. 2008, Khodeir et al. 2014, and Aziz 2012). Table 4 represents the top twenty risk factors concluded from this study and its comparison with the top attributes in the other researches.

Table 4. Ranking of the 20 most significant risk factors in Egypt and their correspondings in the other researches in Egypt

Rank	Author	[7]	[31]	[21]	[8]	[9]
	Year	2008	2012	2013	2013	2014
1	Ineffective planning of project		11	10	5	
2	Insufficient project scheduling			3	5	
3	Owner's financial problems		5		6	
4	Fluctuations in cost/currency		9			1
5	Delay in material delivery	5				
6	Delays in contractor's progress payment (of completed work) by owner	2		1	1	

(continued)

Table 4. (*continued*)

Rank	Author	[7]	[31]	[21]	[8]	[9]
	Year	2008	2012	2013	2013	2014
7	Lack of the contractor's commitment to safety rules and regulations					
8	Poor site management and supervision	6		5		
9	Financing by contractor during construction	1				
10	Delay in approval of completed work by client					
11	Slow decision making by owner	7				
12	Variation orders/changes of scope by the owner during construction	3	14		2	
13	Strikes and revolutions					6
14	Inadequate experience of consultant	4	19	12		
15	Change in material prices or price escalation due to monopoly		10			
16	Insufficient data collection and survey before design		15			
17	Unqualified workforce			19	10	
18	Shortage of construction materials, equipment and labors in market			3	8	
19	Delay in settlement of contractor's claim by the owner					
20	Conflicts between contractors and sub-contractors					

It is observed that only three factors have no ranking in the studies, and the other 17 attributes considered being one of the most important attributes affecting construction projects in Egypt. This result emphasized the current research findings.

6 Suggested Risk Response Strategies

In the third part of the questionnaire, the experts suggest suitable risk response strategies for the identified key risks. The results depend on the most likely suggestion from the 73 respondents, Table 2. The strategies include negative risk responses, such as avoidance, transfer, sharing, mitigation and passive acceptance as illustrated in Fig. 4.

Fig. 4. Risk response strategies (Cooper 2005)

7 Factor Analysis

Factor analysis is a powerful statistical technique that aims at providing greater insight among numerous correlated, but seemingly unrelated variables into a much fewer underlying factors (Doloi 2009). In order to evaluate the adequacy of the survey data for factor analysis, Kaiser–Meyer–Olkin (KMO) test and Bartlett's test were conducted, Field 2005. The value of KMO represents the ratio of the squared correlation between variables to the squared partial correlation between variables. It varies from 0 to 1. A value close to 1 indicates that the pattern of correlations is relatively compact and hence factor analysis should give distinct and reliable results, Field 2005. A minimum value of 0.5 has been suggested, Kaiser 1974. In this research of the 77 attributes, a total of 40 attributes was selected based on the highest risk factors extracted from the descriptive statistics analysis depending on the two equations. After deleting the repetitive attributes, only 20 attributes remained. The KMO value of the selected 20 variables is 0.634 which is deemed good for this study. In order to test the null hypothesis that the correlation matrix is an identical matrix, the principal component analysis requires the probability associated with Bartlett's Test to be less than the level of significance (Zhang 2005). The probability associated with the Bartlett test is < 0.001, which satisfies this requirement as shown in Table 5.

Table 5. KMO and Bartlett's test

Kaiser-Meyer-Olkin	Measure of sampling adequacy	0.634
Bartlett's test	Approx. Chi-Square	413.152
	df	190
	Sig.	.000

7.1 Extracted Components

The principal components analysis, PCA, is adopted to reduce the highly correlated project attributes into a smaller number of risk factors. As stated above, the factor analysis was performed on the selected 20 attributes and principal components (factors) were extracted as shown in Table 7. These principal components were extracted by specifying the minimum initial Eigen value of 1.0. A screen plot to graph the Eigen values against the number of total components was generated in the data reduction process, Field 2005. The first six components, which cumulatively explain 72.1% of the total variances as shown in Table 6, have been kept and other components that accounted for less than 27.9% were dropped. The variables that were grouped into factors were given new headings according to its relationship meaning:

- Factor 1: Lack of experience
- Factor 2: Lack of owners' commitment
- Factor 3: Lack of clarity in project scope
- Factor 4: Egyptian economic crises
- Factor 5: Lack of contractor's commitment
- Factor 6: Inefficient site management

Table 6. Factor analysis-component extracted

ID	Descriptions of factors and the attributes	Factor loading	Variance explained
Factor 1: Lack of experience			*15.792%*
R33	Inadequate experience of consultant	.838	
R36	Poor site management and supervision	.730	
R5	Lack of design team experience in construction projects	.719	
R25	Delay in settlement of contractor's claim by the owner	.655	
R39	Ineffective planning of project	.591	
R77	Strikes and revolutions	.553	
Factor 2:Lack of owners' commitment			*14.495%*
R59	Unqualified workforce	.821	
R23	Delays in contractor's progress payment (of completed work) by owner	.774	
R22	Owner's financial problems	.713	
R27	Slow decision making by owner	.697	
Factor 3: Lack of clarity in project scope			*10.869%*
R28	Variation orders/changes of scope by the owner during construction	.820	
R13	Delay in the project scheduling	.543	
Factor 4: Egyptian economic crises			*10.595%*
R55	Change in material prices or price escalation due to monopoly	.791	
R68	Fluctuations in cost/currency	.774	
Factor 5: Lack of contractors commitment			*10.203%*
R13	Delay in the project scheduling	.533	
R14	Lack of the contractor's commitment to safety rules and regulations	.533	
Factor 6: Inefficient site management			*10.146%*
R69	Delay in approval of completed work by client	.769	
R53	Delay in material delivery	.727	
R72	Insufficient data collection and survey before design	.572	
R70	Financing by contractor during construction	.557	

8 Mathematical Validity of Factor Analysis

8.1 Validity

Validity analysis examines whether what is expected to be measured is truly measured, Zhang 2005, i.e. whether the attributes grouped under a certain factor (or component) in the data reduction process collectively explain the same measure within the target dimensions. If the attributes truly explain the measure of the factor identified in the factor analysis, they should significantly correlate with one another. By taking the Pearson correlation (r), it can be interpreted the amount by which the two variables affect one another. Using the SPSS, the Pearson bivariate correlation analysis was performed to examine the relationships between the measured attributes within all the six extracted factors as listed in Table 7. Table 7 shows the correlations between the attributes in factors 1 to 6. Pearson bivariate correlations are greater than 0.4 in most of the cases among different attributes in all the factors. From these results, factors formed in factor analysis contain attributes which are related.

Table 7. Correlation matrix for the attributes

Factor 1

	R33	R36	R5	R25	R39	R77
R33	1	.611	.516	.506	.425	.354
R36	.611	1	.439	.665	.457	.159
R5	.516	.439	1	.430	.301	.293
R25	.506	.665	.430	1	.574	.214
R39	.425	.457	.301	.574	1	.114
R77	.354	.159	.293	.214	.114	1

Factor 2

	R59	R23	R22	R27
R59	1	.573	.398	.507
R23	.573	1	.460	.458
R22	.398	.460	1	.510
R27	.507	.458	.510	1

Factor 3

	R28	R13
R28	1	.489
R13	.489	1

Factor 4

	R55	R68
R55	1	.701
R68	.701	1

Factor 5

	R13	R14
R13	1	.532
R14	.532	1

Factor 6

	R69	R53	R72	R70
R69	1	.293	.428	.425
R53	.293	1	.254	.415
R72	.428	.254	1	.477
R70	.425	.415	.477	1

8.2 Reliability

Reliability is an important measure to ensure the consistency of the construct over time (i.e. consistency of measured attributes and scale), Cronbach's alpha test was performed on entire data as well as attributes in each factor which are shown in Table 8. The value of $C\alpha$ could be anywhere in the range of 0 to 1, where a higher value denotes the greater internal consistency and vice versa.

The value of $C\alpha$ is inflated by a large number of variables, so there is no set interpretation as to what is an acceptable limit, Zhang 2005. However, a rule of thumb applies to most situations with the following ranges: $C\alpha > 0.9$ denotes excellent, $0.9 > C\alpha > 0.8$ as good, $0.8 > C\alpha > 0.7$ as acceptable, $0.7 > C\alpha > 0.6$ as questionable, $0.6 > C\alpha > 0.5$ as poor and $0.5 > C\alpha$ denotes unacceptable, Doloi 2009. The value of $C\alpha$ for all attributes calculated is 0.859 which is considered to be good.

Table 8. Reliability Cronbach's alpha for the attributes.

Attributes Cronbach's alpha (Cα)	Attributes Cronbach's alpha (Cα)
Attributes in factor 1	0.801
Attributes in factor 2	0.774
Attributes in factor 3	0.652
Attributes in factor 4	0.821
Attributes in factor 5	0.692
Attributes in factor 6	0.712
All attributes selected for factor analysis	0.859

Table 8 shows the values of the Cronbach's alpha (Cα) for the attributes under each factor and the results indicate a good overall reliability and internal consistency of the measured attributes in the analysis.

9 Ranking of Risk Factors According to AHP Technique

In order to adapt the AHP methodology, the first step is to arrange the elements of the decision problem in the form of a hierarchy. Figure 5 summarizes the concequence of conducting AHP Technique. A top down approach has been adopted in formulating the AHP model for this research. A hierarchy that consists of four levels, and descends from the general to the more particular was developed. The top level is the overall goal of the decision, followed by the decision criteria which impact the goal directly in the second level. The sub-criteria level comes next against the alternatives to be evaluated at the lowest level. The goal of this decision problem is the ranking of the risk factors affecting construction projects in Egypt and assigning the top twenty risk factors. This objective can be achieved by considering four strategic criteria, namely project life cycle, project parties, resources and external related factors, which form the second level in the hierarchy. The third level of the hierarchy contains the sub-criteria which were already before. The criteria and sub-criteria used in the hierarchy can be assessed using the AHP approach of pairwise comparison of elements in each level with respect to every parent element located one level above. Local priorities result directly from pairwise comparisons of the sub-criteria with respect to the criteria, while global priorities result from the multiplication of criteria and sub-criteria priorities. For instance, a set of global priority weights is produced for each of the sub-criteria by multiplying local weights of the sub-criteria with weights of all the parent elements above it. The local priorities define a share of a given decision-making element in reaching the goal at the upper level, where the global priorities of a given level represent in turn the share of each element in reaching the main goal, which is the ranking of the top twenty risk factors affecting construction projects in Egypt. After building the hierarchy, next is the pairwise comparisons phase. Hence, from the hierarchy shown in Fig. 6, there are 92 sets of pairwise comparison matrices, one for the criteria with respect to the goal, 14 for the sub-criteria with respect to each criterion, and 77 for the risk factors with respect to each of the sub-criteria. Table 9 illustrates the ranking of the top risk factors according to global weight using AHP technique. Figure 7 shows the ranking of the categories according to AHP assessment.

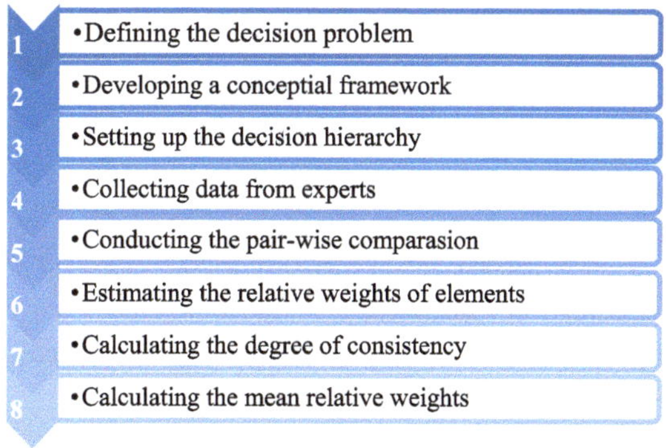

Fig. 5. AHP- processes.

10 Ranking of Risk Factors Using Crystal Ball Simulation Technique

Crystal Ball is an easy-to-use simulation program that helps to analyse the risks and uncertainties associated with Microsoft Excel spread sheet models. To use Crystal Ball, the following steps must be performed as shown in Fig. 8. Table 10 shows priority list of the top twenty project risks according to 90% confidence level. The ranking is according to the two formulas; [P * C] and [P + C− (PC)]. Using the two formulas gives the same results; expect the two risk factors; Force Majeure (R64) and conflicts between contractors and sub-contractors (R32) are added according to the second formula. The output charts and statistics data developed by crystal ball software are depending on the three probability distribution; uniform, triangular and custom distribution. The results show that using the different charts for assumption, are almost consistent and give the same top twenty project risks with different priorities.

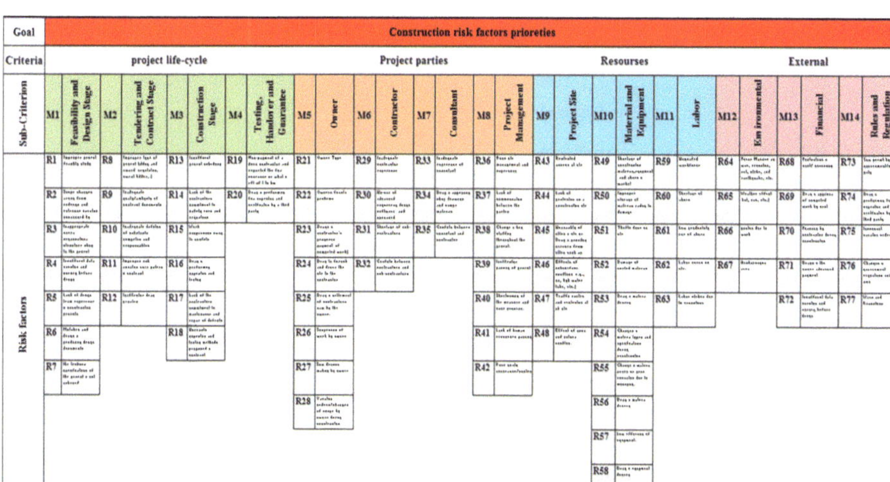

Fig. 6. AHP hierarchy for risk factors affecting construction projects in Egypt

Table 9. Top twenty risk factors using AHP assessment

Rank	ID	Description	Global weight
1	R22	Owner's financial problems	0.0614
2	R39	Ineffective planning of project	0.0564
3	R33	Inadequate experience of consultant	0.0528
4	R36	Poor site management and supervision	0.0506
5	R68	Fluctuations in cost/currency	0.0498
6	R23	Delays in contractor's progress payment (of completed work) by owner	0.0429
7	R40	Shortcoming of the measure and value process	0.0351
8	R13	Insufficient project scheduling	0.0308
9	R69	Delay in approval of completed work by client	0.0296
10	R35	Conflicts between consultant and contractor	0.0291
11	R27	Slow decision making by owner	0.0246
12	R14	Lack of the contractor's commitment to safety rules and regulations	0.0233
13	R1	Improper project feasibility study	0.0229
14	R5	Lack of design team experience in construction projects	0.0224
15	R70	Financing by contractor during construction	0.0210
16	R77	Loss due to Egyptian revolutions	0.0205
17	R28	Variation orders/changes of scope by owner during construction	0.0204
18	R42	Poor quality assurance/control	0.0192
19	R34	Delay in approving shop drawings and sample materials	0.0161
20	R74	Delay in performing final inspection and certification by a third party	0.0156

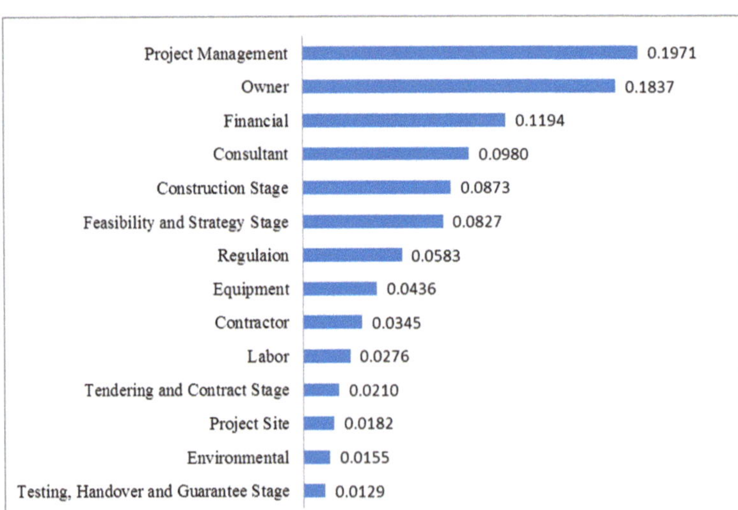

Fig. 7. Ranking of the categories according to AHP assessment

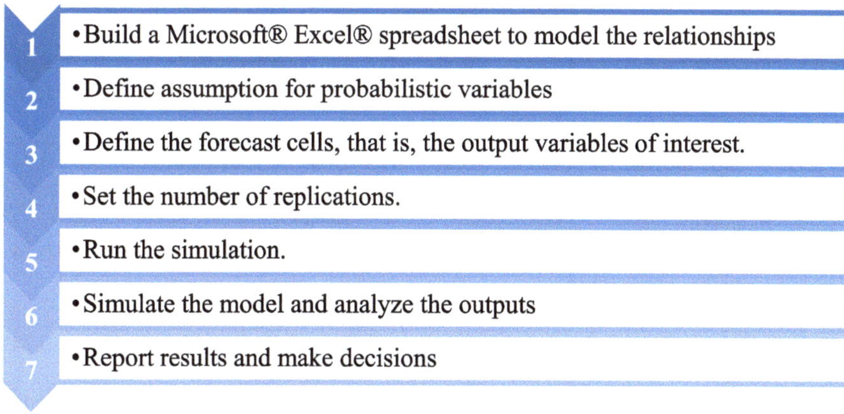

Fig. 8. Model steps development in Crystal Ball.

Table 10. Priority list of project risks (90% confidence level)

No.	ID	Forecast formula [P*C]	Risk factor	ID	Forecast formula [P + C – (PC)]	Risk factor
		Risk description			Risk description	
1	R39	Ineffective planning of project	0.41	R39	Ineffective planning of project	0.83
2	R22	Owner's financial problems	0.4	R22	Owner's financial problems	0.81
3	R13	Insufficient project scheduling	0.36	R14	Lack of the contractor's commitment to safety rules and regulations	0.81
4	R14	Lack of the contractor's commitment to safety rules and regulations	0.35	R13	Insufficient project scheduling	0.8
5	R23	Delays in contractor's progress payment (of completed work) by owner	0.34	R36	Poor site management and supervision	0.8
6	R36	Poor site management and supervision	0.34	R28	Variation orders/changes of scope by owner during construction	0.78
7	R53	Delay in material delivery	0.33	R68	Fluctuations in cost/currency	0.77
8	R68	Fluctuations in cost/currency	0.33	R55	Change in material prices or price escalation due to monopoly	0.77
9	R70	Financing by contractor during construction	0.31	R23	Delays in contractor's progress payment (of completed work) by owner	0.76
10	R77	Loss due to Egyptian Revolutions	0.31	R77	Loss due to Egyptian Revolutions	0.76

(*continued*)

Table 10. (*continued*)

No.	ID	Forecast formula [P*C] Risk description	Risk factor	ID	Forecast formula [P + C – (PC)] Risk description	Risk factor
11	R55	Change in material prices or price escalation due to monopoly	0.31	R25	Delay in settlement of contractor's claim by the owner	0.74
12	R28	Variation orders/changes of scope by owner during construction	0.3	R53	Delay in material delivery	0.74
13	R27	Slow decision making by owner	0.3	R46	Effects of subsurface conditions	0.74
14	R69	Delay in approval of completed work by client	0.29	R33	Inadequate experience of consultant	0.73
15	R33	Inadequate experience of consultant	0.29	R69	Delay in approval of completed work by client	0.73
16	R25	Delay in settlement of contractor's claim by the owner	0.29	R70	Financing by contractor during construction	0.73
17	R46	Effects of subsurface conditions	0.29	R64	Force Majeure	0.72
18	R72	Insufficient data collection and survey before design	0.28	R27	Slow decision making by owner	0.72
19	R5	Shortage of construction materials, equipment and labours in market	0.28	R72	Design errors/incomplete made by designers	0.72
20	R59	Unqualified workforce	0.28	R32	Conflicts between contractors and sub-contractors	0.72

11 Comparasion of Using Different Risk Assessment Techniques

Comparing the top 20 risk factors affecting construction projects in Egypt from the result of using different risk assessment techniques are summarises in Table 11. Thirteen factors from the twenty risk factors are existing in the three risk assessment techniques.

Table 11. Top 20 risk factors comparison using different risk assessment techniques

No.	Descriptive statistical analysis		Analytic Hierarchy Process (AHP) analysis		Crystal Ball software	
	ID	Description	ID	Description	ID	Description
1	R39	Ineffective planning of project	R22	Owner's financial problems	R39	Ineffective planning of project
2	R13	Insufficient project scheduling	R39	Ineffective planning of project	R22	Owner's financial problems
3	R22	Owner's financial problems	R33	Inadequate experience of consultant	R14	Lack of the contractor's commitment to safety rules and regulations
4	R68	Fluctuations in cost/currency	R36	Poor site management and supervision	R13	Insufficient project scheduling
5	R53	Delay in material delivery	R68	Fluctuations in cost/currency	R36	Poor site management and supervision
6	R23	Delays in contractor's progress payment (of completed work) by owner	R23	Delays in contractor's progress payment (of completed work) by owner	R28	Variation orders/changes of scope by owner during construction
7	R14	Lack of the contractor's commitment to safety rules and regulations	R40	Shortcoming of the measure and value process	R68	Fluctuations in cost/currency
8	R36	Poor site management and supervision	R13	Insufficient project scheduling	R55	Change in material prices or price escalation due to monopoly
9	R70	Financing by contractor during construction	R69	Delay in approval of completed work by client	R23	Delays in contractor's progress payment (of completed work) by owner
10	R69	Delay in approval of completed work by client	R35	Conflicts between consultant and contractor	R77	Loss due to Egyptian Revolutions
11	R27	Slow decision making by owner	R27	Slow decision making by owner	R25	Delay in settlement of contractor's claim by the owner
12	R28	Variation orders/changes of scope by owner during construction	R14	Lack of the contractor's commitment to safety rules and regulations	R53	Delay in material delivery
13	R77	Loss due to Egyptian revolutions	R1	Improper project feasibility study	R46	Effects of subsurface conditions (e.g., soil, high water table, etc.)

(*continued*)

Table 11. (*continued*)

No.	Descriptive statistical analysis		Analytic Hierarchy Process (AHP) analysis		Crystal Ball software	
	ID	Description	ID	Description	ID	Description
14	R33	Inadequate experience of consultant	R5	Lack of design team experience in construction projects	R33	Inadequate experience of consultant
15	R55	Change in material prices or price escalation due to monopoly	R70	Financing by contractor during construction	R69	Delay in approval of completed work by client
16	R72	Design errors/incomplete made by designers	R77	Loss due to Egyptian revolutions	R70	Financing by contractor during construction
17	R59	Unqualified workforce	R28	Variation orders/changes of scope by owner during construction	R64	Force Majeure
18	R49	Shortage of construction materials, equipment and labors in market	R42	Poor quality assurance/control	R27	Slow decision making by owner
19	R25	Delay in settlement of contractor's claim by the owner	R34	Delay in approving shop drawings and sample materials	R72	Design errors/incomplete made by designers
20	R32	Conflicts between contractors and sub-contractors	R74	Delay in performing final inspection and certification by a third party	R32	Conflicts between contractors and sub-contractors

12 Case Studies

Three case studies from construction projects in Egypt are used to analyse and quantify the impact of risks factors using the information collected from the construction professionals. Experts were asked for their opinion about causes of cost variation and the percentage of impact to enable prediction of actual cost of such project before completion. A total of 35 out of 77 critical risk factors were identified as major influencing risk factors on the construction project. Experts gave every risk factor; out of the 77 risk factors; influencing cost overrun factor on the construction project Percentages of each observed risk factors. The Impact *(d_j)* of critical risk factors associated with construction projects were used to analyse and identify the possible project cost variation in order compare different risk assessment techniques. The project Cost Variation Coefficient ***(CVC)*** and the Predicted Actual Cost ***(PAC)*** of the three case studies are presented in

12 due to the use of statistical analysis, AHP and Crystal Ball techniques and compared with the actual cost of the projects. From previous analysis of collected data from construction projects field, the planner can predict approximately the construction actual cost of any new construction project before construction using the following equations (Aziz 2013).

$$CVC = 1 + \frac{\sum_{J=5}^{J=1}(dj \times ERIIj)}{\sum_{J=5}^{J=1}(ERIIj)} \tag{3}$$

$$PAC = VC \times TBC \tag{4}$$

Where;

- ***CVC***: is the project Cost Variation Coefficient;
- ***ERII$_J$*** (%): is the Equivalent weighted average percentage of Relative Importance Index per Risk Factor;
- ***dj:*** is the percentage of each Risk Factor impact that ranged between (0.00–1.00).
- ***PAV:*** is the Predicted Actual Cost at completion before construction of the studied project;
- ***TBC:*** is the Total Budgeted Cost before construction of the studied project.

From studying and analysing the previous projects, it was found that for Case Study (1); there is a variation in cost as actual cost increased from the budgeted cost with 9.26% and the predicted actual cost increased from the budgeted cost by 6.76% to 10.52%. Similarily, the Case Study (2) results revealed that there is a variation in cost as actual cost increased from the budgeted cost with 147.67% and the predicted actual cost increased from the budgeted cost by 143.84% to 154.59%. For Case Study (3); there is a variation in cost as actual cost increased from the budgeted cost with 10% and the predicted actual cost increased from the budgeted cost by 9.89% to 11.70%. These differences in cost variation ranges resulted from using different risk assessment techniques. The comparison results found that the prediction of actual cost using different techniques was different from the views of the three case studies. The results presented in Table 12 revealed that using different techniques give predicted actual cost almost consistent with max variance from the actual cost less than 8% which is accepted.

Table 12. Summary of case studies' results

No.	Assessment technique			Case study (1) Initial cost = 540,000,000 L.E		Case study (2) Initial cost = 15,000,000 L.E		Case study (3) Initial cost = 43,000,000 L.E	
				CVC %	PAC	CVC %	PAC	CVC %	PAC
				9.26	590,000,000	146.7	37,000,000	10%	47,300,000
1	Statistical Analysis	(P * C)	PERT	7.4	580,016,341	147.4	37,110,253	10.0	47,315,251
2			Mean	6.9	577,030,137	144.5	36,668,471	9.9	47,253,120

(continued)

Table 12. (*continued*)

No.	Assessment technique			Case study (1) Initial cost = 540,000,000 L.E		Case study (2) Initial cost = 15,000,000 L.E		Case study (3) Initial cost = 43,000,000 L.E	
				CVC %	PAC	CVC %	PAC	CVC %	PAC
3		(P + C) − (PC)	PERT	7.6		147.5		10.5	47,502,372
4			Mean	6.9	577,423,968	144.8	36,717,300	10.0	47,306,168
5	AHP	(P * C)	PERT	10.5	596,823,728	152.4	37,853,463	11.1	47,779,950
6			Mean	9.7	592,560,969	149.3	37,400,595	11.7	48,031,261
7		(P + C) − (PC)	PERT	10.1	594,399,258	154.6	38,187,996	11.4	47,914,375
8			Mean	10.5	596,823,728	152.4	37,853,463	11.1	47,779,950
9	Crystal Ball		Custom	7.4%	580,000,637	145.9	36,888,180	10.4	47,471,821
10			Normal	7.4%	580,000,637	145.9	36,888,180	10.4	47,471,821
11			Triangular	6.8%	576,478,287	143.8	36,562,187	9.9	47,274,356
Min. and Max CVC %				6.8%	10.5%	143%	154.6%	9.9	11.7

13 Discussion of the Critical Risk Factors

13.1 Factor 1: Lack of Experience

Inadequate experience of the contractor is due to the lack of site management skills of the client. The inexperienced contractor may not be able to cope up with the progress of work or may not understand the complexity of the project leading to misinterpretation and confusion. Inadequate experience of contractor in turn leads to improper management of site and thus cause time overruns. The Poor site management and supervision clearly highlights the lack of coordination between various bureaucratic hierarchies involved in Egyptian construction industry. Efficient site management and effective supervision is one of the vital factors for achieving success in Egyptian projects. Improper planning during the bidding stage is one important reason of risk which is generally overlooked in analysis of construction delays (Doloi 2009). In most of the construction companies, people who apply for bid are different from who actually execute the work. Hence there is a tendency that people who apply for bid tend to be a bit overly optimistic and don't envisage for various practical contingencies.

13.2 Factor 2: Lack of Owners' Commitment

In case of the unavailability of workforce with the required skill set and hiring of unskilled labor is inevitable, they must be trained properly before putting them at work. Delay in payment to the contractor, which in turn causes financial difficulties. Financial difficulties of contractors have reportedly been one of the important reasons for delay in construction projects (Assaf and Al-Hejji 2006). Slow decision from the owner is due to the lack of proper coordination between owner and consultant or owner and contractor. This occurs when a contractor or consultant fails to make the client understand the time significance of the decision to be taken or the owner's decision is not communicated properly to concerning parties.

13.3 Factor 3: Lack of Clarity in Project Scope

Variation orders/changes of scope clearly come under lack of clarity of project scope by owner and designer. Change in scope of work and variation orders at a later stage delay the project completion due to change in quantities and change in project schedule. It may further delay the project due to unavailability of appropriate spare resources to the contractors. In fact, it results in a complete drain out of the contractor's resources and reduce his capability to follow the time plan. Insufficient scheduling is due to lack of clarity in project scope. Though the unrealistic schedule not only causes time overrun, it also compels contractors for compromising quality of construction leading to mistakes and reworks in construction activities.

13.4 Factor 4: Egyptian Economic Crises

Change in material prices or price escalation due to monopoly and fluctuations in cost/currency are mainly the results of the Egyptian economic crises. It causes increase in all of the project activity cost and cause time overrun and May causes the project to be stopped.

13.5 Factor 5: Lack of Contractor's Commitment

Insufficient scheduling in this factor is due to lack of coordination between client and contractor about the realistic difficulties at the site. Site accidents due to lack of safety measures are due to lack of commitment from both client and contractor towards the project. Site accidents not only harm individuals and consume time, but also it is also observed that productivity of labor reduces significantly after an accident. Time is also wasted in attending to accidents and replacing the injured person by a person with lesser or irrelevant skills. This then relates to the efforts required for training and development. These can be avoided if the client and contractor commit to appropriate safety measures adopted on the site.

13.6 Factor 6: Inefficient Site Management

Delay in approval of completed work by client occurs due to lack of communication between contractor and the approval authority. Lack of communication can be mutual, i.e. either the client is unaware of the completed work or order of approved work is not communicated back to the contractor. Delay in material delivery by vendors shows inefficient site management and the lack of commitment in terms of contractor's procurement planning prior to the construction phase of the project. Ignorance of the lead time for material delivery by the vendors potentially results in material shortage, which has reportedly been one of the significant causes of schedule delay across construction projects.

14 Conclusions

The objective of this research was to identify the main causes of risk that affect construction projects in Egypt. A literature review was conducted to identify the risk factors stipulated in the literature. 73 responses were conducted to identify the most appropriate risk attributes. A compiled list of 77 causes was obtained and subjected to further quantitative evaluation in a questionnaire survey to confirm the causes and identify the most important causes of project risk. The most important causes identified by the survey, and based on statistical analysis, AHP and Simulation techniques were: ineffective planning of the project; delay in the project scheduling; owner's financial problems; fluctuations in cost/ currency; delay in material delivery and delays in contractor's progress payment (of completed work) by owner.

Project management; Owners' related factors; financial and consultants' related factors are the most important categories affecting construction projects

In addition to identifying the top major key risks, the paper also suggests a group of risk response strategies that suit each of the identified key risks. These risks and their equivalent response strategies have been identified to be added. Factor analysis of responses on the project attributes has extracted critical factors; lack of experience; lack of owners' commitment; lack of clarity in project scope; Egyptian economic crises; lack of contractor's commitment; and Inefficient site management are the six critical factors. Validity analysis and reliability analysis confirm the quality of the questionnaire survey and the soundness of the factor analysis.

15 Data Availabilty Statement and Limitations

- All data generated or analyzed during the study are included in the submitted article or supplemental materials files. Data are covering the last decade of the Egyptian construction market and still valid. This research investigated construction projects in general and made no differentiation between international companies joining either Egyptian private sector companies or Egyptian public sector companies. There is a need to investigate such differences in-depth because the two types of Egyptian company have different behaviours and organisational structures. The study focuses on the construction projects in general and the data were collected with this in mind. Therefore determination and ranking the risk factors are generated based on all types of Egyptian construction projects and doesn't consider other types of construction projects such as building construction projects; construction wastewater projects; and road construction projects.

References

Smith, N.J.: Managing Risk in Construction Projects. Blackwell, Oxford (1999)
Rowe, W.D.: An Anatomy of Risk. Wiley, New York (1977)
Fong, S.W.: Risk management. Cost Eng. **25**, 12–16 (1987)
AS/NZS 4360, Australian/New Zealand Standard on Risk Management. Standards, Australia and Standards, New Zealand (2004)
Amer WH. Analysis and evaluation of delays in construction projects in Egypt. MSc. Thesis, Zagazig University, Egypt (1994)
Abd El-Razek, M.E., Bassioni, H.A., Mobarak, A.M.: Causes of delay in building construction projects in Egypt. J. Constr. Eng. Manag. **134**(11), 831–841 (2008)
Marzouk, M.M., El-Rasas, T.I.: Analyzing delay causes in egyptian construction projects. J. Adv. Res. **5**(1), 49–55 (2014)
Khodeir, L.M., Mohamed, A.H.: Identifying the latest risk probabilities affecting construction projects in Egypt according to political and economic variables. From January 2011 to January 2013. HBRC J. **11**(1), 129–135 (2015)
Marzouk, M., El-Dokhmasey, A., El-Said, M.: Assessing construction engineering-related delays: Egyptian perspective. J. Prof. Issusse Eng. Educ. Pract. **134**(3), 315–326 (2008)
Al-Khalil, M., Al-Ghafly, M.: Important causes of delay in public utility projects in Saudi Arabia. J. Constr. Manag. Econ. **17**(5), 647–655 (1999)
Assaf, S., Al-Khalil, M., Hazmi, M.A.: Delay causes in large building construction projects. J. Manag. Eng. ASCE **11**(2), 45–50 (1995)
Assaf, S., Al-Hejji, S.: Delay causes in large construction projects. Int. J. Proj. Manag. **24**(4), 349–357 (2006)
Faridi, A.S., El-Sayegh, S.M.: Significant factors causing delay in the UAE construction industry. Constr. Manag. Econ. **24**(11), 1167–1176 (2006)
Cooper, D.F.: Project Risk Management Guidelines. Wiley, West Sussix (2005)
Aziz, R.F.: Ranking of delay factors in construction projects after egyptian revolution. Alex. Eng. J. **52**(3), 387–406 (2013a)
Doloi, H., et al.: Analyzing factors affecting delays in Indian construction projects. Int. J. Proj. Manag. **30**(4), 479–489 (2012)
Gündüz, M., Nielsen, Y., Özdemir, M.: Quantification of delay factors using the relative importance index method for construction projects in Turkey. J. Manag. Eng. **29**(2), 133–139 (2013). https://doi.org/10.1061/(asce)me.1943-5479.0000129
Doloi, H.: Analysis of pre-qualification criteria in contractor selection and their impacts on project success. Constr. Manag. Econ. **27**(12), 1245–1263 (2009). https://doi.org/10.1080/01446190903394541
Field, A.: Discovering Statistics Using SPSS. Sage, London (2005)
Kaiser, H.F.: An index of factorial simplicity. Psychometrical **39**, 31–36 (1974)

Zhang, X.: Concessionaire's financial capability in developing build-operate-transfer type infrastructure projects. J. Constr. Eng. Manag. **131**(10), 1054–1064 (2005). https://doi.org/10.1061/(ASCE)0733-9364(2005)131:10(1054)

Aziz, R.F.: Factors causing cost variation for constructing wastewater projects in Egypt. Alex. Eng. J. **52**(1), 51–66 (2013b). https://doi.org/10.1016/j.aej.2012.11.004

RIAM; A Developed Risk Impact Assessment Model for Risk Factors Affecting Large Construction Projects

Ahmed Mohammed Abdelalim(✉)

Faculty of Engineering at Mataria, Helwan University,
P.O. Box 11718, Cairo, Egypt
dr.ahmed.abdelalim@gmail.com

Abstract. Cost saving and time performance are usually essential to all parties who are involved in a construction project, i.e. owner, contractor, subcontractor, etc. The main causes of disputes in construction projects involve delay and failure to complete the work in the specified cost and time frame. The delivery time of a project is a key factor for the owner in terms of cost as much as it is for the contractor. Unexpected increase in cost and delays in construction projects is caused by owner, contractor and environments. In which several types of risk factors may occur concurrently. The effect of cost overrun and schedule overrun do not only influence the construction industry, but the overall economy as well. Even though construction project increasing in cost and schedule have received extensive attention of researchers, but because of continuous changes and development in the field, the study considered of increasing demand to the construction industry, in addition to risk strategy and plan analysis. In order to meet the deadline of a project and due to the complex nature of construction projects, cost and scheduling should be flexible enough to accommodate changes without negatively affecting the overall project cost and duration. As such, the objectives of the presented research in this paper are to identify, study, and assess the effect of the factors that affect cost and time contingency. Data are collected from the biggest sixteen construction companies in Egypt. The collected data, output charts and analysis spreadsheets will be used for development of computerized built model with the identification abbreviation "Risk Impact Assessment Model"; ***RIAM***.

Keywords: Risk assessment · Cost overrun · Schedule overrun
Likelihood · Probability · Scale · Modeling

1 Introduction

Risk management has become an essential requirement for construction projects. The risk management process includes Hazard identification, Risk assessment and Risk control. Risk is assessed by Qualitative Methods and Quantitative Methods. Risk management is the systematic process of identifying, analyzing, and responding to project risk, it includes maximizing the probability and consequences of positive attributes and minimizing the probability and consequences of attributes adverse to

M. Shehata and F. Rodrigues (Eds.): GeoMEast 2018, SUCI, pp. 155–178, 2019.
https://doi.org/10.1007/978-3-030-01905-1_8

project objectives. Project Risk is an uncertain event or condition that, if occurs, has a positive or negative effect on a project's objectives. Components of risk are an event that may or may not happen, the probability of the occurrence of that event and the impact of the occurrence of that event. There are many sources of uncertainty in construction projects, which include the performance of construction parties, resources availability, environmental conditions, involvement of other parties, contractual relations, etc. As a result of these sources, construction projects may face problems that cause delay(s) in the project completion time [1].

The key success indicators of construction management system(s) include completing the project with cost and time, within the planned budget and duration, and within the required quality, safety, and environmental limits. These goals are interrelated where each of them is affected and affected by the others. An accurate cost estimating and scheduling should be sought in order to meet the overall budget and time deadlines of a project. Time contingency is used to guarantee the completion time of either an activity or a project [2]. Due to the unique nature of construction projects, cost overrun and schedule overrun uncertainty are essential for true budget and scheduling, which should be flexible enough to accommodate changes without negatively affecting the overall cost and duration. It is also essential to allocate a contingency value to both cost and time [3].

Yet, there are situations where there could be delays in activities, whether they are within the critical path or not, which result in a delay in the overall project duration. These delays will consequently have a negative impact on the quality, budget, and might be the safety of a project. Therefore, estimating cost and time contingencies are seen as a prime factor in achieving a successful construction project. Although several industrial sectors developed and used software for estimating time and cost contingencies in order to minimize delays and avoid being over budget, yet limited efforts are reported in the literature in the area of predicting time contingency in construction projects. The overall objectives of the presented research in this paper are: 1. to identify and study the factors that affect cost overrun and schedule overrun; 2. to develop a probability distribution chart for likelihood, cost impact and schedule impact; and 3. to quantify the risk assessment impact on cost and schedule. See Figs. 1, 2 and 3. Strategies for Negative

Rank	Descriptor	Description	Probability
Very high	Almost Certain	Even chance	> 50%
High	Likely	One in every 4 projects	> 25%
Moderate	Possible	One in every 10 projects	>10%
Low	Unlikely	One in every 20 projects	>5%
Very low	Rare	Less than 1 in every 20 projects	< 5%

Fig. 1. Probability matrix

Rank	Schedule	Cost	Safety	Quality
Very high	>3 months	>$10 million	Fatality	>10 %
High	2-3 months	$ 5-10 million	Severe injury	5-10 %
Moderate	1-2 months	$ 2-5 million	Medical treatment	3-5 %
Low	2-4 weeks	$1-2 million	First Aid	1-3 %
Very low	< 2 weeks	< $1 million	No injury	<1 %

Fig. 2. Impact matrix

Fig. 3. Strategies for negative risks

Risks or Threats are Avoid, Transfer, Mitigate or Accept. On the other hand Strategies for Positive Risks or Opportunities are Exploit, Enhance or Accept.

2 Background

Delays have an adverse impact on project success in terms of time, cost, quality and safety [1]. Time-delays and cost overruns are among the most common phenomena in the construction industry [4]. Therefore, planners and schedulers have used time contingency to guarantee the completion time of either an activity or a project [2]. The easiest and safest way to build a time contingency is to extend the project end date to a point where there is a comfortable amount of positive float, which may not be cost effective or acceptable to the client. However, it might not also be acceptable to proceed on a project with a zero float plan [5]. There is no standard definition of contingency in which it could imply different meanings to estimators, contractors, and owners' organizations [6]. Contingency is probably the most misunderstood, misinterpreted, and misapplied word in project execution [7]. It is an amount of money or time (or other resources) added to the base estimated amount to achieve a specific confidence level or allow for changes where experience shows obligation [8]. It can also be defined as the budget that is set aside to cope with uncertainties during construction [3] or the amount of money/time needed above the estimate to reduce the risk of overruns of project objectives to an acceptable level within the organization [9]. [10] Identified two major categories of contingency for construction projects:

1. Design contingency—it addresses the changes during the design process for factors, such as incomplete scope definition and inaccuracy of estimating methods as well as data [11].
2. Construction contingency—it addresses the changes during a construction process. Under a traditional procurement arrangement, the contract typically contains a variation clause(s) to allow for changes and provide a mechanism for determining and valuing variations.

There are many definitions for contingency in the literature; most of them focused on cost contingency. Contingency has different meanings to different people. Despite its importance, estimating time contingency was not thoroughly addressed in the literature. Prior to reviewing the estimating methods for project contingency, there are different attributes that affect contingency. [12] Identified 55 factors affecting the performance of project schedule. They observed seven factors that have the most significant impact on schedule outcome and divide them into two main categories. The first category included factors that encompassed the capability to improve performance level, such as owner's competence as well as commitment and conflict among project members. The second included factors that tended to retain the schedule at its existing level, such as coordination among project members, lack of knowledge and skills for the project managers, hostile socioeconomic environment, and uncertainty in selecting project members/team. In addition, [3] indicated that the effect of change orders increased the original cost and schedule since they modified the original contract. [13] Presented a methodology that incorporated network analysis and duration uncertainty in the project time analysis. They studied the effect of various factors on time contingency using a linear equation. There are many factors that impact time contingency in which it is not only important to identify these factors, but assess their impact on the project duration as well. Most engineers, planners, and agencies relied on their experience to estimate cost and time contingencies [3, 13, 14, 15]. The contractor's contingency was represented as a fixed percentage of the contract value [16] or as a percentage of total project cost or duration [3]. The completion date of a project was often missed due to uncertain events in which their impact was difficult to predict because of the uniqueness of construction projects.

Schedule contingency analysis is the process of identifying and evaluating contingency factors, present or anticipated, and determining both the probability and impact of identified contingency factors [17]. It is a preliminary step in establishing a schedule and time control strategy, which is intended to increase the probability of desired outcome while minimizing risk factors [12].

3 Factors Affecting Cost Overrun and Schedule Overrun

Based on literature and the opinion of practitioners/Expert through fifty nine questionnaires, several imperative factors that affect cost and time contingency are identified and studied. They are divided into four major criteria: (A) Site conditions, (B) Resources, (C) Project parties and (D) Project features related factors. Table 1 shows the detail of these main criteria and their factors. Site conditions include environmental, Subsurface and Site location. Resources include Labor, Equipment and Material. On the other hand, project parties cover Owner, Engineering and Design, Contractor and Project management. In addition Project features cover Financial, Political and schedule sub-criteria. The detail attributes/risk factors related to each sub-criterion are shown in Table 1. It is quite clear that the identified criteria and factors effectively contribute to the uncertainty in construction project cost and scheduling, which in turns, impact the assessment of cost and schedule overruns. In the present research, these factors are considered in predicting project budget and time contingencies.

Table 1. Factors affecting cost and schedule overruns

Criteria	Sub-criterion	Attributes/risk factor
Site conditions	Environmental	Earthquake
		Precipitation/flood
		Unpredicted Weather conditions
		Pollution
	Subsurface	Unexpected Surface conditions
		Archeological survey done
		Geo-technical investigation
	Site location	Construction area (rural/urban)
		Access conditions
		On-site congestion
		Delay in permits and licenses
		Security requirements
		Safety regulation
		Differing site conditions
Resources	Labor	Labor skills level
		Labor availability
		Drop in Labor productivity
		Labor accidents
		Human resource planning
		Working hours restrictions
	Equipment	Equipment quality
		Equipment breakdown
		Equipment maintenance
		Equipment malfunctions
	Material	Material delivery
		Material storage
		Material theft and damage
		Material procurement
		Nonconforming material
		Material monopoly
		Nominated vendors
Project parties	Owner	Owner type
		Management strategy
		Organization structure
		Work/labor permits
		On-site access
	Engineering and design	Team experience
		Project goal
		Complexity of design
		Ad-hoc consultants
		A design error

(*continued*)

Table 1. (*continued*)

Criteria	Sub-criterion	Attributes/risk factor
Project parties cont.	Contractor	Contractor pre-qualified
		New technology
		Defective work
		Rework
		Number of subcontractors
		Contractor Reputation
		Nominated sub-contractors
		Number of current projects
	Project management	Management experience
		Owner quality assurance
		Scope definition
		Quality control process
		Type of contract
		Availability of variations
Project features	Financial	Type of Funds
		Fluctuation in prices
		Invoices delay
		Change in currency rate
		Owner financial capacity
		Progress payment
		Rate of interest
		Tax rate
		Foreign currency
		Project size
	Political	Bribery and corruption
		Wars and revolutions
		Military coup
		Changes in laws and regulations
	Schedule	Fast track schedule
		Project duration

4 Research Methodology

Figure 4 shows the detailed steps utilized to perform the various activities of the present research. Factors that affect cost and schedule overruns are identified and discussed using literature review and expert opinion. A questionnaire survey is conducted to collect the impact of each factor. The research methodology is performed using probability distribution developed by crystal ball software. The collected data through questionnaires will be used to illustrate the minimum, mean, maximum and standard deviation values.

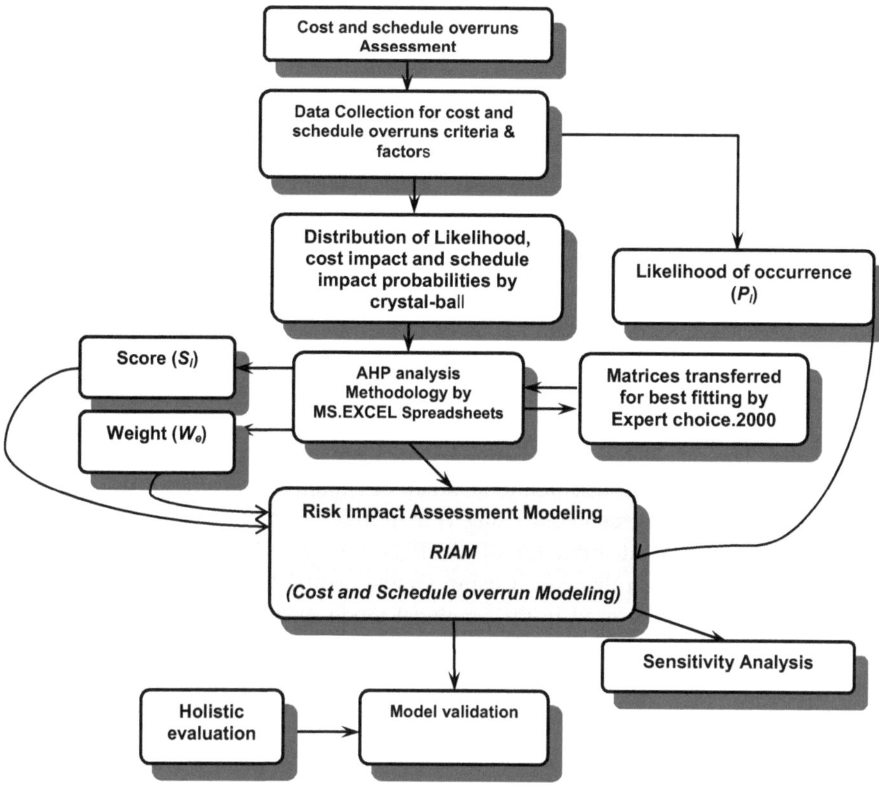

Fig. 4. Research methodology flow chart

The output charts developed by crystal ball software will be divided into three stages, the first one illustrates the probability distribution per each attribute likelihood [Pi], the second stage will concern the cost impact probability distribution either for attributes which will be defined as [weight Wi], or concerning the sub-criterion itself, which will be defined as [Score Si], furthermore the third stage will reflect the schedule impact probability distribution either for attributes/risk factor or for the sub-criterions themselves as described previously. These data will be used by AHP based simulation modeling, which will be described in detail in another paper. As illustrated in Fig. 4, the upper steps will be done in this paper, on the other hand the lower part of the flow chart includes AHP analysis will be illustrated in detail in another paper which can be considered as a complementary part of the present paper, [18].

5 Development of Data and Charts

5.1 Introduction

Cost and schedule overruns analysis are defined in this research as the process of identifying and evaluating contingency factors, present or anticipated, and determining the cost and schedule overruns indexes. In order to assess these indexes, some steps are utilized to perform the intended analysis. These steps are discussed further in the following sections. The questionnaires have been developed to evaluate the probability of attributes in addition to attributes cost impact and schedule impact, these questionnaires cover a lot of attributes/risk factors from literature to allocate various categories as per environmental and financial effects which had covered by 65 factors and increased up to 70 factors by practitioners advice, These added factors are equipment malfunctions, ad-hoc consultants, availability of variations, owner financial capacity and tax rate. The data have been mentioned in Excel spreadsheets to determine the min., max., mean values in addition to standard deviation, the evaluated weights have been transformed into percentages using (PMBOK® Guide- 5th edition) chapter-11 scales. The probability distribution of likelihood of occurrence, cost impact and schedule impact has been developed using Oracle crystal ball Release 11.1.2.3 software, these charts will be used in the graphical mode mentioned in the development model as discussed later, and on the other hand the user can use the numerical mode simply. Table 1 and Figs. 5, 6, 7 illustrated the set of attributes. The spreadsheets will be used to present the heat maps and tornado charts as mentioned in the paper.

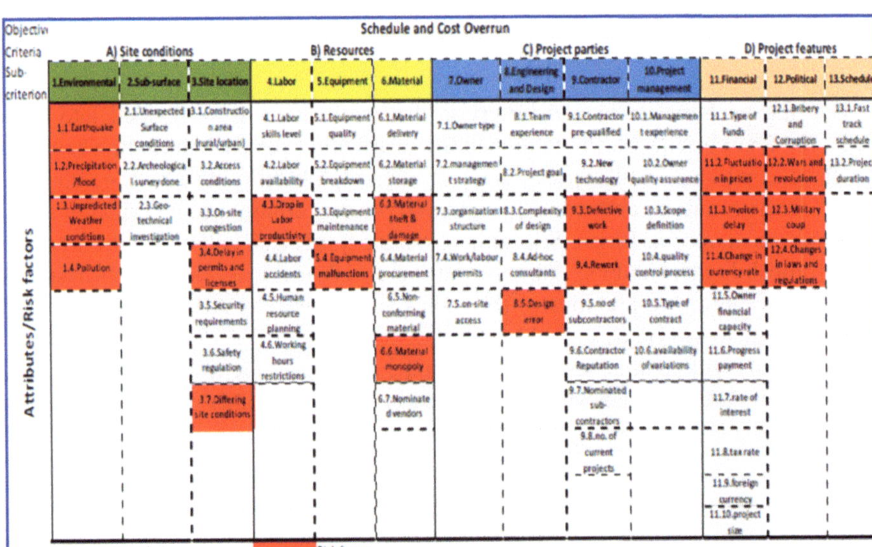

Fig. 5. Summary of risk attributes.

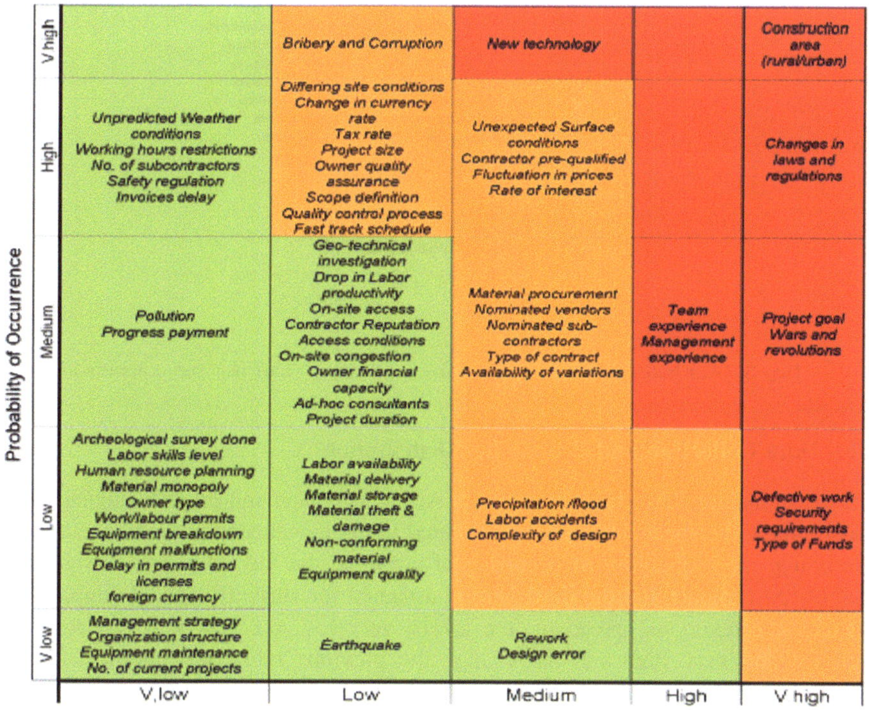

Fig. 6. Heat map concerning attributes cost impact.

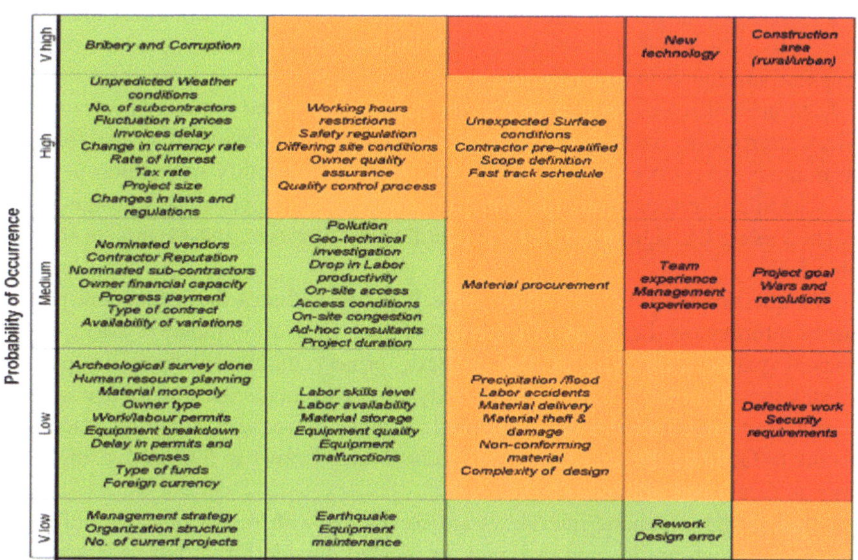

Fig. 7. Heat map concerning attributes schedule impact.

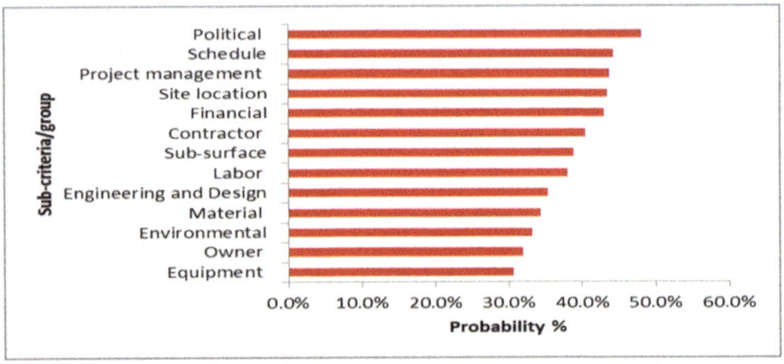

Fig. 8. Attributes probability of occurrence tornado chart (for sub-criterions).

5.2 Data Collection and Probability Distribution

The questionnaires have been prepared in Arabic language and English language to be simply filled; the sheets have been distributed in three evaluated portions, the first one concerning the probability of occurring of attribute, the second portion assigned for attribute cost impact and the third portion assigned for attribute schedule impact. The evaluation category is spread to five levels, very low = 1, low = 2, moderate = 3, high = 4 and very high = 5. To change the previous ranks into numerical scales (PMBOK® Guide) – Fifth Edition chapter-11 will be utilized, MS. Excel 2010 has been used to develop spreadsheets to present the collection data and calculate the mean, minimum and maximum values in addition to the standard divisions, Oracle® Crystal Ball Release 11.1.2.3 software will be used for representing the probability distribution for all attributes as same as the probability distribution of sub-criterions themselves, referred to (PMBOK® Guide) some distributions are frequently used as per Beta, triangular and lognormal distribution.

Data are collected through one mean questionnaire and structured interviews to collect information on case study projects. A survey is conducted in the form of a questionnaire to collect projects risk factors and its probability in addition to its cost and schedule overruns from the construction firms in Egypt. The questionnaire is designed using the significant factors identified by literature and experts as shown in Table 1.

The questionnaire includes three parts. Part one includes the respondent personal general information, i.e. years of experience, total value of completed works, type of company partnership, type of the projects, project duration. Otherwise part two includes the identified attributes probability of occurrence, cost impact and schedule overrun. Part three includes any additional attributes have been added by practitioner. In order to facilitate the answers of reviewers, a scale from 1 to 5 is used, using a scale range as mentioned above.

Physical and telephone interviews are conducted with senior managers of sixteen companies, which are located in Egypt, or located outside Egypt but have some projects here. The surveyed companies have experience history ranges from 15 to 60 years in the construction industry.

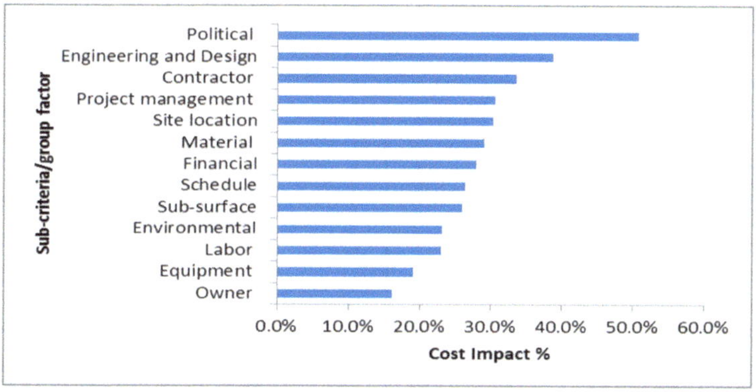

Fig. 9. Tornado chart for sub-criterions cost impact.

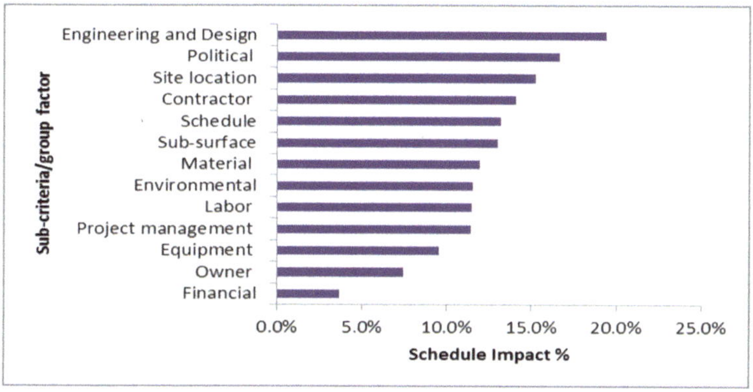

Fig. 10. Tornado chart for sub-criteria schedule impact.

The said companies work in variety of projects, such as pipelines, tanks, refineries, infrastructure, silos …etc. The variety of companies' type is taken into consideration as per national/international huge companies, construction firms, consultants and insurance group. 38 feedbacks have been received, 5 have been rejected due to shortage data, the type of contact engineers are 3 consultants, 4 project managers, 6 construction managers, 4 planners, 7 cost estimators, 1 contract administrator, 1 insurance engineer and 7 site engineers. The budget values of projects that run by the interviewed managers range from $25 up to $500 Million US dollars and the durations range from 6 months up to 3 years. Three case study projects are selected to test the robustness of developed model.

Some figures concerning tornado charts and probability distribution have been illustrated as samples, due to limited space all charts and figures can't be illustrated.

The sample charts mentioned in Figs. 11, 12, 13, 14, 15, 16, 17, 18, 19 and 20 will be used as a part of model structure named RIAM which has been built by the same author and described in another paper titled "Quantitative Risk Assessment in Construction Projects Using AHP-ANP Simulation Based Techniques".

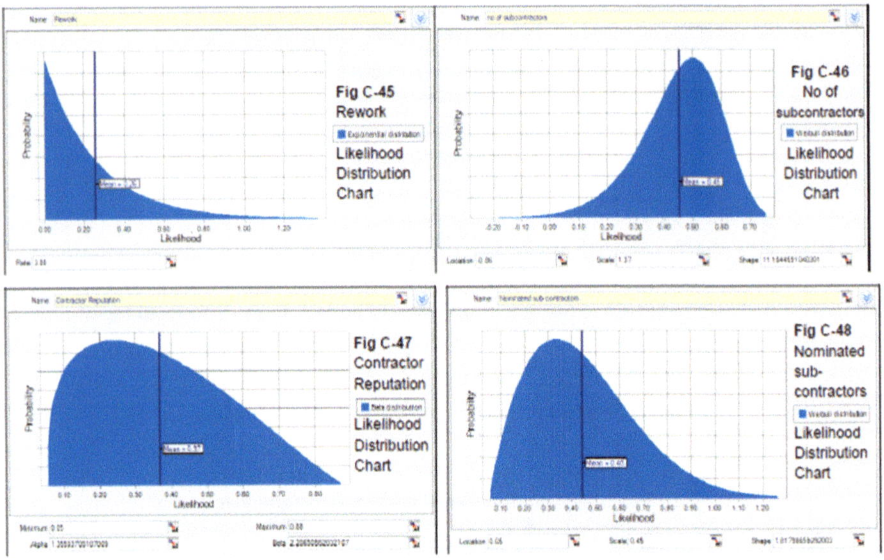

Fig. 11. Likelihood probability distribution charts for some contractor's related factors.

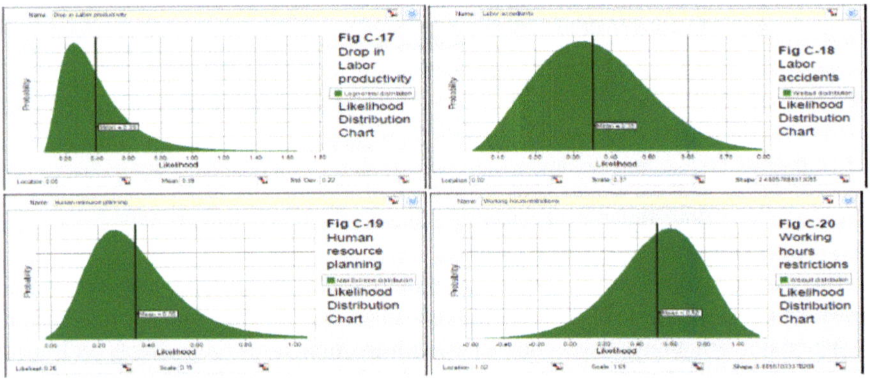

Fig. 12. Likelihood probability distribution charts for some labors' related factors.

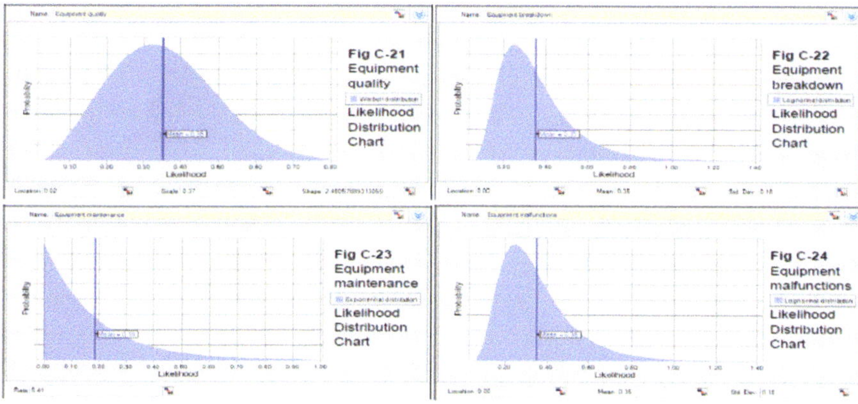

Fig. 13. Likelihood probability distribution charts for some equipment related factors.

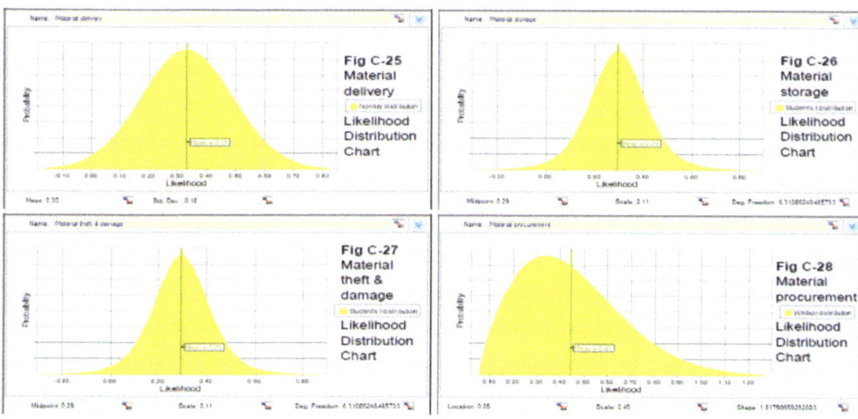

Fig. 14. Likelihood probability distribution charts for some materials related factors.

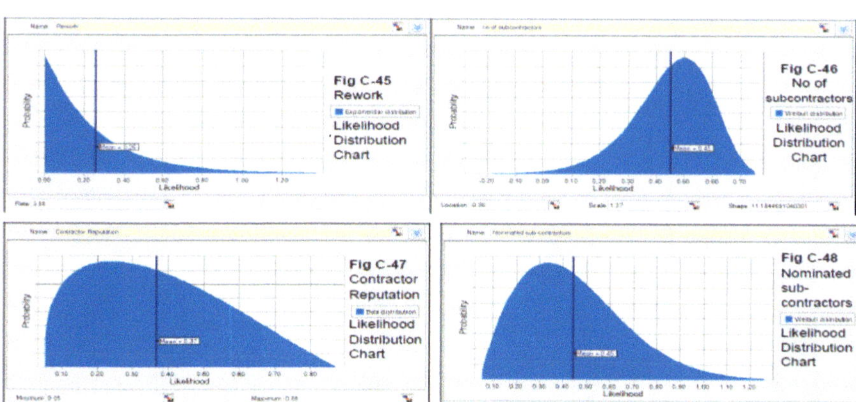

Fig. 15. Likelihood probability distribution charts for some subcontractors' related factors.

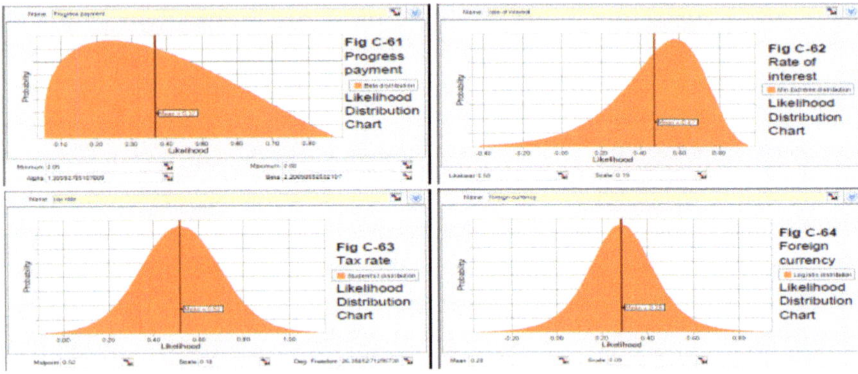

Fig. 16. Likelihood probability distribution charts for some financial related factors.

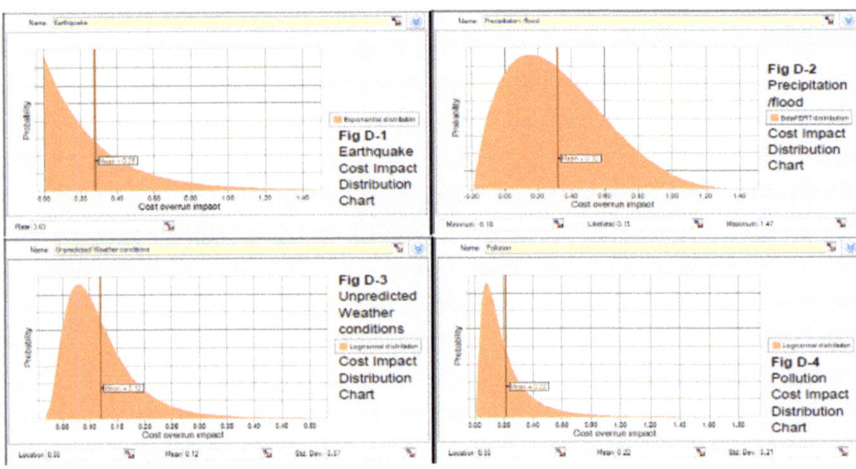

Fig. 17. Cost impact probability distribution charts for some environmental related factors.

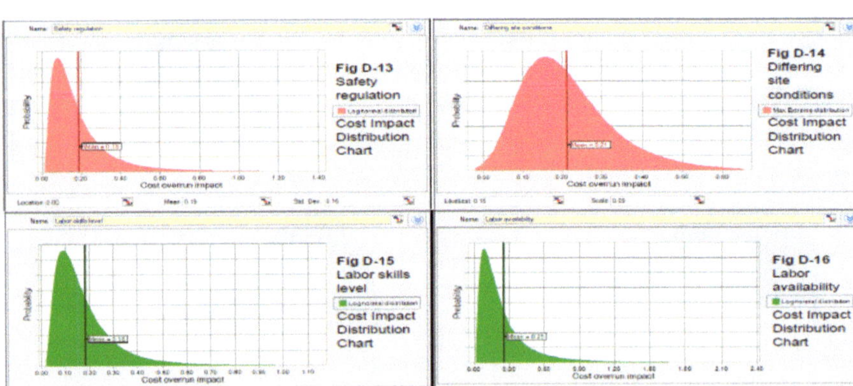

Fig. 18. Cost impact probability distribution charts for some labors'/site related factors.

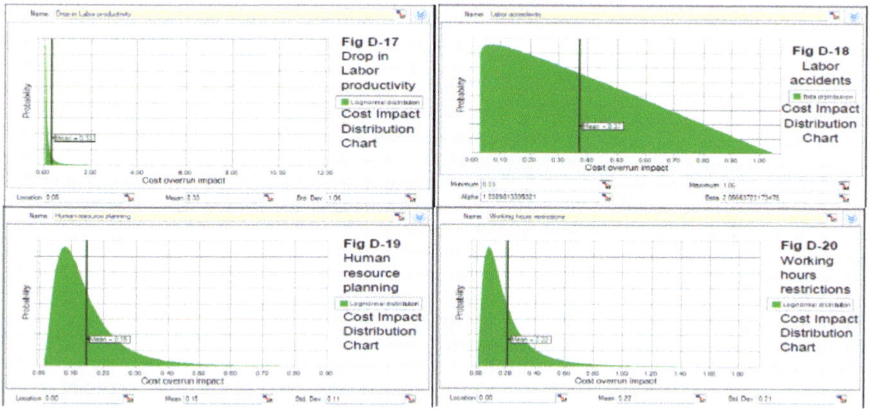

Fig. 19. Cost impact probability distribution charts for other labors' related factors.

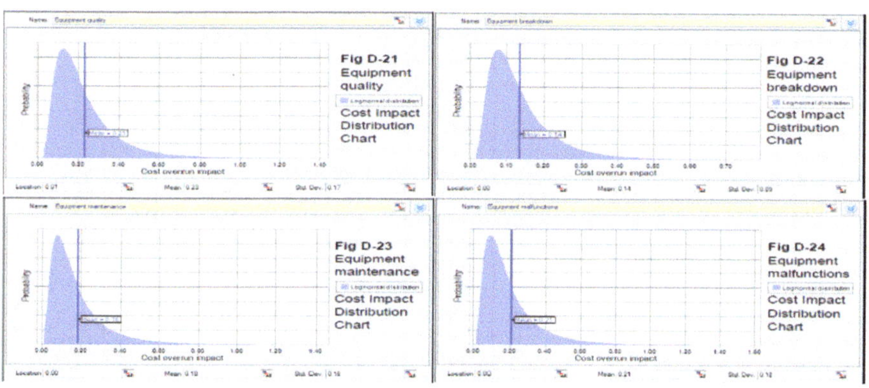

Fig. 20. Cost impact probability distribution charts for some equipment's related factors.

6 Implementation of Cost and Schedule Overruns Assessment Model

The above explained steps are utilized in implementing the developed model and performing the intended analysis, Table 2 will brief the collected data using mean values for probability of occurrence, score of sub-criterion and weight of attributes, this table is repeated another two times, one based on minimum values and the other one based on maximum values. These data will be illustrated to calculate the optimistic and the pessimistic overruns to help the top management to decide the right decision for cost and time contingency percentages.

For risk response, if the attribute impact will be avoided, the proposed value will be taken equal to zero, if the action will be accepted, the proposed value will be taken as same as its original value, if the action is mitigated, the proposed value will be reduced to its half value, if it's transferred, 12% only of its initial value will be taken.

Table 2. Probability of occurrence, local weight of attributes impact and score of sub-criteria impact using mean values

Sub-criterion cost score/ schedule score (S_i)	Attributes/risk factor	Attribute Probability of occurrence (P_i)	Attribute cost impact local weight (W_i)	Proposed mean cost overrun% $S_i \times W_i \times P_i$	Cost Risk response	Planned mean cost overrun %	Attribute schedule impact local weight (W_i)	Proposed mean schedule overrun% $S_i \times W_i \times P_i$	Schedule Risk response	Planned mean schedule overrun %
Material	Material delivery	33.03	8.90	0.16	transfer	0.02	6.50	0.11	accept	0.11
5.4%/5%	Material storage	29.39	17.60	0.28	accept	0.28	13.10	0.19	avoid	0.00
	Material theft & damage	29.39	9.80	0.16	transfer	0.02	6.50	0.10	accept	0.10
	Material procurement	43.94	2.50	0.06	mitigate	0.03	3.30	0.07	accept	0.07
	Non-conforming material	29.39	9.80	0.16	avoid	0.00	6.50	0.10	accept	0.10
	Material monopoly	31.21	49.00	0.83	avoid	0.00	39.30	0.61	avoid	0.00
	Nominated vendors	43.94	2.40	0.06	avoid	0.00	24.80	0.54	avoid	0.00
Owner	Owner type	33.03	42.00	3.15	avoid	0.00	46.20	2.33	avoid	0.00
22.7%/15.3%	management strategy	25.76	14.00	0.82	accept	0.82	11.50	0.45	avoid	0.00
	organization structure	25.76	14.00	0.82	accept	0.82	11.50	0.45	avoid	0.00
	Work/labour permits	33.03	28.00	2.10	avoid	0.00	23.10	1.17	accept	1.17
	on-site access	42.12	2.00	0.19	mitigate	0.10	7.70	0.50	mitigate	0.25
Engineering	Team experience	40.30	10.80	0.09	avoid	0.00	13.30	0.10	accept	0.10
and Design	Project goal	36.67	2.70	0.02	mitigate	0.01	6.70	0.05	mitigate	0.02
2%/1.9%	Complexity of design	33.03	15.10	0.10	avoid	0.00	26.70	0.17	avoid	0.00
	Ad-hoc consultants	40.91	60.50	0.50	avoid	0.00	40.00	0.31	avoid	0.00
	Design error	25.76	10.80	0.06	transfer	0.01	13.30	0.07	accept	0.07
Contractor	Contractor pre-qualified	51.82	5.20	0.06	avoid	0.00	6.50	0.14	avoid	0.00
2.4%/4.1%	New technology	59.70	5.20	0.07	avoid	0.00	6.50	0.16	avoid	0.00
	Defective work	34.85	1.30	0.01	accept	0.01	2.20	0.03	accept	0.03
	Rework	25.76	5.20	0.03	accept	0.03	6.50	0.07	accept	0.07
	no of subcontractors	45.15	26.10	0.28	accept	0.28	11.10	0.21	avoid	0.00
	Contractor Reputation	36.67	23.90	0.21	avoid	0.00	30.90	0.46	avoid	0.00
	Nominated sub-contractors	43.94	6.80	0.07	avoid	0.00	25.90	0.47	avoid	0.00
	no. of current projects	25.76	26.10	0.16	avoid	0.00	10.30	0.11	avoid	0.00
Project	Management experience	40.30	2.40	0.03	avoid	0.00	1.60	0.05	accept	0.05
Environmental	Earthquake	16.67	9.00	0.17	avoid	0.00	14.30	0.19	avoid	0.00
11.3%/8.1%	Precipitation /flood	29.39	3.30	0.11	avoid	0.00	6.80	0.16	avoid	0.00
	Unpredicted Weather conditions	46.36	76.70	4.02	mitigate	2.01	59.10	2.22	mitigate	1.11
	Pollution	40.30	10.90	0.50	accept	0.50	19.80	0.65	accept	0.65
Sub-surface	Unexpected Surface conditions	48.18	14.90	0.62	accept	0.62	15.90	0.38	accept	0.38
8.6%/5%	Archeological survey done	27.58	74.50	1.77	mitigate	0.88	63.60	0.88	avoid	0.00
	Geo-technical investigation	40.91	10.60	0.37	mitigate	0.19	20.60	0.42	avoid	0.00
Site location	Construction area (rural/urban)	57.27	1.40	0.03	mitigate	0.01	3.30	0.08	mitigate	0.04
3.2%/4.1%	Access conditions	40.30	7.20	0.09	mitigate	0.05	10.10	0.17	mitigate	0.08
	On-site congestion	42.12	7.20	0.10	mitigate	0.05	11.10	0.19	mitigate	0.10
	Delay in permits and licenses	33.03	36.90	0.39	accept	0.39	35.50	0.48	accept	0.48
	Security requirements	34.85	3.20	0.04	accept	0.04	5.60	0.08	avoid	0.00
	Safety regulation	50.00	28.80	0.46	accept	0.46	22.40	0.46	avoid	0.00
	Differing site conditions	46.36	15.20	0.23	accept	0.23	12.00	0.23	avoid	0.00
Labor	Labor skills level	34.85	29.40	1.24	accept	1.24	25.50	0.72	avoid	0.00
12.1%/8.1%	Labor availability	33.03	6.80	0.27	accept	0.27	6.40	0.17	avoid	0.00
	Drop in Labor productivity	38.48	8.90	0.41	mitigate	0.21	12.80	0.40	mitigate	0.20
	Labor accidents	34.85	1.80	0.08	transfer	0.01	4.30	0.12	accept	0.12
	Human resource planning	34.85	44.30	1.87	avoid	0.00	38.30	1.08	avoid	0.00
	Working hours restrictions	52.42	8.90	0.56	avoid	0.00	12.80	0.54	accept	0.54
Equipment	Equipment quality	34.85	10.00	0.60	avoid	0.00	14.30	0.57	avoid	0.00
17.1%/11.5%	Equipment breakdown	34.85	50.20	2.99	mitigate	1.50	42.90	1.72	mitigate	0.86
	Equipment maintenance	18.48	29.80	0.94	mitigate	0.47	28.60	0.61	mitigate	0.30
	Equipment malfunctions	34.85	10.00	0.60	transfer	0.07	14.30	0.57	accept	0.57
management	Owner quality assurance	46.36	31.20	0.42	accept	0.42	5.80	0.22	avoid	0.00
2.9%/8.1%	Scope definition	46.36	10.40	0.14	avoid	0.00	49.40	1.86	avoid	0.00
	quality control process	46.36	40.50	0.54	accept	0.54	5.80	0.22	avoid	0.00
	Type of contract	38.48	7.40	0.08	accept	0.08	23.00	0.72	avoid	0.00
	availability of variations	44.55	8.10	0.10	avoid	0.00	14.40	0.52	avoid	0.00
Financial	Type of Funds	29.39	0.70	0.01	transfer	0.00	5.90	0.37	transfer	0.04
4.1%/21.6%	Fluctuation in prices	48.18	1.70	0.03	accept	0.03	11.80	1.23	avoid	0.00
	Invoices delay	50.00	18.70	0.38	accept	0.38	5.90	0.64	avoid	0.00
	Change in currency rate	51.82	3.30	0.07	accept	0.07	11.80	1.32	avoid	0.00
	Owner financial capacity	36.67	8.60	0.13	accept	0.13	11.80	0.93	avoid	0.00
	Progress payment	36.67	10.40	0.16	accept	0.16	11.80	0.93	avoid	0.00
	rate of interest	48.18	1.60	0.03	accept	0.03	11.80	1.23	avoid	0.00
	tax rate	51.82	3.30	0.07	mitigate	0.04	11.80	1.32	avoid	0.00
	foreign currency	29.39	46.30	0.56	mitigate	0.28	5.90	0.37	avoid	0.00
	project size	48.18	5.50	0.11	mitigate	0.05	11.80	1.23	avoid	0.00
Political	Bribery and Corruption	63.33	77.10	0.29	accept	0.29	56.60	0.90	avoid	0.00
0.6%/2.5%	Wars and revolutions	42.12	9.20	0.02	avoid	0.00	11.80	0.12	avoid	0.00
	Military coup	33.03	4.60	0.01	avoid	0.00	5.90	0.05	avoid	0.00
	Changes in laws and regulations	53.64	9.20	0.03	transfer	0.00	25.80	0.35	avoid	0.00
Schedule	Fast track schedule	51.82	25.00	1.00	avoid	0.00	33.30	0.78	accept	0.78
7.7%/4.5%	Project duration	36.67	75.00	2.12	accept	2.12	66.70	1.10	accept	1.10
100%	Summation			35.2%		16.2%		38.5%		9.5%

7 Implementation of the Proposed Model; Riam

This SECTION presents the steps for developing a computerized system (RIAM) to quantify the risks associated with schedule and cost concerning construction projects. The system will be developed by Microsoft® Visual Basic ™ programming language and Microsoft Access. The RIAM - model will be run through three software, Microsoft Excel and Crystal ball in addition to Expert Choice 2000.

- This model is designed to be flexible and easy to use. It does not require previous knowledge of the above mentioned programs. Thus RIAM-model overcomes the deficiencies of previous risk analysis and quantification tools in being complicated, non-flexible, and time consuming. A computerized system (RIAM) will be used for risk assessment and risk strategy (plan) for project attributes as mentioned in Fig. 4 above. Figure 21a illustrates the front page of running the model.
- As per Fig. 21b, the user will insert brief of the project which will be assessed. The data can be allocated either by writing or using the arrows attached.

7.1 User Interfaces and Data Entry

As per Fig. 21c, when the mentioned screen is opened, the user will choose each attribute by ticking the enclosed arrow, a sub - screen will be opened to choose the attribute probability, cost impact and schedule impact. The user can choose the minimum, mean or maximum values easily by either pressing the indicating button or write the fraction in the related field. Taken into confederation that (a) if the user insert any number out of range, it will be rejected; (b) the default values are equal to mean values.

- As referred in Fig. 21d, the lower part of the previous screen will illustrate the user choices through tables of data. In addition, the user can use the right buttons in order to restore the values to their mean values furthermore to transfer data in order to start the analytical hierarchy process AHP for building the relationship matrices.
- As previous, the user can allocate the fractions either numerically or through the graphical solution by illustrating the charts attached with each part of the form presented in Fig. 21d.The normal distribution of probability will be appeared as per Fig. 21e, three buttons are appended with each attribute, one will illustrate the probability normal distribution of attribute likelihood and its button in the upper part, the second chart will consider the probability of cost impact related to its attribute, the button mentioned in the middle portion, on the other hand the schedule impact probability distribution chart will be appeared through the lower button.
- To utilize the allocated data by the user and building the matrices in order to develop the AHP technique, the user will transfer data using the related button as illustrated in Fig. 21c above; RIAM-model will start the analysis and the calculation to form the relationship matrices either between the attributes or between the sub-criterions themselves. With respect of the matrices, RIAM-model will develop the

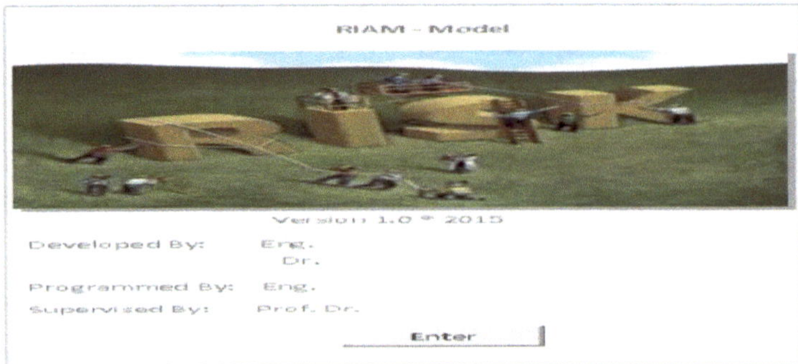

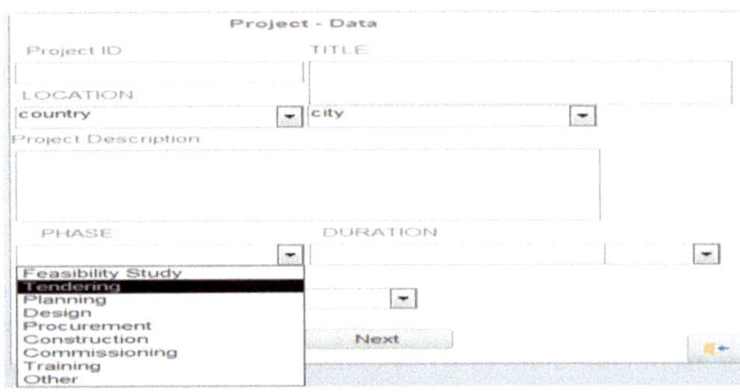

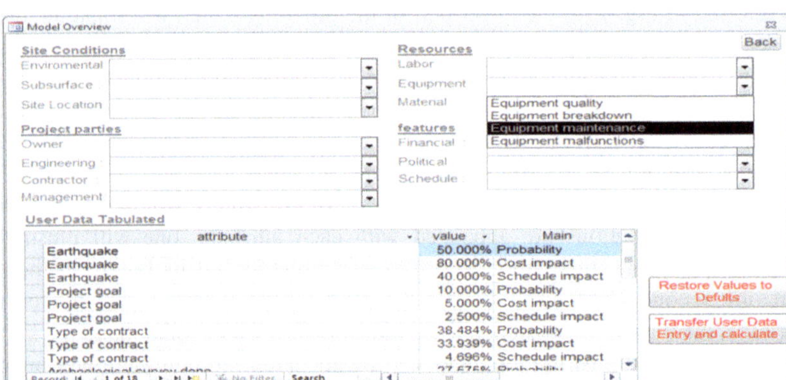

Fig. 21. **a.** Model information screen. **b.** Model information screen. **c.** Model main screen. **d.** User data entry. **e.** The developed charts through graphical solution. **f.** The matrices main screen. **g.** The matrices before best fitting. **h.** The matrices after best fitting. **i.** Risk response screen

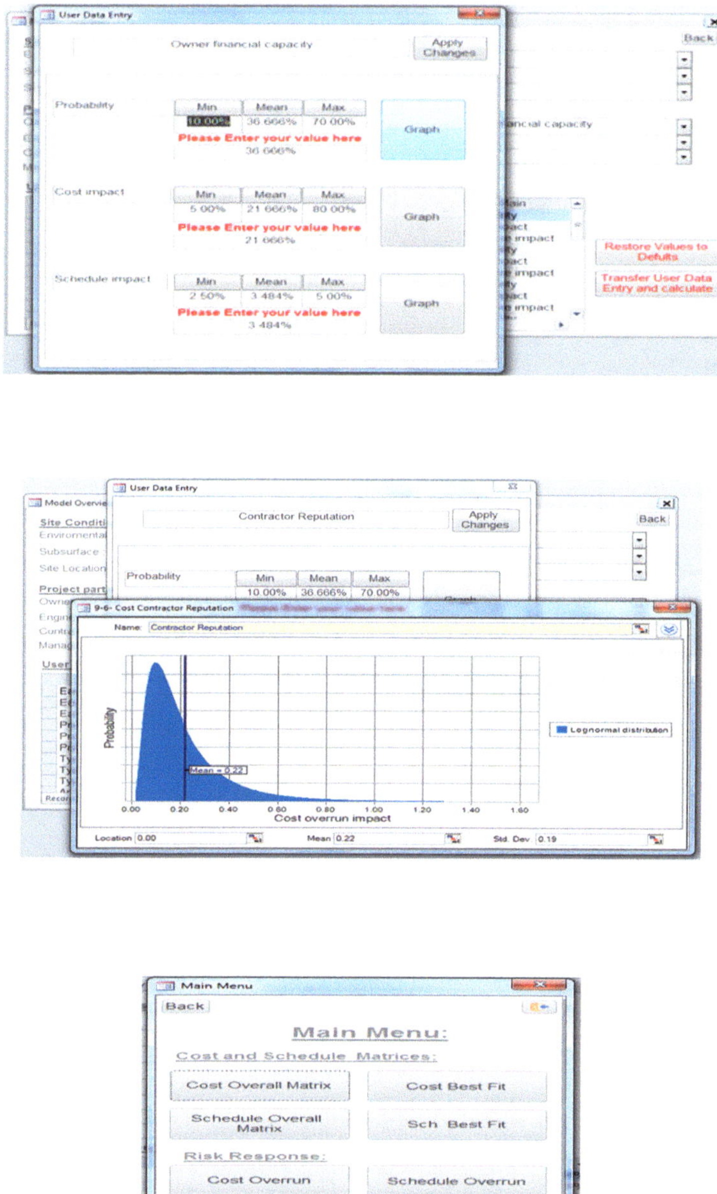

Fig. 21. *(continued)*

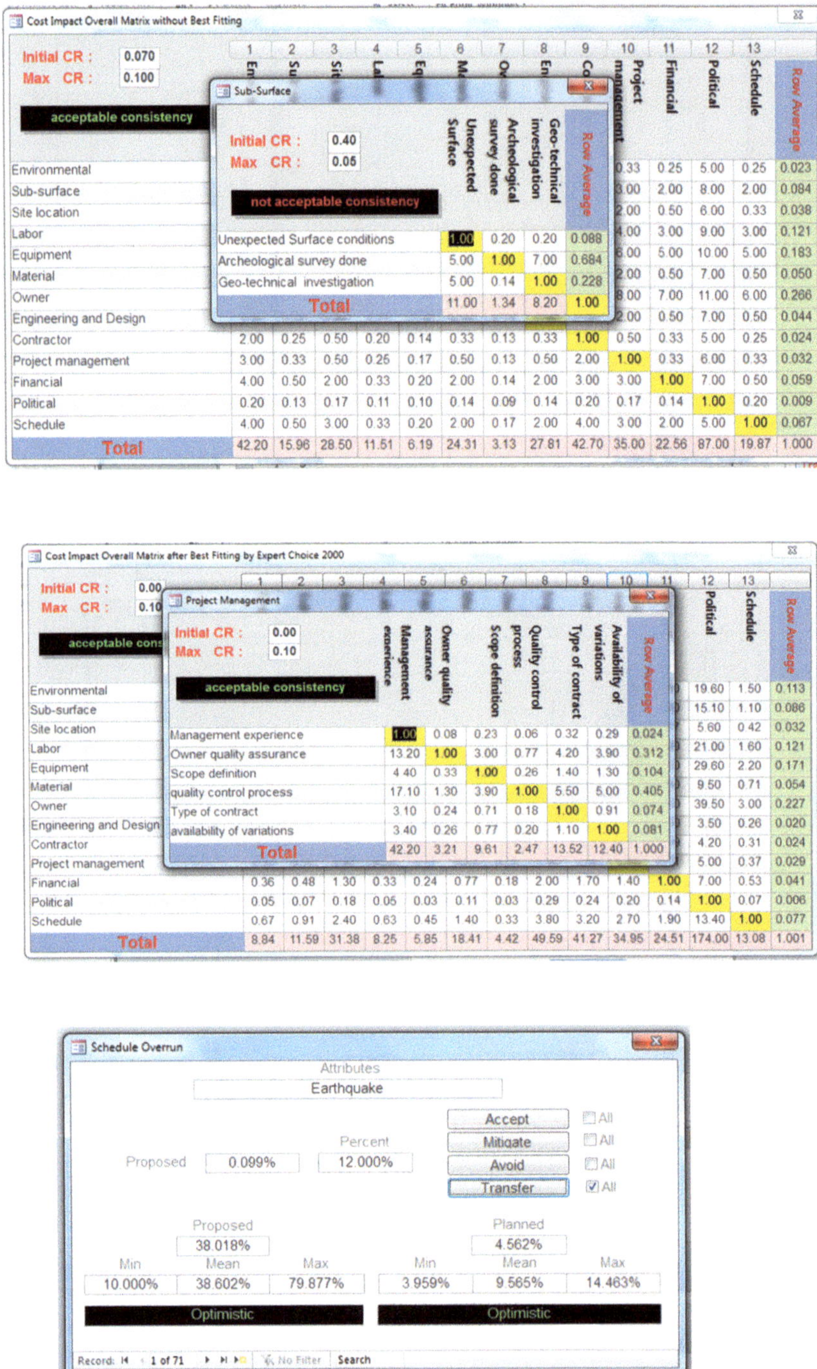

Fig. 21. *(continued)*

row average, the totals, consistency ratio and the maximum consistency. In addition, the model will evaluate the matrix inconsistency if it acceptable, the font color will be green without animation, if it is not acceptable, it will be red color and appeared with animation. Figure 21f will illustrate the main entrance of matrices either for cost impact or concerning schedule impact.

- The user can start his choice by pressing on the overall matrices buttons to present the matrices without best fitting as per the data entry, the overall matrix will be appeared as Fig. 21g and the user can review each sub-criterion matrix by ticking its button above
- In order to enhance matrices inconsistency to reach the optimum value (closed to zero), the model will apply the best fitting iterations typically as per Expert choice 2000 software procedure. These steps will be applied for all main matrices subjected to main criteria themselves in addition to their sub-criterions matrices either with respect of attributes cost impact or attributes schedule impact.
- Furthermore, the revised inconsistency ratio will be appeared at the top part of the screen as indicated in Fig. 21h, the user can check their values equal to zero.
- As per Fig. 21i, the user can choose the response plan related to each attribute either for cost overrun or schedule overruns, the proposed values will indicate the range of output values without risk response which logically will be equal or more than the planned values that indicate the expected plan to affiliate the project risks.

Taken into consideration that, (a) the developed graphs have been issued by crystal ball software, (b) all calculations and matrices import from Microsoft Excel spreadsheets developed by the author.

7.2 Validation of RIAM Model

The developed model RIAM is accurate enough for applying the analytical hierarchal analysis, therefore the user can utilize either the Microsoft excel spreadsheets developed by the author or the developed model RIAM which in fact depends mainly on the said spreadsheets, the accuracy of RIAM model with respect to Super-Decisions software is less than $\pm$ 3%. In order to verify the precision of predicted cost contingency and time contingency index using the developed model, data are collected from experts and projects database (closeout reports) regarding the actual cost and delays in their previous construction projects. These data include the actual and planned cost in addition to actual and planned finish dates. There are three huge construction projects will be presented in Table 3 to compare with the developed model as follow:

After then a comparison was carried out to illustrate the results depend on Microsoft Excel spread sheets, Expert choice 2000 software and Super-Decisions software (which developed by (Saaty, 1982) as follows in Table 4.

Table 3. Comparison between three actual projects

Description	Project-1	Project-2	Project-3
Name	Tina-Abu Sultan Gas Pipeline	Port Said East Container Phase-2	Al-Arish – Taba Gas Pipeline
Scope of work	Construction only of 32"/60 km steel PL	EPC of terminal and buildings	Construction only of 36"/254 km steel PL
Client	–	–	–
Type of project	Unit rates	FIDIC, Lump sum	Lump sum
Location	Ismailia	Port said	Sinai
Price in US$	9,728,685	164,869,316	45,269,440
Planned cost	7,958,064	147,887,777	35,672,319
Actual cost	8,772,651	156,834,988	38,533,239
Planned start date	26/9/2005	12/9/2009	31/3/2002
Actual start date	14/9/2005	23/9/2009	13/4/2002
Planned duration (months)	12	32.7 (140 weeks)	12
Actual duration (months)	13.2	34.9	12.8

Table 4. The results comparison between different software.

Description	Microsoft excel	Expert choice 2000	Super-decisions	Accuracy %
Proposed mean cost overrun%	35.1%	35.2%	35.0%	99.7151%
Planned mean cost overrun%	16.2%	16.2%	15.9%	98.1481%
Proposed min cost overrun%	10.0%	10.0%	10.0%	100.0000%
Planned min cost overrun%	4.4%	4.3%	4.4%	100.0000%
Proposed max cost overrun%	73.2%	72.1%	72.8%	99.4536%
Planned max cost overrun%	37.6%	36.3%	37.2%	98.9362%
Proposed mean schedule overrun%	38.6%	38.5%	38.6%	100.0000%
Planned mean schedule overrun%	9.6%	9.5%	9.4%	97.9167%
Proposed min schedule overrun%	10.0%	10.0%	10.0%	100.0000%
Planned min schedule overrun%	4.0%	4.1%	4.0%	100.0000%
Proposed max schedule overrun%	79.9%	79.7%	79.9%	100.0000%
Planned max schedule overrun%	14.5%	14.2%	14.1%	97.2414%
Optimistic cost overrun %	10.3%	10.3%	10.2%	99.0291%
Pessimistic cost overrun %	26.9%	26.3%	26.6%	98.8848%
Using PERT technique cost overrun%	17.8%	17.6%	17.6%	98.8764%
Optimistic schedule overrun %	6.8%	6.8%	6.7%	98.5294%
Pessimistic schedule overrun %	12.0%	11.8%	11.8%	98.3333%
Using PERT technique schedule overrun%	9.4%	9.4%	9.3%	98.9362%

8 Conclusions

Estimating cost and scheduling contingencies are major factors in achieving a successful and realistic budget and schedule for construction projects. In the present research, a survey is sent to many construction companies to identify, qualify study, assess, and quantify the factors that affect budget and time contingency. From Figs. 6 and 7 as presented above, there are some attributes located in the red zone, which reflect greater probability and impact effect of such factors, if the cost estimator takes the impact of those attributes as same as their values, the budget and schedule will increase accordingly, and the tenders may be unsuccessful. Therefore, risk response and risk plan shall be studied and controlled to transfer or mitigate the impact of said factors to another party, such as insurance companies, sub-contractor or the client himself, which enhance the cost and schedule contingencies demand. On the other hand the attributes located in the green zone have little probability and impact, these factors may be neglected or avoid. Applying these results in the region of study, from Figs. 8, 9 and 10 political sub-criteria can be considered as the most probable and impacted factor, on the other hand owner sub-criterion can be considered the lowest probable and impacted factor.

Finally, a developer model is designed to be flexible and easy to use. It does not require previous knowledge of the above mentioned programs. Thus RIAM-model overcomes the deficiencies of previous risk analysis and quantification tools in being complicated, non-flexible, and time consuming. A computerized system (RIAM) will be used for risk assessment and risk strategy (plan) for project attributes as mentioned before.

Acknowledgements. Special thanks are due to Eng. Mohammed Basiouny for his fruitful help in building up the model and Dr. Nawawy, O.A. for facilitating conducting interviews with key participants.

Data Availability Statement

All data generated or analyzed during the study are included in the submitted article or supplemental materials files.

References

1. Faridi, A., El-Sayegh, S.: Significant factors causing delay in the UAE construction industry. J. Constr. Manag. Econ. **24**(11), 1167–1176 (2006)
2. Park, M., Peña-Mora, F.: Reliability buffering for construction projects. J. Constr. Eng. Manag. **130**(5), 626–637 (2004)
3. Touran, A.: Calculation of contingency in construction projects. IEEE Trans. Eng. Manag. **50**(2), 135–140 (2003)
4. Koushki, P., Al-Rashid, K., Kartam, N.: Delays and cost increases in the construction of private residential projects in Kuwait. Constr. Manag. Econ. **23**(3), 285–294 (2005)
5. Moselhi, O.: Applied earned value for project control. In: CIB W-65 Conference Proceedings, Trinidad, Wisconsin (1993)

6. Moselhi, O.: Risk assessment and contingency estimating. Trans. Am. Assoc. Cost Eng., p. 22 (1997)
7. Patrascu, A.: Construction Cost Engineering Handbook. Marcel Dekker, New York (1988)
8. AACE: AACE international's risk management dictionary. Cost Eng. **42**(4), 28–31 (2000)
9. PMI: A Gide to the Project Management of Knowledge (PMBOK Guide), 5th edn. Project Management Institute, Pennsylvania (2013)
10. Treasury, H.: CUP Guidance: No. 41—Managing Risk and Contingency for Works Projects. Central Unit on Procurement, London (1993)
11. Clark, F., Lorenzoni, A.: Applied Cost Engineering. CRC Press, Boca Raton (1996)
12. Iyer, K., Jha, K.: Critical factors affecting schedule performance: evidence from Indian construction projects. J. Constr. Eng. Manag. **132**(8), 871–881 (2006)
13. Boskers, N., AbouRizk, S.: Modeling scheduling uncertainty in capital construction projects. In: Proceeding of the Winter Simulation Conference, pp. 1500–1507 (2005)
14. Barraza, G.: Probabilistic estimation and allocation of project time contingency. J. Constr. Eng. Manag. **137**(4), 259–265 (2011)
15. Brinkkemper, S., Lyytinen, K., Welke, R.: Method Engineering: Principles of Method Construction and Tool Support. Kluwer Academic Publishers, Massachusetts (1996)
16. Smith, G., Bohn, C.: Small to medium contractor contingency and assumption of risk. J. Constr. Eng. Manag. **125**(2), 101–108 (1999)
17. Sukumaran, P., Bayraktar, M.E., Hong, T.H., Hastak, M.: Model for analysis of factors affecting construction schedule in highway work zones. J. Transp. Eng. **132**(6), 508–517 (2006)
18. Abdelalim A.M., O.M. Nawawy, Basiouny: Identification and assessment of risk factors Affecting construction projects. HBRC J. (2015)

Perception of Consultants and Contractors to Performance Factors of Construction Projects

Ayodeji E. Oke(✉), Clinton O. Aigbavboa, and Khosi Mohapeloa

Department of Construction Management and Quantity Surveying, Faculty of Engineering and the Built Environment, University of Johannesburg, Johannesburg, South Africa
emayok@gmail.com, kakanyof@gmail.com, caigbavboa@uj.ac.za

Abstract. Due to various problems and challenges of construction industry especially in the production and delivery of sustainable infrastructure, performance of construction projects has become an issue of concern not only to contractors and professionals but other stakeholders including clients, project sponsors and end-users. This is due to complexity of construction process, innovation in the use of raw materials and machinery, new construction management techniques and more importantly, change in the demand of clients. Thus, this research evaluates performance indicators for construction projects from the perspective of consultants and contractors. This is to improve construction project performance with a view to enhancing the quest of construction industry of contributing positively to the economy and overall growth of the nation. Using quantitative approach, data were collected through close-ended questionnaires administered on a group of learned consultants and contractors with sufficient knowledge and experience of the construction industry. From the analysed data, there is a significant difference in the perceptions of consultants and contractors regarding factors affecting performance of construction projects. To consultants, general time factors and conformance to specification are the most important performance indicator for project success while from contractors' view, cost factors including delay in payment from owner to contractor is a major critical success factor. The two groups unilaterally agreed that bringing about the best planning and scheduling technique before and during construction would assist in improving performance of construction projects in the country.

Keywords: Construction performance · Key performance indicators (KPIs)
Performance management · Critical success factors (CSFs)
Performance improvement

1 Introduction

Infrastructural developments are becoming more complex due to increase in size, scope and technology requirement. However, if construction projects are accomplished within contract time, budget and quality to the satisfaction of stakeholders, then the

M. Shehata and F. Rodrigues (Eds.): GeoMEast 2018, SUCI, pp. 179–189, 2019.
https://doi.org/10.1007/978-3-030-01905-1_9

project is considered successful. In their quest to produce projects to satisfaction of clients, understanding factors to measure projects success have been a major concern to construction projects practitioners, professionals and concerned stakeholders (Parmenter 2010). Organisations around the world are failing in their expected performance and project delivery, whether they are operating as multinationals, government parastatals or agencies, or small local entities.

Performance management is a complex holistic system that arose out of a combination of performance appraisals and measurement systems (Furnham 2004). With this holistic approach, performance management enhance effective measurement systems that enable executives and stakeholders to take a comprehensive view of the entire system. More so, with the integrated perspective provided by strategic measurement, projects executives and administrators are positioned to understand and forecast how actions taken in one aspect of the system can affect performance of another system. According to Brown and Armstrong (1999), performance management are all the processes involved in a system to improve total performance of organisation or process and it can be measured using the tool called Key Performance Indicators (KPIs). Masilamani (2005) defined KPIs as a set of factors acting as relative measure of the performance of an organization or process. In view of the KPIs, Auma (2014) noted that the construction industry is have underperformed compared to other industries.

Performance of construction projects are affected by various internal and external factors that could lead to the success or failure of such project. However, there are different views to indices for measuring performance of project success depending on the interest of the party in consideration. There are numerous parties involved in construction of projects including clients, owners, sponsors, financiers, clients' representative also known as consultants, statutory bodies, etc. Hence, this study examines perception of two of the major stakeholders, that is, consultants and contractors to factors affecting performance of construction projects.

2 Overreview of Construction Project Performance

Traditionally, performance or success of projects are measured in term of cost, time and quality. Over the years, due to complexity and changing nature of the various industries and varying demands of clients, attention has shifted to such measures as energy efficiency, stakeholders' satisfaction, projects sustainability, etc.

Performance is the degree of achievement of certain milestone, effort or undertaking depending on the prescribed and planned goals or objectives that formed the project parameters (Chitkara 2005). In the current construction industry that is highly competitive and characterised by uncertain business environment, clients of construction projects, who are the major stakeholder, are demanding timely delivery of their project (Costa et al. 2004). This is related to early confirmation of design, early start of construction projects, certainty of performance in terms of acceptable delivery indices specific to the project, value for money for their capital and investment as well as minimal exposure to risk.

Performance of construction projects affects economic development of every countries especially the developing ones (Olatunji et al. 2016). This is because of the contribution of construction industry to the economy of a country. Success of any construction project depends mainly on success of approved and acceptable performance indicators for such project. This implies that success or delivery indices to measure performance vary from one project to the other. Reichelt (1999) remarked that the important structures underlying the dynamics of project performance are work accomplishment structure, effects of feedback on productivity and work quality as well as effects from upstream to downstream phases. More so, Ugwu and Haupt (2007) stated that an adequate awareness and ample knowledge of performance indicators are required for archiving managerial goals of an organization.

3 Factors Affecting Performance of Construction Project

Construction project development may be impaired without a good knowledge and management of the factors influencing the performance of such projects (Akanni et al. 2015). Dissanayaka and Kumaraswamy (1999) concluded that one of the principal factors for construction industry's poor performance is the inappropriateness of adopted or chosen procurement system. Furthermore, Ng et al. (2002) identified the main performance criteria of construction projects as financial stability, standard of quality; project duration; health and safety; relationship among clients, consultants, contractors, sub-contractors and other stakeholders; amount of subcontracting; claim and contractual disputes; and management capabilities. In addition, Akogbei et al. (2015) identified project size as a key project performance factor.

3.1 Cost Performance Factors of Construction Projects

According to Mbamali et al. (2005) a project is successful when it is completed within budgeted and planned cost, specified quality, stipulated time and delivered safely to the satisfaction of clients and other stakeholders. Msani (2011) stated that setting up cost objectives and indicators is driven mainly by the process and individuals involved in the process. The level of experience, skills, training, knowledge and expertise possessed by project team members shouldered with the responsibility of costing is a major factors for the process. The cost also depends on the process adopted at arriving at the cost and procedure that will be used to execute the project. For instance, transporting raw materials from source by railway carriages or trucks would influence the cost of such materials, and thereby, affect overall cost of the project.

3.2 Time Performance Factors of Construction Projects

Jouini et al. (2004) stated that managing time in the procurement, engineering and construction of projects is a key competing factor among innovative firms. It is believed that customers who consider time as a valuable performance indicator and resource will be keen on timely delivery of their projects and sometimes, their demand influence contractors to improve their time performance. Walker and Shen (2002) found that time performance of construction projects is directly proportional to design team performance. More so, Aiyetan (2010) emphasized that where there are challenges in getting to the site, as a result of bad roads, narrowness of the road or long distances between the storage and usage point, construction speed are likely be negatively affected. According to Toor and Oguniana (2008), this eventually causes delays in completion of construction projects.

According to Yakubu and Sun (2009), change in design at various times is the most influential factor inhibiting timely delivery of construction projects in the United Kingdom. Proverb and Holt (2000) observed that construction methods adopted for the procurement of a project significantly affect construction time performance. More so, Koushki and Kartam (2004) concluded that projects experience delays due to late delivery of materials, late selection and approval, type of construction materials and their availability at either local or international market.

3.3 Quality Performance Factors of Construction Projects

Quality of a project is influenced by general project characteristics, project participants, contractual arrangements and interactive processes among the participants (Iyer and Jha 2006). Chan et al. (2004) stated that factors affecting quality performance of construction projects are: effectiveness of design team leader and project manager; project complexity, size and scope; nature and competency of client; environmental factor; client's requirement and emphasis on quality, cost and time; project management actions; procurement and contractual method and nature of project.

Ling et al (2009) submitted that important practices for quality management are to control the quality of contract document, quality of response to perceived variations and extent of changes to signed contract. According to Aiyetan (2010), planning factors to be considered for project quality management are analysis of construction methods; analysis of materials and general resource movement to and within site; analysis of work and activities sequencing to achieve and maintain workflow; monitoring and updating of plans to reflect work status; responding to problems and taking advantage of present opportunities; effective coordination of project resources, and development of appropriate and effective organisational structure to maintain workflow.

3.4 Health and Safety Performance Factors of Construction Projects

Most of the factors affecting cost, time and quality performance of projects were found to affect other measures of performance such as sustainability, health and safety, satisfaction, etc. Mc Donald and Hrymak (2002) investigated the factors that influence safety behaviour and compliance with safety requirements on construction sites. It was revealed that the most significant safety compliance factor is the presence or absence of a safety representative for the project. The presence of this representative is strongly related to the effectiveness of response to audits and reported hazards on construction sites. This pattern of relationships suggests that presence of safety representatives affect project health and safety performance.

4 Research Methodology

Qualitative research approach was adopted for the study due to the nature of the research. Survey method was adopted to collect information from respondents through distribution of close-ended questionnaires. The target population are contractors and professionals acting as consultants or clients' representatives in Gauteng region, South Africa. These include architects, quantity surveyors, civil engineers, project managers, construction managers and construction project managers. Using convenient sampling method, questionnaire were designed and administered on 60 of these respondents to gather information about factors affecting the performance of construction projects. Prior to the actual data collection, pilot test was conducted on selected professionals in the construction industry. Base on their comments and observations, adjustments were made to the questionnaire before the final distribution.

Information relevant to the study was provided on the cover letter attached to the questionnaire. It was stated that data supplied will strictly be used for academic purpose. More so, respondents can complete the instrument at their convenience and also withdraw from the survey at any time. Confidentiality of the respondents was ensured by excluding questions that may reveal their identity. The major questions relating to the study were asked using 5-point Likert scale (5 = strongly agree; 4 = agree; 3 = average/neutral; 2 = disagree; and 1 = strongly disagree) from which Mean Item Score (MIS) for individual factors for both consultants and contractors were computed. The data were coded and analysed using Statistical Package for Social Science (SPSS 21) software. The MIS was used to evaluate and rank the identified variables in descending order. More so, Spearman' rho was employed to examine significant relationship between the opinions of contractors and consultants concerning factors affecting project performance using SPSS software.

5 Findings and Discussion

5.1 Background Information of Respondents

Forty-Four questionnaires were retrieved and found worthy of further analysis. For years of experience, 34% reported to have experiences varying between 1 and 5 years, 27% are between 6 and 10 years, 11% are between 16 and 20 years and 9% are 20 years and above. On the number of projects respondents have executed in the last five years, 25% executed 1 to 5 projects, 19% executed 6 to 10 projects, 27% executed 11 to 20 projects, 9% executed 21 to 30 projects while 20% executed more than 30 projects. On the worth of projects executed in the last 5 years, none of the respondents were involved in project worth less than R 13 million. 36% have executed projects worth between R 13 million and R 130 million while the remaining 59% have executed projects worth more than R 130 million. Currently, 34% of the professionals are involved in 1 to 5 projects, while 27%, 11%, 14% and 14% are involved in 3 to 4, 5 to 6, 7 to 8 and more than 8 projects respectively. This indicate that the respondents have the necessary knowledge and experience to supply adequate information require for this study.

5.2 Construction Project Performance Factors

Overall, the general opinions of contractors and consultants in Table 1, indicates that conformance to specification, cash flow of project and planned time for construction projects are the most important factors to measure project performance. Using a benchmark of 4.00 mean score, other important factor are quality assessment system in organization, percentage of orders delivered late, availability of personals with high experience and qualification, delay because of closures and materials shortage and availability of resources as planned through project duration. All the factors have MIS score of above 3.50 indicating that they are all important performance indicators. However, the least important factors were found to be medical fitness, project overtime cost and project design cost.

Table 1. Performance indicators for construction project performance

Factors	Contractors			Consultants			Average		
	MIS	SD	Rank	MIS	SD	Rank	MIS	SD	Rank
Conformance to specification	3.60	0.71	18	4.46	7.56	1	4.36	8.35	1
Cash flow of project	3.40	0.71	23	4.28	8.26	2	4.18	8.67	2
Planned time for project construction	3.80	1.22	14	4.21	9.93	3	4.16	10.30	3
Quality assessment system in organization	3.20	1.73	30	4.15	10.03	4	4.05	8.04	4
Percentage of orders delivered late	3.60	0.71	17	4.10	7.60	5	4.05	8.67	5
Availability of personals with high experience and qualification	4.00	1.00	6	4.05	8.17	7	4.05	9.73	6
Delay because of closures and materials shortage	3.80	1.22	12	4.05	7.69	6	4.02	7.40	7
Availability of resources as planned through project duration	4.00	1.00	5	4.03	6.65	8	4.02	9.23	8
Time needed to rectify defects variation orders	3.80	1.22	11	4.00	6.38	9	3.98	6.69	9
Material and equipment cost	3.80	1.22	9	4.00	8.47	11	3.98	9.50	10
Project labour cost	3.80	1.22	10	3.97	7.16	14	3.95	7.26	11
Profit rate of project	3.60	1.41	20	3.97	6.06	12	3.93	7.26	12
Escalation of material prices	4.00	1.22	8	3.92	8.11	19	3.93	7.56	13
Liquidity of organization	3.60	1.22	19	3.97	6.76	13	3.93	7.79	14
Cost control system	3.60	1.73	21	3.95	6.72	16	3.91	6.57	15
Quality training/meeting	4.00	1.22	7	3.90	6.38	21	3.91	7.73	16
Quality of equipments	3.00	1.00	34	4.00	7.46	10	3.89	7.01	17
Participation of managerial levels with decision making	3.60	1.73	22	3.90	5.89	20	3.86	6.61	18
Waste rate of materials	3.60	0.71	16	3.90	6.87	22	3.86	7.46	19
Training and development	3.40	0.71	25	3.92	8.02	18	3.84	6.42	20

(continued)

Table 1. (*continued*)

Factors	Contractors			Consultants			Average		
	MIS	SD	Rank	MIS	SD	Rank	MIS	SD	Rank
Market share of organization	3.20	0.71	27	3.92	6.14	17	3.84	8.04	21
Time needed to implement variation orders	4.00	1.00	4	3.82	7.22	26	3.84	8.47	22
Delay in claim approval	4.20	1.73	2	3.79	5.89	28	3.84	8.67	23
Delay in payment from owner to contractor	4.20	1.00	1	3.71	6.10	33	3.82	6.42	24
Site preparation time	3.00	0.00	31	3.87	6.26	23	3.77	6.76	25
Appointment of health and safety officers or managers	3.00	1.00	33	3.95	6.69	15	3.77	7.50	26
Regular project budget update	3.20	1.73	29	3.82	6.87	24	3.75	6.30	27
Regular visits of the Health and Safety officer	3.40	1.00	26	3.79	5.85	27	3.75	6.69	28
Cost of rework	4.00	1.00	3	3.72	5.50	32	3.75	7.98	29
Cost of variation orders	3.40	0.71	24	3.79	6.94	29	3.75	8.79	30
Regular toolbox talk meetings onsite	3.00	1.00	32	3.82	7.22	25	3.73	7.46	31
Overhead percentage of project	3.20	1.00	28	3.77	5.36	30	3.70	8.59	32
Project design cost	2.80	1.73	35	3.77	7.26	31	3.66	6.14	33
Project overtime cost	3.80	1.22	13	3.62	5.40	34	3.64	6.38	34
Medical fitness	3.80	1.22	15	3.59	7.85	35	3.61	8.17	35

MIS mean item score, SD standard deviation, R rank

In order to examine the relationship in the opinions of the two groups of respondents, Spearman' rho was employed. The analysis revealed a correlation coefficient of −0.21 and significance (2-tailed) value of 0.903 indicating that the relationship is not significant at both 5% and 10% level. This implies that there is difference in the opinion of contractors and consultants concerning factors affecting project performance. The general opinions of the two groups are closely related to the opinions of consultants. However, contractors were of the opinion that delays in payment from owner to contractors and delays in claim approval are the important performance factors with MIS value of above 4.00. The least important factor is project design cost with MIS of 2.80.

5.3 Performance Improvement Strategies for Construction Projects

For improvement in the performance of construction projects, several measures were analysed. These measures include project planning and scheduling; meeting for monitoring, updating and controlling of project progress; request for sub-contractors to submit activity schedule; provision of incentives for workers' motivation; presence of quantity surveyors in project team; safety meetings; and site visit by safety officers.

As for project planning and scheduling, 64% use the bar or Gantt chart, 16% use the line of balance method, 18% use the network method and 2% use other methods.

However, on frequency of project team meeting for discussion of monitoring, updating and controlling progress of work, 86% meet weekly, 11% daily and only 2% meet monthly, this shows that most meetings to monitor progress and implement necessary measures to enhance performance are undertaken weekly. Most construction firms require sub-contractors or suppliers to submit their detail activity schedule in advance in order to adjust their actual schedule weekly (82%), 9% require it daily while the remaining 9% never ask for it. This indicates that most respondents require their sub-contractors or suppliers to submit their activity schedule weekly but the challenge lies with the 9% that never request for the document.

The most adopted incentives and reward packages to motivate personnel in the quest to enhance project performance are through bonuses (39%), 37% increase salaries while 23% reduce the knock-off time. 80% of project teams organises meetings for safety issues weekly, 11% organise the meeting monthly while 9% organise it daily. More so, 80% of the projects are always visited by safety officer while 18% and 2% are visited often and sometimes. About 75% of project team always have a quantity surveyors who is responsible for dealing with cost control of construction projects. However, 14%, 5%, 5% and 2% often, sometimes, rarely and never observes this practice respectively.

5.4 Discussion of Findings

Conformance to specification is the most important factor for measuring performance of construction projects followed by cash flow of the project. This is supported by Shaban (2008) where it was stated that construction professionals such as contractors and consultants are averagely satisfied with performance of their project due mainly to non-conformance to specification. It is therefore necessary for construction professionals to devote more attention to enforcement of specification for construction projects in their quest for improved performance of construction projects. Another factor affecting project performance of construction projects is cashflow, which is related to cost planning, control and management. Oladinrin et al. (2016) noted that healthcare projects are abandoned due to lack of proper cost planning by contractors.

Planned time for construction projects and quality assessment system were also indicated as important factors. In support, Shaban (2008) observed that if planned time is not implemented appropriately, the project will suffer from delays and disputes between the owner and other parties. More so, Akogbei et al. (2015) noted that most of the problems in construction are related to planning issues. From contractors view, financial issues relating to payment and claims are major factors. In agreement, Sweis et al. (2014) noted that contractors' top three project performance indicators are too many change orders from owners, contractors' financial difficulties and owners' financial constraints.

For performance improvement, majority of construction firms adopt Bar Chart for project planning and scheduling. This according to Alsakini et al. (2004) is because it is quicker to design and easier to understand. More so, formal meeting for monitoring, updating and controlling project progress are mostly done weekly. Navon (2005) noted that such issues are usually discussed at site meetings which are usually organised weekly for most construction projects. About 9% of construction firms rarely require subcontractors to submit detail activity schedule. Against this practice, Thomas and Wentao (2006) noted that reports from sub-contractors and suppliers for the provision of various construction services and materials are necessary to monitor their progress and identify areas for immediate attention or improvement. This will also help the subcontractors and suppliers as Errasti et al. (2007) stated that, subcontractors in the infrastructure sector undergo tremendous difficulty in delivering their services to time and cost.

Most of the construction firms supply bonuses and increase salaries to motivate their employees. However, Chan and Kumaraswamy (2002) noted that better training and motivation are major requirements to improve performance of construction projects. The study also revealed that majority of the firms engage the service of quantity surveyors in dealing with cost control. Chan (2009) stated that quantity surveyors can control the cost of projects to stay within the budget and help to achieve the objective of the project. On the issue of safety, it was observed that safety meetings are usually organised monthly and safety officers always visit most construction sites. Mc Donald and Hrymak (2002) observed that the presence of a site safety representative enhances a higher level of compliance with safety. It was recommended that all sites should have a safety representative and their role and functions should be reinforced as part of the safety management system. In support, El-Nagari et al. (2015) noted that safety training and planning is one the necessary but less utilised factors by contractors.

6 Conclusion and Recommendations

Factors affecting the performance of construction projects are conformance to specification, cash flow of project, planned time for project construction and quality assessment system in organization. Compliance is the most crucial thing in construction and this is where consultants and contractors both need to comprehend original intention of their clients. For contractors, delays in payment and claim approval are basic. It is therefore essential that professionals shouldered with the responsibilities of processing contractors' payment, that is architect and quantity surveyors, improve on document preparation and approval process for prompt payment of contractors. In order to provide value for money for construction clients, owners and financiers, stakeholders in the industry must ensure and improve on the following performance improvement measures: project planning and scheduling; formal meeting of project team to monitor, update and control progress; submission of detail activity schedule by sub-contractors or suppliers; and provision of necessary incentives to motivate workers and thereby improve project performance.

References

Aiyetan, A.O.: Influences on construction project delivery time, Ph.D. thesis, Nelson Mandela Metropolitan University. contentpro.seals.ac.za/. Accessed 15 Sep 2014

Akanni, P.O., Oke, A.E., Akpomiemie, O.A.: Impact of environmental factors on building project performance in Delta State, Nigeria. Hous. Build. Natl. Res. Cent. J. **11**(1), 91–97 (2015)

Akogbei, R.T., Feng, X., Zhou, J.: Project performance evaluation based on time-cost-design capacity and plant utilization. J. Archit. Civ. Eng. **2**(2), 1–9 (2015)
Alsakini, W., Wikstrum, K., Kiiras, J.: Proactive schedule management of industrial Turnkey projects in developing countries. Int. J. Project Manag. **22**(1), 75–85 (2004)
Auma, E.: Factors affecting the performance of construction projects in Kenya: a survey of low-rise buildings in Nairobi Central Business District. Int. J. Bus. Manag. **2**(10), 115–140 (2014)
Brown, D., Armstrong, M.: Paying for contribution: real performance-related pay strategies. Kogan, London (1999)
Chan, D.W.M., Kumaraswamy, M.: Compressing construction durations: lessons learned from Hong Kong building projects. Int. J. Project Manag. **20**(1), 23–35 (2002)
Chan, K.W: Current roles and responsibilities of quantity surveyors (2009). http://chankinwo-dissertation-cn3307.blogspot.com/. Accessed 15 Sep 2014
Chan, A.P., Scott, D., Chan, A.P.: Factors affecting the success of a construction project. J. Constr. Eng. Manag. **130**(1), 153–155 (2004)
Chitkara, K.K.: Project management: planning, scheduling and controlling. McGraw Hill, New Delhi (2005)
Dissanayaka, S.M., Kumaraswamy, M.: Comparing contributors to time and cost performance in building projects. Build. Environ. **34**(1), 31–42 (1999)
El-Nagari, R., Hosny, H., Askari, H.S.: Development of a safety performance index for construction projects in Egypt. Am. J. Civ. Eng. Archit. **3**(5), 182–192 (2015)
Errasti, A.R., Roger, B., Aitor, O., Javier, S.: A process for developing partnerships with subcontractors in the construction industry: an empirical study. Int. J. Proj. Manag. **25**(3), 250–256 (2007)
Furnham, A.: Performance management systems. Eur. Bus. J. **16**(2), 83–94 (2004)
Iyer, K.C., Jha, K.N.: Critical factors affecting quality performance in construction projects. Int. J. Project Manag. **17**(9), 1155–1170 (2006)
Jouini, S., Mahmoud, B., Christophe, M., Gilles, G.: Time-to-market vs. time-to-delivery; managing speed in engineering, procurement and construction projects. Int. J. Project Manag. **22**(5), 359–367 (2004)
Koushki, P.A., Kartam, N.: Impact of construction materials on project time and cost in Kuwait. Eng. Constr. Manag. Econ. J. **11**(2), 126–132 (2004)
Ling, F.Y.Y., Low, S.P., Wang, S.Q., Lim, H.H.: Key project management practices affecting Singaporean firms: project performance in China. Int. J. Project Manag. **27**(1), 59–71 (2009)
Masilamani, M.: Key Performance Indicators: Paper Presented at National Conference on Key Performance Indicators (KPis), 19–22 September, Crown Princess Hotel, Kuala Lumpur (2005)
Mbamali, I., Aiyetan, A.O., Kehinde, J.O.: Building design for buildability: an investigation into the current practice in Nigeria. Build. Environ. **40**(9), 1267–1274 (2005)
Mc Donald, N., Hrymak,.V.: Safety behaviour in the construction sector. Available online http://arrow.dit.ie/cgi/ (2002). Accessed 15 Sep 2014
Msani, T.A.: Critical success factors influencing project success in the Durban construction industry, South Africa, Masters degree thesis, Department of Entrepreneurial Studies and Management, Durban University of Technology (2011). http://ir.dut.ac.za:8080/bitstream/handle/. Accessed 15 Sep 2014
Navon, R.: Automated project performance control of construction projects. Autom. Constr. **14**(4), 467–476 (2005)
Ng, S.T., Palaneeswaran, E., Kumaraswamy, M.: A dynamic e-reporting system for contractor's performance appraisal. Adv. Eng. Softw. **33**(6), 339–349 (2002)

Oladinrin, T.O., Oke, A.E., Abidoye, R.B, Koleowo, N.A.: Assessment of the critical success factors for healthcare project delivery in Ondo state, Nigeria. In: Ebohon, O.J., Ayeni, D.A., Egbu, C.O., Omole, F.K. (eds.) Proceeding of Joint International Conference on 21st Century Human Habitat: Issues, Sustainability and Development, 21–24 March, Federal University of Technology, Akure, Nigeria, pp. 311–318 (2016)

Olatunji, S.O., Oke, A.E., Aghimien, D.O., Saidu, S.A.: Effect of construction project performance on economic development of Nigeria. J. Econo. Sustain. Dev. **7**(12), 142–149 (2016)

Parmenter, D.: Key Performance Indicators, Developing, Implementing and Using Winning KPIs. Wiley, Canada (2010)

Reichelt, K.L.J.: The dynamic of project performance benchmarking the drivers of cost and schedule overrun. Eur. Manag. J. **17**(2), 135–150 (1999)

Shaban, S.A.: Factors Affecting the Performance of Construction Projects in the Gaza Strip. The Islamic University of Gaza, Palestine (2008)

Sweis, R.J., Bisharat, S.M., Bisharat, L., Sweis, G.: Factors affecting contractor performance on public construction projects. Life Sci. J. **11**(4), 28–39 (2014)

Thomas, S., Wentao, L.: A parallel bargaining protocol for automated sourcing of construction suppliers. Autom. Constr. **15**(5), 365–373 (2006)

Toor, S.R., Oguniana, S.O.: Problems causing delays in major construction projects in Thailand. Constr. Manag. Econ. **26**(4), 395–408 (2008)

Ugwu, O.O., Haupt, T.C.: Key performance indicators and assessment methods for infrastructure sustainability: a South African construction industry perspective. Build. Environ. **42**(6), 665–680 (2007)

Yakubu, O., Sun, M.: Cost and time control of construction projects: a survey of contractors and consultants in the UK. Constr. Inf. Q. **11**(2), 53–59 (2009)

Utilization of Project Management Tools for Construction Project Success

Sihle Gogela, Ayodeji E. Oke(✉), and Clinton O. Aigbavboa

Department of Construction Management and Quantity Surveying, Faculty of Engineering and the Built Environment, University of Johannesburg, Johannesburg, South Africa
sihlelisonke.gogela@gmail.com, emayok@gmail.com, caigbavboa@uj.ac.za

Abstract. Due to increasing complexity of construction projects and the need to deliver projects to the satisfaction of clients with changing demand, various tools, techniques and methods have been introduced while some are gaining popularity in the construction industry. One of the existing tools are various project management techniques. The purpose of this study is to explore available project management tools and techniques that can be utilized to enhance construction project performance with a view to achieving value for money and ensuring satisfaction of construction clients. The study adopted quantitative research approach and a survey questionnaire was employed as an instrument of collecting data from relevant construction professionals with adequate knowledge and experience of the construction industry and project management tools. The study was limited to Gauteng and Mpumalanga Provinces in South Africa The survey revealed that application of value management or value engineering, reliable feasibility study, usage of computerized integrated software, application of life cycle costing and utilization of emerging technology and development are the five important project management tools for effective project delivery. It can be concluded that client satisfaction can be achieved by the application and full usage of aforementioned project management tools during the preconstruction, construction and post-construction stages of infrastructural projects.

Keywords: Construction stakeholders · Clients' satisfaction
Project performance · Value for money

1 Introduction

Construction projects are continuously experiencing cost overruns, subsequently costing the client more money than the initial estimated project amount. Construction Industry Development Board (CIDB) (2015) revealed that Clients were neutral or dissatisfied with the performance of contractors on 18% of the projects surveyed in 2015 – which reflects a visible increase in dissatisfaction over the period 2013 to 2015. Further, clients were neutral (i.e. neither satisfied nor dissatisfied) or dissatisfied on 17% of the projects with the overall performance of the consulting team employed. Client satisfaction shows a visible decrease over the surveys from 2012 to 2015, namely from 87% in 2012 to 83% in 2015. Client satisfaction is the level of satisfaction

M. Shehata and F. Rodrigues (Eds.): GeoMEast 2018, SUCI, pp. 190–195, 2019.
https://doi.org/10.1007/978-3-030-01905-1_10

of the client on the contractor's performance on the project planning and execution. Consultants and Contractors survival in the industry depends on level of feedback in which they received from their clients. Consultants and Contractors need to provide value for money to the client, as many clients are no longer awarding contracts on the lowest tender price, but also include the performance of the contractor on past contracts. Clients in construction want their projects delivered on time, on budget, free from defects, efficiently, right the first time, safely and by profitable companies. This study therefore examines the importance of various project management tools and techniques in the achievement of project performance with a view to ensuring satisfaction of clients and other stakeholders.

2 Review of Project Management Tools

2.1 Value Management

During the pre-tender stage, the consultants, over and above have to ensure that the project procurement is successful, also have an obligation to achieve client satisfaction. The client's objective is to make an initial investment into a project that will at the end inject good returns. At the strategic level of organization planning, Value Management (VM) or Value Engineering (VE) has a role in the design of program intended to deliver strategic benefits (Thiry 2010). The operational level use of VM continues to develop, because when effectively used, it can deal with problems related to identifying areas of poor value, maximize the use of human resources and develop innovative ways to perform key activities, Stewart (2010). Jay and Bowen (2015) believes that effective use of VM or VE can achieve innovative solutions to product and process design problems as well as addressing design issues that impact the entire organization and its value chain. He further adds that further research need to be done to better understand how to apply the method of VM or VE in less traditional processes such as strategic planning and direction setting. VM or VE is not a cost reduction exercise, but rather a useful mechanism that can be applied to mitigate the design and planning issues during the pre-tender stage of a project.

2.2 Life Cycle Costing

Swaffield and McDonald (2008) research investigated on attitudes and options of quantity surveyors working for design and construction contractors regarding the imperative use of life cycle costing within private finance initiative projects. The findings of this research identified that design and contractors' quantity surveyors generally consider life cycle costing when procuring new products for Private Finance Initiative (PFI) projects. Most often, considering the time constraints and meeting of deadlines during a project procurement, Life Cycle Costing (LCC) is not used as a cost reduction mechanism. Consequently, research found that failure of effective use of LCC has financial risk implications. Moreover, there is a likelihood of the maintenance cost increasing in the future, compared to the original budget for maintenance cost. With the subsequent findings in research, effective use of LCC can result on cost reduction in future costs. This will mean that the client can gain value for money which is the main objective for project consultants. This research study believes LCC can only be affective if implemented during the pre-tender as a cost reduction tool in order to innovatively solve design specifications and achieve client satisfaction.

2.3 Total Quality Management System

According to a study on total quality management and quality assurance by Jabnoun (2002), Total Quality Management control process emphases on customer satisfaction and continuous improvement. Total Quality Management (TQM) encourages employees to constantly check and improve quality. TQM also stresses the role of management in providing and altering the necessary input of structure, skills, systems and shared values, in order to meet customer or clients varying demands. The results from the study of Bryde and Robinson (2007) concluded that those in organizations with a TQM program in place are more customer-focused in their project management practices than those in organizations with no TQM program. The benefits of TQM from the viewpoint of the individual company, the strategic implementation of TQM include the following:

- Survival in the increasing competitive world
- Better service provided to clients
- Enhancement of the organization's shareholder value
- Improvement of the overall quality and safety of our facilities
- Reduced project durations and costs
- Better utilization of the talents of its people

Quality management is an essential component in the successful management of construction projects; critical success factors include the support of senior management, appropriate leadership styles, cultivation of employee's enthusiasm and participation, and open communication and feedback. The most appropriate strategy for effective TQM in the construction industry is to establish effective management practices.

The existing literature materials reveals that underestimating is caused by underestimating risk, overestimating opportunities, inadequate estimation methods and skills, reliance of weak information and deliberate scope creep and division of project. As of recent, there is an increasing number of unsatisfied clients due to projects exceeding the initial estimates that clients used as a decision making tool. Arguable, initial estimates are becoming unreliable and not credible enough for clients to use as a decision making tool. The findings showed that costs are more often overestimated than underestimated during the pre-tender stage. This should not come as a surprise as contractor's aim is to maximize profit while project consultants' aim is to secure project commitment against the clients funding. Even though, the standard bills of quantities method are an efficient method of estimation, it needs to be utilized and supported with the expertise and experience of the estimator, which is seen to have a significant impact on the estimate accuracy.

3 Research Methodology

This study adopted a quantitative research approach and the respondents were development experts in the Gauteng and Mpumalanga Provinces in South Africa. In a bid to be able to generalize the findings of this study, all forms of projects were considered. This implies that the findings of this study are applicable to building, civil and industrial engineering projects. It is also applicable to residential, commercial and other forms of construction projects. As per the delimitation the respondents, these comprises of design, cost and project management team within the construction industry, these include architects, engineers, quantity surveyors, construction managers and project managers. Convenience sampling method was adopted due to its appropriateness in reaching high number of respondents with ease. Progression of a survey according to the reviewed literature materials were utilized as instruments of controlling data from the respondents. The survey was expressed in English and an introductory letter was given to advise the respondents on the purpose of the study. Questions were identified with the primary research study objective utilizing a 5-poiny Likert scale. A sum of 65 questionnaires were circulated but 40 were recovered and found usable for analysis. Mean Item Score (MIS) and Standard Deviation (SD) were employed in the analysis and ranking of importance of the identified project management tools.

4 Findings and Discussion

Table 1 shows the calculated mean scores (MIS), standard deviation (SD) as well as ranking (R) of the identified tools, it could be observed that the most dominant project management tools and techniques include: application of value management or value engineering (MIS = 4.93, SD = 0.27, R = 1), reliable feasibility study (MIS = 4.93, SD = 0.35, R = 2), usage of computerized integrated software (MIS = 4.78, SD = 0.42, R = 3), application of life cycle costing (MIS = 4.63, SD = 0.59, R = 4), utilization of emerging technology and development (MIS = 4.25, SD = 0.44, R = 5),

Table 1. Project management tools for project performance

Client satisfaction: value for money	MIS	SD	R
Application of value management or value engineering	4.93	0.27	1
Reliable feasibility study	4.93	0.35	2
Usage of computerized integrated software	4.78	0.42	3
Application of life cycle costing	4.63	0.59	4
Utilization of emerging technology and development	4.25	0.44	5
Application of total quality management system	4.00	0.64	6
Application of lean construction	3.98	0.66	7
Creating synergy amongst team members	3.90	0.50	8
Application of front end loading	3.65	0.86	9
Optimize supply chain management	3.38	0.59	10

SD = Standard deviation; MIS = Mean item score; R = Rank

application of Total Quality Management System (MIS = 4.00, SD = 0.64, R = 6), application of lean construction (MIS = 3.98, SD = 0.66, R = 7), creating synergy amongst team members (MIS = 3.90, SD = 0.50, R = 8), application of front end loading (MIS = 3.65, SD = 0.86, R = 9), supply chain management (MIS = 3.38, SD = 0.59, R = 10).

These findings revealed a similar level of importance as Thiry (2010) who added that at the strategic level of organization planning, Value Management (VM) or Value Engineering (VE) plays a vital role in the design of program intended to deliver strategic benefits. The operational level use of VM continues to develop, because when effectively used, it can deal with problems related to identifying areas of poor value, maximize the use of human resources and develop innovative ways to perform key activities, Stewart (2010). Jay and Bowen (2015) believes that effective use of VM or VE can achieve innovative solutions to product and process design problems as well as addressing design issues that impact the entire organization and its value chain. He further adds that further research need to be done to better understand how to apply the method of VM or VE in less traditional processes such as strategic planning and direction setting. VM or VE is not a cost reduction exercise, but rather a useful mechanism that can be applied to mitigate the design and planning issues during the pre-tender stage of a project. Swaffield and McDonald (2008) noted that failure of effective use of LLC has financial risk implications. Moreover, there is a likelihood of the maintenance cost increasing in the future, compared to the original budget for maintenance cost. With the subsequent findings in research, effective use of LLC can result on cost reduction in future costs. This will mean that the client can gain value for money which is the main objective for project consultants. This research study believes LLC can only be affective if implemented during the pre-tender as a cost reduction tool in order to innovatively solve design specifications and achieve client satisfaction. This study also reveals that there is no significant agreement on the application of Total Quality Management Systems in facilitating client satisfaction based on Jabnoun (2002).

5 Conclusion and Recommendations

Client satisfaction is the level of satisfaction of the client on the contractor's performance on the project planning and execution. Consultants and Contractors survival in the industry depends on level of feedback in which they received from their clients. Consultants and Contractors need to provide value for money to the client, as many clients are no longer awarding contracts on the lowest tender price, but also include the performance of the contractor on past contracts. Clients in construction want their projects delivered on time, on budget, free from defects, efficiently, right the first time, safely and by profitable companies. The application of Value Management or Value Engineering, reliable Feasibility Study, usage of Computerized Integrated Software, application of Life Cycle Costing and Utilization of Emerging Technology and Development were the most important project management tools for the achievement of project delivery. Therefore, client satisfaction can be achieved by the application and full usage of aforementioned project management tools in the execution of construction projects. Further to this study, there is a need to examine the influence of each of the identified project management tools on project performance, especially in the area of providing value for money for construction clients.

References

Bryde, D.J., Robinson, L.: The relationship between total quality management and the focus of project management practices. TQM Mag. **19**(1), 50–61 (2007)

Construction Industry Development Board: The CIDB construction industry indicators (2015). Summary Results: http://www.cidb.org.za/publications/Documents/. Accessed Aug 2016

Jabnoun, N.: Control processes for total quality management and quality assurance. Work Study **5**(4), 182–190 (2002)

Jay, C.I., Bowen, P.A.: Value management and innovation. J. Eng. Des. Technol. **13**(1), 123–143 (2015)

Stewart, R.B.: Value Optimization for Project and Performance Management. Wiley, Hoboken (2010)

Swaffield, L.M., McDonald, A.M.: The contractor's use of life cycle costing on PFI projects. Eng. Constr. Archit. Manag. **15**(2), 132–148 (2008)

Thiry, M.: Program Management. Gower, Farnham, Surrey (2010)

Text and Information Analytics for Fully Automated Energy Code Checking

Peng Zhou(✉) and Nora El-Gohary

University of Illinois at Urbana-Champaign, Urbana, IL, USA
{pzhou6, gohary}@illinois.edu

Abstract. Energy compliance checking aims to check the compliance of design information embedded in building information models (BIMs) with applicable building energy codes. To fully automate the compliance checking process – without the need to manually code the rules in a compliance checking software, there is a need to automatically process and understand the text in the codes, extract the relevant requirements from the text, and match the extracted requirements to the information in the BIMs. This is challenging because of the text complexities of the codes, including longer provisions, requirement exceptions, hierarchically-complex sentence structures, and different terminologies and levels of detail compared to BIMs. To address these challenges, a set of text and information analytics methods were proposed and implemented in an energy compliance checking prototype. The prototype includes four main modules: text classification, information extraction, semantic information alignment, and compliance checking and reporting. For testing the prototype, a BIM of an educational building was checked for compliance with three energy codes – the 2012 International Energy Conservation Code, the 2013 Building Energy Efficiency Standards (known as the California Energy Code), and the Ontario Building Code Supplementary Standard SB-10. A promising performance of 91.7% recall and 84.6% precision in noncompliance detection was achieved. This study could benefit different participants in the domain of architecture, engineering and construction, by streamlining the energy code compliance process in a fully-automated manner. Future research may test the proposed methods and prototype on other energy topics (e.g., fenestration) and different types of documents (e.g., contract specifications). The challenges for developing a compliance checking method that is fully-automated and generalized across different types of documents could also be studied.

1 Introduction

Building energy codes are developed to help building projects achieve energy efficiency and sustainability. A + ∂ Energy compliance checking aims to check the compliance of design information embedded in building information models (BIMs) with applicable building energy codes. Because manual code checking is time and cost intensive, several research efforts have been undertaken to automate the code compliance checking process. A variety of semi-automated and automated code compliance checking methods have been developed and implemented in recent years [e.g., Zhang and El-Gohary (2017), Ilal and Günaydin (2017), Solihin and Eastman (2016),

M. Shehata and F. Rodrigues (Eds.): GeoMEast 2018, SUCI, pp. 196–208, 2019.
https://doi.org/10.1007/978-3-030-01905-1_11

Beach et al. (2015), Pauwels et al. (2011)]. Such methods have achieved different levels of automation, expressivity, representativeness, accuracy, and efficiency. Semi-automated approaches [e.g., Ilal and Günaydin (2017), Solihin and Eastman (2016), Beach et al. (2015), Pauwels et al. (2011)] use different representations and methods to model regulatory requirements in a computer-processable format. But, they all require a degree of manual effort to achieve that. Fully-automated approaches are, on the other hand, limited in number; primarily because of the challenge in enabling computers to automatically understand and process the text in regulatory documents, and automatically represent it in a computer-processable format. An example of such efforts is the work by Zhang and El-Gohary (2017), which used semantic natural language processing (NLP) technique to automatically extract requirements from building codes and represent the extracted requirements in a computer-interpretable representation.

Despite the importance of all previous efforts, they all share one common limitation. They are unable to deal with complex requirements such as those in energy codes. Fully automated checking of the compliance of BIMs with energy codes is challenging because of the text complexities of the energy codes, including (1) longer provisions: longer provisions are likely to contain semantically complex requirements; (2) requirement exceptions: requirements may have multiple exceptions; (3) hierarchically-complex sentence structures: one super-level provision may contain multiple sub-levels of provisions, and one sub-level provision may contain multiple requirements; and (4) different terminologies and levels of detail compared to BIMs: such terminology and detail gaps require complex linking and matching of the right text entities to the right BIM entities.

To address these challenges, a set of text and information analytics methods were proposed and implemented in an energy compliance checking prototype. The prototype includes four main modules: (1) text classification, which aims to filter out irrelevant text in the codes to avoid unnecessary processing effort and potential errors, thereby improving the efficiency and performance of the checking; (2) information extraction, which aims to extract energy requirements from the classified text and extract relevant design information from the BIMs; (3) semantic information alignment, which aims to match the extracted energy requirements and the extracted design information to the same data representation so that they can "speak the same language"; and (4) compliance checking and reporting, which performs the check automatically and reports the checking results. The prototype was tested in checking the compliance of the BIM of an educational building project with thermal insulation requirements from three energy codes. The remainder of this paper presents the proposed methods and the preliminary experimental results.

2 Background and State of the Art in Automated Code Checking

A large number of research efforts have been undertaken in the past five decades to automate the code checking process. Automated code checking (ACC) requires two essential elements: (1) code requirement representation: representing the code requirements in the form of computer-interpretable rules, and (2) code-BIM representation matching: matching the representations in the codes/rules to those in the BIMs. As such, existing ACC efforts can be categorized into three groups, according to the degree of manual effort involved in code requirement representation and code-BIM representation matching – intensive effort, substantial effort, and no or minimal effort. First, the majority of existing ACC efforts require intensive manual effort. For example, Goel and Fenves (1969) manually encoded building structural design requirements into decision tables. Pauwels et al. (2011) manually encoded building acoustic performance requirements as SPARQL Protocol and RDF Query Language (SPARQL) queries and manually developed a mapping ruleset to match the SPARQL-represented requirements to the RDF-represented design information. Solihin and Eastman (2016) manually modeled the general building design requirements into conceptual graphs, in which BIM terminologies were used to encode the requirements for matching to BIMs. These intensive manual efforts are time-consuming and costly. Second, a few ACC efforts require substantial manual effort. For example, the RASE methodology [e.g., Beach et al. (2015), Ilal and Günaydin (2017)] requires semi-automatic encoding of the requirements by manually annotating the code sentences using four tags – "Requirement, Applicability, Selection, Exception" – and manually matching to the BIMs. Such semi-automated approaches are still labor-intensive and hard to scale up. Third, a very limited number of efforts require no or minimal manual effort. For example, Zhang and El-Gohary (2017) proposed a fully-automated semantic NLP-based approach to automatically extract requirements from building codes, and used a semi-automated machine learning approach to match the regulatory information to the Industry Foundation Classes (IFC)-represented design information. While this work reached a level of full automation in building code requirement extraction and representation, it has not been tested on energy codes – which contain more challenging text complexities (i.e., longer provisions, exceptions, hierarchically-complex sentence structures and different terminologies and levels of detail compared to BIMs). It also requires manual effort – even if little or minimal – in semi-automated code-BIM matching.

3 Proposed Methods and Prototype for Fully Automated Energy Compliance Checking

To address the challenging text complexities for fully automated energy compliance checking, a set of text and information analytics methods are proposed and implemented in an energy compliance checking prototype. The prototype includes four main modules: text classification, information extraction, semantic information alignment, and compliance checking and reporting.

3.1 Text Classification

Text classification aims to filter out irrelevant text in energy codes to improve the efficiency and performance of information extraction and alignment. The classification is conducted on the most granular subheading level – the clause level. One energy code contains many clauses and one clause may contain multiple requirements/exceptions. A regular expressions-based text splitting method is used to automatically split an energy code into individual clauses, in which a clause is recognized based on the text feature patterns of clause headings. The split clauses are then classified. An ontology-based text classification method (Zhou and El-Gohary 2016) is used for the classification. The classification methodology includes four main steps: semantic modeling, text preprocessing, deep learning, and label assignment. A topic hierarchy is developed to identify and model the labels used in text classification, and a subontology is built for each topic (label) in the hierarchy to assist in semantic text classification. Text processing techniques – including tokenization, stemming, and stopword removal – are used to preprocess each clause prior to classification. A deep learning technique is used to measure the semantic similarity between each clause and each subontology. An empirical method is used to analyze the degrees of semantic similarities and accordingly assign zero, one, or multiple labels to a clause. The zero-label clauses are then filtered out. For the details of the ontology-based text classification method, the readers are referred to Zhou and El-Gohary (2016). Examples of a relevant clause and an irrelevant clause are shown in Table 1.

Table 1. Examples of a relevant clause and an irrelevant clause

Relevant clause	Irrelevant clause
(a) Roof/Ceiling Insulation. The opaque portions of the roof/ceiling that separates conditioned spaces from unconditioned spaces or ambient air shall meet the applicable requirements of Items 1 and 2 below: 1. Metal Building - The weighted average U-factor of the roof assembly shall not exceed 0.098 2. Wood Framed and Others - The weighted average U-factor of the roof assembly shall not exceed 0.075	(g) Administrative requirements. Administrative requirements relating to permit requirements, enforcement by the Commission, locally adopted energy standards, interpretations, claims of exemption, approved calculation methods, rights of appeal, and certification and labeling requirements of fenestration products and roofing products are specified in California Code of Regulations, Title 24, Part 1, Sections 10-101 to 10-114

3.2 Information Extraction

Information extraction (IE) includes: (1) regulatory IE, which aims to automatically extract requirements from the classified clauses and represent these requirements in a computer-interpretable rule format; and (2) BIM IE, which aims to automatically extract design information from BIMs for further alignment to the requirement rules.

Regulatory Information Extraction

For regulatory IE, the energy requirements are automatically extracted using an ontology-based IE method (Zhou and El-Gohary 2017). The proposed method uses a rule-based approach and includes four primary steps. First, the classified clauses are preprocessed using common text processing techniques including tokenization, sentence splitting, and morphological analysis. Second, text features, including syntactic features (e.g., part-of-speech tags) and semantic features (e.g., ontology concepts), are selected for developing extraction rules. Third, nine target semantic information elements (SIEs) are used to represent a requirement/exception, including "Subject", "Subject Restriction", "Compliance Checking Attribute", "Deontic Operator Indicator", "Quantitative Relation", "Comparative Relation", "Quantity Value", "Quantity Unit/Reference", and "Quantity Restriction". Fourth, a set of extraction rules are developed by analyzing and defining the text feature patterns to extract the target SIEs. For the details of the proposed method, the readers are referred to Zhou and El-Gohary (2017). To deal with the challenging text complexities including long provisions, requirement exceptions, and hierarchically-complex sentence structures, two text and information analytics methods are used (Zhou and El-Gohary 2017). First, a domain-specific text splitting and stitching method is used to deal with hierarchically-complex sentence structures and exceptions. The exceptions (if any) are split from the provisions (i.e., clauses), and the provisions/exceptions with hierarchically-complex sentence structures are automatically split to the lowest level. The conjunctive or disjunctive relationships among the multiple requirements within a provision are used, at later step (by the compliance checking and reporting module), for reasoning about the compliance with the whole provision. Second, a cascaded extraction method is used to extract target information from text with long provisions, hierarchically-complex sentence structures, and exceptions. Text with such complexities usually contains complex SIEs (e.g., "Subject Restriction" and "Quantity Restriction"), where a complex SIE may be composed of multiple simple SIEs (e.g., "Subject", "Compliance Checking Attribute", "Quantity Value"). Instead of extracting the complex SIEs directly, the cascaded extraction method breaks down such a challenging task to several simple extraction tasks; the simple SIEs are first extracted and the complex SIEs are then extracted using the extracted simple SIEs. An example of extracted requirement in the SIEs representation is shown in Table 2.

BIM Information Extraction

For BIM IE, the design information is extracted in three steps. First, the Revit model is exported into an IFC data file using the Revit built-in IFC exporter. Second, an EXPRESS-based data extraction method is used to automatically extract the relevant design information from the IFC data file into an intermediate format, where the relevant design information refers to the instances of a specific set of IFC entities (e.g., IfcProduct, IfcMaterial, IfcRelationship, IfcSimpleProperty, IfcSIUnit). Third, a data post-processing method is used to transform the design information in the intermediate format to the final computer-interpretable representation, in which the design information is ready for use in the semantic information alignment module. An example of extracted BIM design information is shown in Table 3.

3.3 Semantic Information Alignment

Semantic information alignment aims to align the representation of the design information to the representation of the regulatory requirements so that they can "speak the same language" – align design information instances to regulatory concepts (i.e., find instance-concept matches). A fully automated semantic information alignment method is proposed and used. The proposed method includes three primary sub-methods:

Table 2. Example of an extracted energy regulatory requirement

Subject	Subject restriction	Compliance checking attribute	Deontic operator indicator	Quantitative relation	Comparative relation	Quantity value	Quantity Unit/Reference	Quantity restriction
Metal building roof	N/A	Weighted average U-factor	Shall	N/A	Not exceed	0.098	N/A	N/A

Table 3. Example of extracted BIM design information

Key	Value	Example ({Key: Value})
IfcRoof instance ID	List of properties of the Key IfcRoof instance, each property is represented in: (Property set name, Property instance ID, Property name, Value, Unit)	{#5327028: Pset_RoofCommon, #5327243, Thermal Transmittance, 0.748031496062992, {{null, kelvin, '−1'},{kilo, gram, '1'},{null, second, '−3'}}}
IfcRoof instance ID	The entity attribute values of the Key IfcRoof instance	{#5327028: Basic Roof:EPDM – 4 1/2″ – lsf 8:9839086, Basic Roof: EPDM – 4 1/2″ – lsf 8, 9839086, notdefined}

concept matching, semantic similarity analysis, and graph search and network modeling.

Concept Matching

Concept matching aims to automatically recognize all candidate instance-concept matches. The buildingSmart Data Dictionary (bSDD) and a domain ontology are used to capture domain knowledge to support the recognition of candidate matches. The matching involves the matching of design information instances (e.g., "Basic Roof: EPDM – 4 1/2" – lsf 8:9839086") to regulatory concepts (e.g., "metal building roof"). For example, two methods are used for concept matching: (1) the bSDD is searched to find perfectly matching IFC entities (e.g., searching "roof" in the bSDD finds "IfcRoof" and "IfcRoofType"). The instances of these matching IFC entities become candidate matching instances; and (2) a given object concept (e.g., "metal building roof") is decomposed into two parts based on the ontology: a core concept (e.g., "roof") that carries the primary information, and a secondary part (e.g., "metal building") that carries further descriptive information to characterize the core concept. The bSDD is then searched to find matching IFC entities to the core concept (e.g., "roof"), and the instances of those matching IFC entity parents become candidate design information instances.

Semantic Similarity Analysis

Semantic similarity analysis aims to assess the semantic similarity between candidate instance-concept matches, and accordingly select the final matches. The semantic similarity analysis includes three steps: semantic similarity scoring, semantic similarity weighting and ranking, and semantic similarity threshold analysis. For scoring, the instance-concept semantic similarity is calculated by aggregating the term-to-term semantic similarities, which refer to the semantic similarity between each term in the instance descriptors (e.g., instance property value "Roofing – Standing Seam Metal Panel") and each term in the concept descriptors (e.g., concept name like "metal building roof"). A deep learning technique is used to calculate the term-to-term semantic similarities. For semantic similarity weighting and ranking, a set of weighting schemes are proposed and used to capture the factors that may impact the term-to-term semantic similarities and the instance-concept semantic similarities. The weighted instance-concept semantic similarities of all instances for a concept are then ranked in decreasing order. For semantic similarity threshold analysis, an empirical way is used to analyze the patterns of weighted instance-concept semantic similarities, and select the final instance-concept matches based on a set of semantic similarity thresholds.

Graph Searching and Network Modeling

A graph-and-network-based method is proposed and used to recognize the set of design information instances (called instance groups) that are related to one regulatory requirement. First, graph searching is used to automatically recognize the relationships that link two IFC instances as a pair (i.e., instance pair). An IFC instance graph is built for graph searching, where each node in the graph refers to a design information instance, and each edge between two nodes refers to an instance of the IfcRelationship entity. A graph searching algorithm is then used to search the path(s) between two design information instances, and it is assumed that the shortest path (if any) determines that those two instances can be linked as an instance pair. Second, network modeling is used to model concept groups (concepts in one regulatory requirement) and the associated design information instances into a set of networks, where each network models a concept group and a set of instance groups. An instance group is created by grouping and linking the created instance pairs, if those instance pairs have one-to-one correspondence to the concept pairs and can be linked in the same way as those concept pairs in the concept group. An example of design information that were matched to a regulatory requirement is illustrated in Fig. 1.

Semantic Information Element (SIE)	Energy Requirement	Matched Design Information
Subject	metal building roof	#5327028= IFCROOF('05YY7Kwq155h2aebeszJ1e',#42,'**Basic Roof:EPDM - 4 1/2" - lsf 8:9839086**',$,'Basic Roof:EPDM - 4 1/2" - lsf 8',#5327025,$,'9839086',.NOTDEFINED.);
Compliance checking attribute	weighted average U-factor	#5327243= IFCPROPERTYSINGLEVALUE('**ThermalTransmittance**',$,IFCTHERMALTRANSMITTANCEMEASURE(0.748031496062992),$);
Comparative relation[a]	not exceed	N/A
Quantity value	0.098	#5327243= IFCPROPERTYSINGLEVALUE('ThermalTransmittance',$,IFCTHERMALTRANSMITTANCEMEASURE(**0.748031496062992**),$);

[a]SIE that is only used for representation of regulatory information.

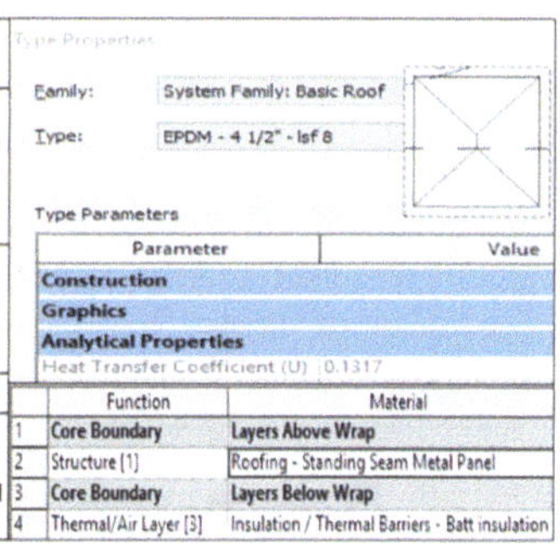

Fig. 1. An example of matched design information to a regulatory requirement

3.4 Compliance Checking and Reporting

Compliance checking and reporting aims to perform the check automatically and report the checking results. After semantic information alignment, the requirements are represented as logic rules, and the matched design information instances are represented as logic facts. The logic representations proposed in Zhang and El-Gohary (2016) are adapted and used to represent the logic rules and logic facts. An example of a logic rule and related logic facts, in B-prolog language, is shown in Fig. 2. The built-in reasoner of B-prolog is used to automatically reason about the compliance of the logic facts to the logic rules, and report whether the design is compliant or noncompliant as well as the violations(s) if noncompliant.

Logic Rule

Primary Logic Clause

```
compliance_weighted_average_u_factor_of_metal_building_roof(Weighted_average_u_f
actor,Metal_building_roof) :- metal_building_roof(Metal_building_roof),
weighted_average_u_factor(Weighted_average_u_factor),
relation(Metal_building_roof,Weighted_average_u_factor),
less_than_or_equal(Weighted_average_u_factor, quantity(0.556501987507, {{null,
kelvin, '-1'},{kilo, gram, '1'},{null, second, '-3'}})).
```

Secondary Logic Clause 1

```
check_a_roofceiling_insulation(X) :- (metal_building_roof(Metal_building_roof),
weighted_average_u_factor(X), relation(Metal_building_roof,X)) ->
check_weighted_average_u_factor_of_metal_building_roof(X,Metal_building_roof);true.
```

Secondary Logic Clause 2

```
check_weighted_average_u_factor_of_metal_building_roof(X,Y) :-
compliance_weighted_average_u_factor_of_metal_building_roof(X,Y) ->
(writeln((X,of,Y,is,compliant,with,'(a),RoofCeiling
Insulation',requirement1_1)));(writeln((X,of,Y,is,noncompliant,with,'(a),RoofCeiling
Insulation',requirement1_1,it,should,be,'less_than_or_equal, 0.098')), false).
```

Logic Facts

```
metal_building_roof('basic_roof_epdm_4_1_2_lsf_8_9839086#5327028').
relation('basic_roof_epdm_4_1_2_lsf_8_9839086#5327028','thermal_transmittance#5327
243'). weighted_average_u_factor('thermal_transmittance#5327243').
get_quantity('thermal_transmittance#5327243', 0.748031496062992, {{null, kelvin, '-
1'},{kilo, gram, '1'},{null, second, '-3'}}).
```

Fig. 2. An example of a logic rule and logic facts in B-prolog language

4 Preliminary Experimental Results and Analysis

The proposed methods and the energy compliance checking prototype were implemented in a Java-based platform. A gold standard was manually built for testing and performance evaluation. The gold standard includes: (1) 20 building thermal insulation requirements from three energy codes – the 2012 International Energy Conservation Code (ICC 2012), the 2013 Building Energy Efficiency Standards (California Energy Commission 2013), and the Ontario Building Code Supplementary Standard SB-10 (Ontario Ministry of Municipal Affairs 2011); and (2) 79 design information sets, which were added to the Revit model of an educational building project in Illinois. The design information sets include both compliant and noncompliant cases, and there are 24 noncompliant design information sets in total. To evaluate the performance, recall and precision in noncompliance detection were used. Recall, here, is defined as the percentage of total number of correctly detected noncompliance instances over the total number of noncompliance instances in the gold standard. Precision, here, is defined as the percentage of total number of correctly detected noncompliance instances over the total number of noncompliance instances detected. For noncompliance detection, recall is more important than precision. Recall errors are more critical because they may result in missing noncompliance instances, while precision errors could be easily double-checked and filtered out by the user of the prototype. The experimental results are summarized in Table 4. A promising performance of 91.7% recall (95% confidence interval [74.2%, 97.7%]) and 84.6% precision (95% confidence interval [66.5%, 93.8%]) in noncompliance detection was achieved, which indicates the effectiveness of the proposed methods. A compliance checking example, using the prototype, is shown in Fig. 3.

Table 4. Preliminary experimental results

Number of noncompliant instances	Result
In gold standard	24
Detected	26
Correctly detected	22
Precision	84.6%
Recall	91.7%

An error analysis was conducted to identify the sources of errors. The analysis shows that the majority of errors originate from semantic information alignment. For example, in some cases, the concept matching methods failed to recognize matches for subjects that have no apparent matching IFC entities (e.g., "radiant panel"), which resulted in false negatives. This indicates that using only apparent knowledge (e.g., concept names) is not sufficient in concept matching. The unapparent knowledge (e.g., superconcept information), captured in semantic knowledge bases (e.g., ontology), may also be used. For the previous example, "radiant panel", the superconcept information, in terms of an energy ontology, may help find the matching IFC entity parent; searching its superconcept "material" in bSDD may find IfcMaterial. In other cases, it failed to recognize matches considering the restrictions, which resulted in false positives. For example, in the following provision, P1, the restriction "in ground contact" was not reflected in the match, which led to false positives – the insulation materials of heated slab floors that did not meet the following absorption rate requirement, but were *not* in ground contact, were found noncompliant. This indicates the challenges in automatically reflecting the meaning of subject restrictions in concept matching, especially for those that contain abstract information (e.g., "in ground contact"). Machine intelligence may be explored to address such issues, but cannot go beyond its limit. Human involvement may still be required in some cases to capture implicit domain knowledge, clarify ambiguities, etc.

- P1: "(g) Insulation Requirements for Heated Slab Floors. Heated slab floors shall be insulated according to the requirements in Table 110.8-A.
  1. Insulation materials in ground contact must:
  B. Have a water absorption rate for the insulation material alone without facings that are no greater than 0.3% when tested in accordance with Test Method A-24 Hour-Immersion of ASTM C272."

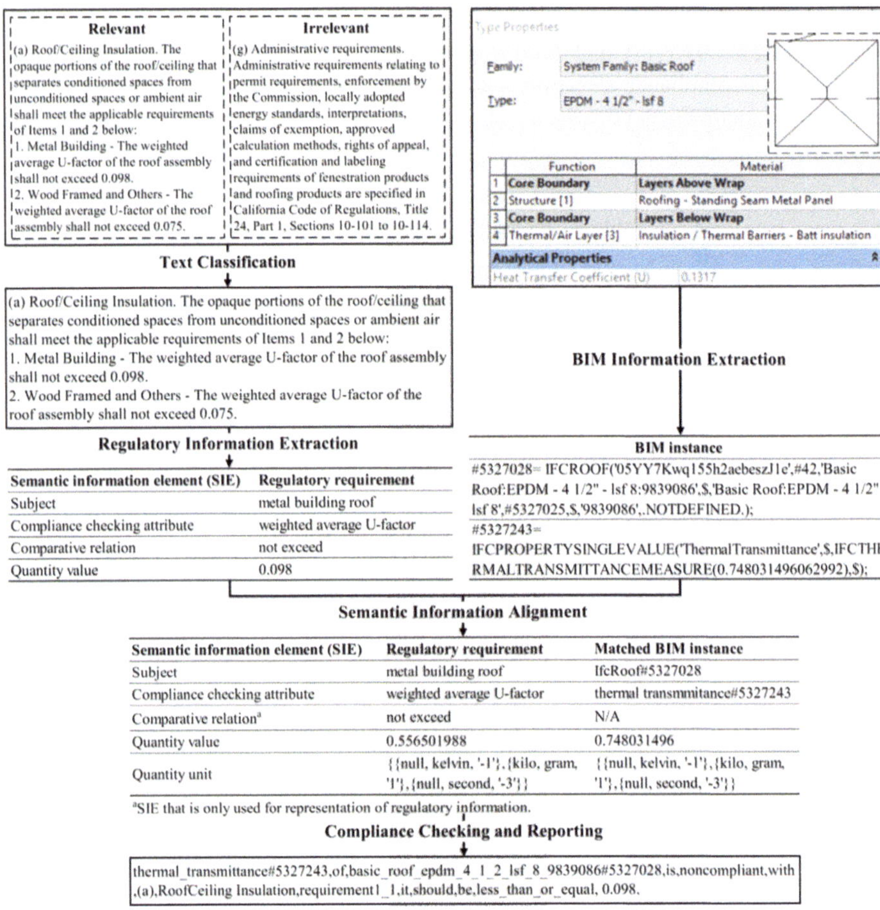

Fig. 3. A compliance checking and reporting example

5 Conclusions

This paper proposed a set of text and information analytics methods for fully automated compliance checking of BIMs with energy codes. The proposed analytics methods aim to address the challenging text complexities in energy codes. The proposed methods were implemented in an energy compliance checking prototype. The prototype was tested in checking the compliance of the BIM of an educational building project with building thermal insulation requirements from three energy codes. The prototype achieved a performance of 91.7% recall and 84.6% precision in noncompliance detection, on the testing data, which is highly promising. An error analysis showed that the majority of errors originate from semantic information alignment. In their future work, the authors will further improve the proposed text and information analytics methods to address these errors, as well as deal with more complex cases in semantic information alignment, including requirements with exceptions, multiple requirements with complex conjunctive/disjunctive relationships within a provision, and requirements in both natural language and tables. The authors will also further test the proposed methods on other energy topics (e.g., fenestration) and other types of regulatory documents (e.g., contract specifications).

Acknowledgments. The authors would like to thank the National Science Foundation (NSF). This material is based upon work supported by NSF under Grant No. 1201170.

References

Beach, T.H., et al.: A rule-based semantic approach for automated regulatory compliance in the construction sector. Expert Syst. Appl. J. (2015). https://doi.org/10.1016/j.eswa.2015.02.029

California Energy Commission: Building Energy Efficiency Standards for Residential and Nonresidential Buildings (2013). http://www.energy.ca.gov/2012publications/CEC-400-2012-004/CEC-400-2012-004-CMF-REV2.pdf

Goel, S.K., Fenves, S.J.: Computer-aided processing of structural design specifications. University of Illinois at Urbana-Champaign (1969). https://www.ideals.illinois.edu/bitstream/handle/2142/14777/SRS-348.pdf?sequence=2

ICC (International Code Council): 2012 International Energy Conservation Code (2012). http://publicecodes.cyberregs.com/icod/iecc/2012/

İlal, S.M., Günaydın, H.M.: Computer representation of building codes for automated compliance checking. Autom. Constr. J. (2017). https://doi.org/10.1016/j.autcon.2017.06.018

Ontario Ministry of Municipal Affairs: Ontario Building Code Supplementary Standard SB-10 (2011). http://www.mah.gov.on.ca/AssetFactory.aspx?did=9227

Pauwels, P., et al.: A semantic rule checking environment for building performance checking. Autom. Constr. J. (2011). https://doi.org/10.1016/j.autcon.2010.11.017

Solihin, W., Eastman, C.: A knowledge representation approach in BIM rule requirement analysis using the conceptual graph. J. Inf. Technol. Constr. J. International Council for Research and Innovation in Building and Construction (CIB) (2016). http://www.itcon.org/2016/24

Zhang, J., El-Gohary, N.: Semantic-based logic representation and reasoning for automated regulatory compliance checking. J. Comput. Civ. Eng. J. ASCE. (2016). https://doi.org/10.1061/(ASCE)CP.1943-5487.0000583

Zhang, J., El-Gohary, N.: Integrating semantic NLP and logic reasoning into a unified system for fully-automated code checking. Autom. Constr. J. (2017). https://doi.org/10.1016/j.autcon.2016.08.027

Zhou, P., El-Gohary, N.: Ontology-based multilabel text classification of construction regulatory documents. J. Comput. Civ. Eng. J. ASCE (2016). https://doi.org/10.1061/(ASCE)CP.1943-5487.0000530
Zhou, P., El-Gohary, N.: Ontology-based automated information extraction from building energy conservation codes. Autom. Constr. J. (2017). https://doi.org/10.1016/j.autcon.2016.09.004

Integrating BIM-Based Simulation Technique for Sustainable Building Design

Ahmed Mohammed Abdelalim(✉) and Yasser Abo.elsaud

Faculty of Engineering at Mataria, Helwan University,
P.O. Box 11718, Cairo, Egypt
dr.ahmed.abdelalim@gmail.com

Abstract. Sustainable development, of which construction is a part, has been recognizably divided into energy use reduction, industrial development, climate change, water conservation, recycling and waste reduction, sustainable procurement of materials, transport strategies and biodiversity. The sustainable building or structure is one that satisfies environmental, social and economic concerns in a balanced way. With the increasing demand for sustainable buildings' design, the whole building energy simulation tools are extensively involved in the design process for saving energy and improving building performance. On the other hand, Building Information Modeling (BIM) is utilized as a simulation technique for various sustainability practices like; building location, orientation, sun- path analysis, thermal properties of materials, heating and cooling loads, daylighting and ventilation. However, sustainability is more concerned with all environmental aspects and the whole building design rather than energy simulation. The aim of this research is to introduce a simplified approach for BIM-based simulation technique as a tool for pre-assessment of sustainable buildings' design. The proposed method helps project development team to conduct the required evaluation of design alternatives according to a specified Green Building Rating system. Moreover, a case study was applied on the rating system Leadership in energy and environmental design (LEED) to investigate the configuration of using BIM in accordance with the LEED rating system and producing an easy to be handled assessment template that measures compliance with sustainable requirements and measures the potential points than the design can achieve.

Keywords: Green buildings · BIM-based sustainability
Building performance analysis · Energy simulation
Building energy performance

1 Introduction

Building Information Modeling, BIM, provides more accurate and integrated documentation, less confusion and rework, this results in saving time, cost and efforts. The use of BIM to provide data for energy performance evaluation and sustainability assessment is defined as "Green BIM". Large design organizations adopt this approach to enable integrated design, construction and maintenance towards Net Zero Energy buildings. Green BIM includes Building Energy Modelling dealing with the project

M. Shehata and F. Rodrigues (Eds.): GeoMEast 2018, SUCI, pp. 209–238, 2019.
https://doi.org/10.1007/978-3-030-01905-1_12

energy performance to optimize energy efficiency during the Project life cycle. During the design phase, project team can ensure that sustainable compliance can be realized beside the technical and economic requirements. Thus, BIM can provide information to determine the credit points assigned to specified levels of sustainability rating systems (Sebastiano Maltese 2016). Moreover, decision making in the early design stage impacts on the performance and cost-effectiveness of the buildings significantly (Ajla Aksamija 2012). Project owner and design team need a third-party verification to certify the project as a sustainable building. In order to deal with an authorized third party and go through the certification process, it takes a lot of time, effort, and cost. The owner and the design team may prefer if they have a prior verification tool for saving design time and cost. In addition, the third party may use the same tool in the integration design process with concerned parties. Meanwhile, the main issue in the whole building design approach is to have an accurate and accessible source of information. Integrated design helps to engage all consultants and stakeholders at early stages of the project. Whether, contractually allowed for this dynamic interface at early stages or not, architects have increasing responsibility to address energy and environmental factors in the conceptual stages of design projects (Quinn 2013). LEED, BREEAM, CESBA and other rating systems, are only a certification to be achieved at the end of the construction. Their strong connection with design tools boosts for a solid integration with BIM. Using the Industry Foundation Classes (IFC) protocol, in combination with defined procedures and tools, will allow for a better data extraction and elaboration, it is providing more reliable results with less effort (Sebastiano Maltese 2016).

This research introduces a simplified approach, which help the concerned parties to take good decisions in early design stage related to sustainable design targets to benefit from the advantages of the information modeling revolution. In addition, the research surveys the specialists' opinions for the best practices and the effects of the BIM-simulating approach on Architecture, Construction and Engineering, AEC market.

2 Research Objectives

The use of building information modeling (BIM) has provided a means of increasing total project quality, providing accurate quantity take-off, improving scheduling, and consequently diminishing total project contingencies and costs. Although BIM is a recent development technique, a lot of research has been conducted in order to further enhance the capabilities of BIM in design and construction. However, there is a little research done on the effect that BIM has on sustainable construction practices (Bynum and Olbina 2013). The main objective of this research is investigating to what extent the integration between BIM and LEED certification process can be helpful. The study attempts to outline a conceptual framework for BIM-LEED integration. This conceptual framework indicates the various BIM plugin tools that can support and enhance the manual process of calculating credits for LEED certification.

The main objectives of the research are:

- Check the ability of BIM-Based simulation tools to measure the credit categories of a specific green building rating system using a simple workflow.

- Develop a simplified template can be utilized by Add-ons to the BIM-based software
- Provide green building pre-assessment tool for the design team in order to save time and efforts of green building certification process.

3 Research Methodology

This research is carried out based on gathering information from past relevant researches as a literature review, for integration between BIM-Based sustainability software and green building design in order to answer the question, how can the BIM software support design process to achieve LEED credits? After literature review, the research compromised an assessment excel sheet that contain LEED credit categories checklist. The purpose of excel sheet is to classify the required data to be extracted from BIM software. A case study was undertaken to achieve the targeted LEED Silver certification from USGBC for the selected Maisalon Hotel project located at Northern Province of Saudi Arabia. By applying the previous prepared checklist in the BIM model to measure the potential green building credits could be achieved under LEED's credit categories. Finally, the research proposes a flowchart for sub-LEED credit categories. Every checked credit category exports the potential credits to a particular column in the final credits sheet either automatically or manually.

4 Literature Review of the Main Parameters

4.1 Sustainable Construction

The terms like; high performance, green, and sustainable construction often are used interchangeably. However, the term "sustainable construction" most comprehensively addresses the ecological, social, and economic issues of a building in the context of its community. In 1994, the *Conseil International du Bâtiment*, CIB defined sustainable construction as "creating and operating a healthy built environment based on resource efficiency and ecological design", and articulated seven principles of *Sustainable Construction* that ideally would inform decision makers during each phase of the design and construction process, continuing throughout the building's entire life cycle (Kibert 2016). The Architectural, Engineering, Construction & Operation (AECO) industry has witnessed assured development through BIM implementation in parallel with constant drive of green building movement in the last two decades. Although the two concepts are independent of each other, BIM and green building collectively are able to overcome present challenges in productivity and (Amarnath CB 2016). Assessment tools help industry stakeholders to evaluate the performance of green buildings. Building information modeling (BIM) enables architecture, engineering, and construction managers to evaluate the performance of green buildings during preconstruction. BIM-based sustainable analysis extracts the data from a building model, which supports the assessment for green building certification (Mohamed Solla 2016). Green building rating systems, which feature their respective criteria and allocation of points and credits; thus, BIM may possibly be integrated with the green building certification process. This factor is a strong reason driving several countries to implement BIM in most construction phases (Azhar 2011). The BIM model assists project stakeholders, including designers, contractors and clients, in calculating, documenting, and measuring the green building rating system scores based on the LEED, BEAM Plus, and Green Star certification. In the future, the design team and construction approaches should meet the LEED, BEAM Plus, and Green Star rating systems for green building accreditation requirements (Azhar 2011).

4.2 Green Building Councils

Green Building Councils are independent, non-profit organizations include organizations working in the building and construction industry. As members of World Green Building Council (World-GBC), they work to develop green building in their own countries, as well as cooperating with other green building councils to achieve environmental, economic and social goals on a larger, global scale. Table 1 shows the World Green Building Council members classification; Figs. 1 and 2 show the World Green Building Councils distribution around the world.

Table 1. World green building council members classification (World-GBC 2017)

Membership	(Prospective)	(Emerging)	(Established)	(Potential future)
GB councils total no. of organizations	**30**	**12** **74**	**32**	33 33

There are currently seventy-four members around the world, each at different stage of their own green building journey. There are three membership levels for Green Building Councils:

- **Established:** a fully developed and operational organization that is running impactful green building programs of work, delivering change on a national level, and embracing best practice governance, accountability and transparency.
- **Emerging:** a membership which has a strong foundation, such as an elected board and staff to manage day-by-day operations. It is expected to progress to Established status within 24 months.
- **Prospective:** an organization at the early stages of developing, but has put in place a comprehensive strategy on how it will operate and advance green building in its country. It is expected to progress to Emerging status within 24 months (World-GBC 2017).

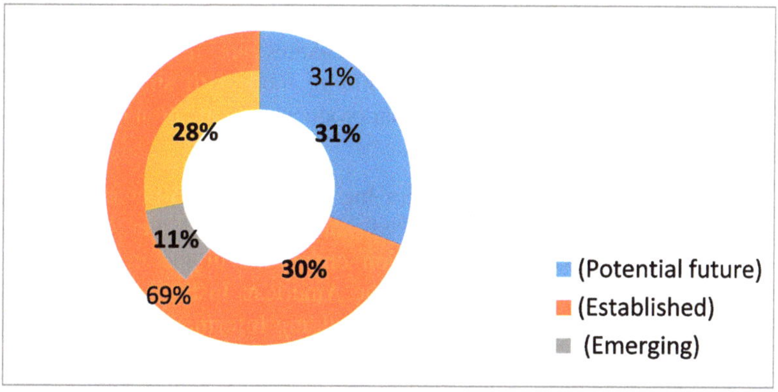

Fig. 1. World-GBC Members percentage distribution (World-GBC 2017)

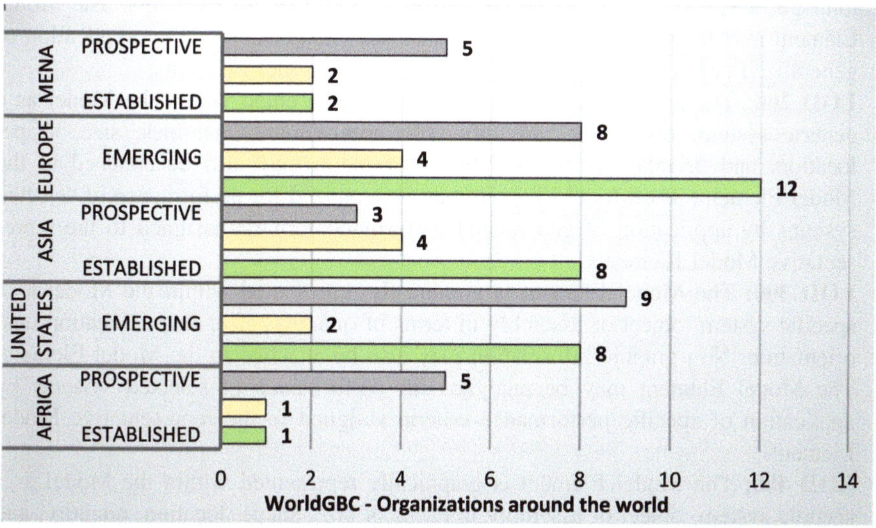

Fig. 2. World-GBC - Organizations distribution around the world

5 Integration of BIM-Based Performance Analysis with Sustainable Design

In order to evaluate and optimize building performance, different analysis cycles should be part of an integrated design process. This challenges the current traditional design paradigm with a performance-based design method for many reasons;

- Traditional Method has deficiencies because; (1) it may include simplified assumptions based on rules-of-thumb, which can be inaccurate; (2) it may force an aesthetic feature without considering performance impacts; and (3) it doesn't provide performance measurement/evaluation of a certain design solution.

- Building Performance-Based Design Method has an ability to estimate the impact of design solution since: (1) performance measures are investigated with actual quantifiable data; (2) it uses detailed building models to simulate, analyze and predict behavior of the system and (3) can produce an evaluation of multiple design alternatives (Ajla Aksamija 2012).

On other hand, it is very important for the design team to determine the proper Level of Development, LOD, and related analysis. The Level of Development (LOD) refers to the amount of information embedded in the model. The widely accepted example of LOD specification is the American Institute of Architects most recent BIM-protocol document, G202–2013, Building Information Modeling Protocol Form (AIA 2013), as follows;

- **LOD 100:** The Model Element may be graphically represented in the Model with a symbol or other generic representation, but does not satisfy the requirements for LOD 200. Information related to the Model Element (i.e. cost per square foot, tonnage of HVAC, etc.) can be derived from other Model Elements. The Model Element may be analyzed based on volume, area and orientation by application of generalized performance criteria assigned to other Model Elements.
- **LOD 200:** The Model Element is graphically represented within the Model as a generic system, object, or assembly with approximate quantities, size, shape, location, and orientation. Non-graphic information may also be attached to the Model Element. The Model Element may be analyzed for performance of selected systems by application of generalized performance criteria assigned to the representative Model Elements.
- **LOD 300:** The Model Element is graphically represented within the Model as a specific system, object or assembly in terms of quantity, size, shape, location, and orientation. Non-graphic information may also be attached to the Model Element. The Model Element may be analyzed for performance of selected systems by application of specific performance criteria assigned to the representative Model Element.
- **LOD 400:** The Model Element is graphically represented within the Model as a specific system, object or assembly in terms of size, shape, location, quantity, and orientation with detailing, fabrication, assembly, and installation information. Non-graphic information may also be attached to the Model Element. The Model Element may be analyzed for performance of systems by application of actual performance criteria assigned to the Model Element.
- **LOD 500:** The Model Element is a field verified representation in terms of size, shape, location, quantity, and orientation. Non-graphic information may also be attached to the Model Elements (AIA 2013).

5.1 Link Between BIM and Energy Simulation

In 2015 M. Senave & S. Boeykens conducted five studies to examine the link between BIM and energy simulation and they conclude that, "a whole arsenal of tools exists to export data from a BIM model in favor of an energy simulation model. Yet automating the transfer to an application is cumbersome, time-consuming and may give rise to errors and misinterpretations. The development of tools to accomplish dynamic simulations by specialized software like "Energy-Plus" through the Industry Foundation Classes (IFC) standard is still in an experimental stage but offers good prospects for the operation of extensive data exchange and analysis. Therefore, there are several promising developments of BIM applications and energy simulation tools (Senave and Boeykens 2015). With BIM, designers can optimize the building design efficiently in the very early stages of the whole process and produce a better solution.

Table 2 summarizes the major contributions of BIM to sustainable building design (Kam-din Wong 2013)

Table 2. Summary of BIM for sustainable design (Kam-din Wong 2013)

BIM Inherency		Functions		Benefits		Sustainable achievements
✚ BIM-based Analysis Tools	➡	3D Model Visualization Clash Detection Compliance with Regulations Energy Analysis; Solar Analysis Thermal Analysis Lighting Design Acoustic Analysis Ventilation and Air Flow Materials/Resource Management	➡	Integrated Project Delivery Design Optimization Better Communication and Coordination More accurate and efficient	➡	Energy Reduction Water Conservation Wastage Lessen IEQ Improvement

5.2 BIM and LEED Credits

Chena and Nguyen developed a framework for the integration of BIM and Web Map Service (WMS) technologies for location and transportation analysis in the green building certifications. Using Autodesk Revit API and Google Maps API as the development tools, this research converted the integration model into the BIM-integrated plugin in Autodesk Revit. The plugin is used to streamline the certification process of site location and transportation analysis in LEED (Leadership in Energy and Environmental Design), one of the most popular and globally recognized green building standards. Most countries are promoting the use of new technology in sustainable construction. If the BIM technology is involved efficiently in LEED certification, the information will be automatically reprocessed without additional manual calculation or using other methods, it can reduce costs of calculation and enhance the rate of green building application in the world (Chena 2017). Azhar demonstrated the ways designers and planners may use BIM for various sustainability analyses in pursuit of LEED® certification. The scope of the research was limited to nonresidential new building projects only. He concluded that, the results of sustainability analyses software can be used to directly, semi-directly or indirectly generate LEED® documentation. Up to 17 LEED® credits and 2 prerequisites may be documented using results generated by BIM-based sustainability software; however; only 5 credits and one prerequisite have been verified in this study so far. BIM-based sustainability software generates results very quickly as compared to the traditional methods. In other words, a building information model can be used partially to run these analyses. This could save substantial time and resources (Azhar 2011). M. Solla and his team verified the BIM application in achieving credits for an identified case study and achieved 55% of LEED credits, i.e. equivalent to the silver level certification. M. Solla presented that, the LEED rating system has a total of 69 credits. Based on an analysis of BIM software and sustainability, 38 credits can be found directly from BIM, whereas 31 credits require supporting documentation that cannot be prepared by BIM software. For example, some credits require a narrative that is submitted along with the necessary computations (Mohmed Solla 2016). Table 3 illustrates the total points that can be achieved with a BIM tool and the examples of the BIM tools. As shown in Table 3, LEED has six criteria with the maximum points. Water efficiency has full points that can be earned using BIM and minimum points achieved from credits of indoor environmental quality. In this case, the project receives a silver certification in the LEED rating system.

Table 3. BIM achievement in LEED rating system (Mohmed Solla 2016)

Criteria	Maximum points	Weighting	BIM points	BIM weighting	BIM tool
Sustainable Sites	14	20.3%	6	8.7%	Revit
Water Efficiency	5	7.2%	5	7.2%	Revit VE
Energy& Atmosphere	17	24.6%	10	14.5%	VE
Materials & Resources	13	18.9%	9	13.1%	Revit
Indoor Environmental Quality	15	21.8%	4	5.8%	Revit VE
Innovation & Design Process	5	7.2%	4	5.7%	Revit VE
Total	69	100%	**38**	55%	

5.3 Database Management System (DBMS)

Database management system is a collection of programs, which can control and manage a database structure. DBMS provides controlled access to the data stored in the database. The use of a computer database can make for efficient management of the data that can be stored, shared and integrated. The database structure itself is stored as a collection of files. The only way to access this information; is by using a database management system (DBMS) (Jalaei 2015).

5.4 Integration and Interoperability

The building life cycle involves organizations with different responsibilities and functions. However, there is no specific BIM tool that can support all the functions needed at all stages of the building life cycle. Exchanging the data between applications is essential to the project team. Model information can be exchanged between two different software tools in four ways:

- Direct links between specific BIM tools;
- Proprietary Exchange File Format;
- Public Level Exchange Formats;
- XML-based exchange formats (Jalaei 2015).

5.5 BIM and GIS Integration

The separation of low energy building design and urban energy planning process is mainly resulted from the use of different information and modeling systems. BIM gives description of the building in deep geometrical and attributive details. Geographic information system (GIS) is the most important tool to describe a digital city in multi-rate, multi-scale and space-time dimensions. Building information models provide a very rich data source for properties about all of the building elements (e.g., identification, maintenance information, and life cycle and condition based information). These properties are inevitable components of any construction project. Moreover, descriptive information (e.g., transportation network, asset locations, etc.) in GIS should be used to model temporary components, locate temporary facilities, reduce transportation and logistics costs, and many other applications (Karan and Irizarry 2015).

6 Questionnaire

As before discussed, a questionnaire was designed and distributed to 150 specialists on Green Building Design and building Information modeling (BIM). The questionnaire targeted sample was selected from 15 countries. The questionnaire focused on international and regional firms which are working in Middle East and North Africa (MENA) region. The main objective of the questionnaire is investigating to what extent the AEC market benefits from BIM-based simulation tools in measuring the performance of green buildings. The questionnaire checked five experience parameters listed below:

- Sustainable Design and common used rating systems
- Building Information Modeling (BIM) and common used programs
- Building Performance Analysis (BPA) and common used methods
- The accuracy of results of simulation techniques
- The future of using BIM-based simulation tools in sustainable design

6.1 The Questionnaire Respondents' Profile

The questionnaire was sent in electronic form to about 150 AEC practitioners and academics. Those were randomly chosen from companies addresses published, Famous design firms in MENA area and business contacts. The response rate was about 80%. Several responses contained invalid answers and were therefore disregarded. Thus, 119 responses were gathered, 95 from the MENA, 15 from Asia-Pacific and 9 from USA & Europe. Table 4 shows the respondents distribution.

Table 4. The questionnaire respondents' location

Location	Responses	
USA	3.36%	4
Asia-Pacific	12.61%	15
Europe	4.20%	5
MENA (Middle East and North Africa)	79.83%	95
	Answered	**119**

6.2 The Questionnaire Results

The results reflect that the sustainable design began to prosper in the last decade in (MENA) region. 54 respondents claimed that they have a practice in sustainable design, all of them using (LEED) rating system in (MENA) region. 91 respondents declared that they are using (BIM) programs in the design process, 75 of them basically use (Autodesk Revit) software for modeling. But also, the questionnaire reflects that, using BIM in environmental studies and energy simulation still in its beginnings, only 10 respondents using BIM in environmental analysis and 23 using BIM in energy performance analysis. Figure 3 shows the potential uses of BIM. Table 5 represents the respondents' answers of the considered environmental studies for simulation.

In addition, the questionnaire results reflect that, design team may still use traditional methods to evaluate building performance even if use simulation tools. 18 respondents stated they are using traditional method, 15 respondents using BIM-based simulation tools and 19 respondents use both methods. This may be due to a lack of confidence in the simulation outputs. Figure 4 shows using of performance analysis methods. Figure 5 illustrates the conducting of building performance analysis during project phases. 19 respondents are conducting building performance analysis in concept design phase, 19 respondents in the schematic design phase, 20 respondents in the development design phase, 11 in tendering phase, 19 in construction phase and 8 respondents are conducting BPA in operation and maintenance phase. Figure 6 reflects the reasons of conducting building performance analysis. 41 respondents are conducting building performance analysis just upon owner's request, 16 respondents said that, the BPA is a part of the company policy as an environmental responsibility. 2 respondents are conducting BPA because it is a mandatory in their country laws (municipality regulations, License requirements… etc.), the remaining respondents skipped this question. Figure 7 illustrates the specialists' evaluation of the benefits

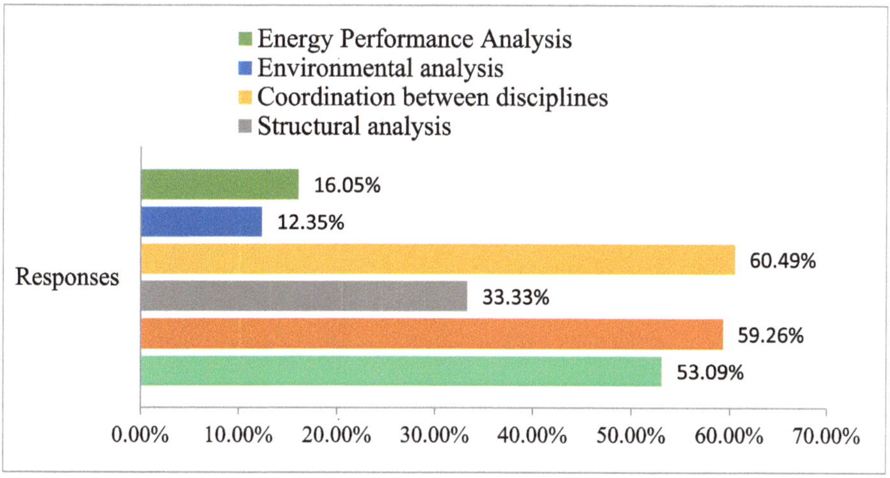

Fig. 3. Purposes of using BIM-questionnaire result.

Table 5. Respondents answers of the considered environmental studies for simulation run

Answer choices	Responses	
Building orientation	75.81%	47
Mass study	51.61%	32
Solar study	40.32%	25
Thermal Properties of envelope materials	50.00%	31
Wind study	51.61%	32
Daylight study	46.77%	29
Glazing ratio	38.71%	24
Potential renewable energy saving	37.10%	23
Other	4.84%	3
	Answered	**62**
	Skipped	**62**

gained from using BIM-based simulation tools and its effect on design outputs. 29 respondents thought it is extremely beneficial, 23 respondents said it is somewhat helpful and 3 respondents said not helpful at all may be because they have not enough experience in this field. Figure 8 illustrates the specialists' evaluation of the simulation outputs accuracy. 10 respondents thought it is excellent, 34 respondents said it is good, 12 considered that it just give a feasible perception and 4 respondents said it's producing confused results, may be because they have not enough experience in this field. Moreover, Fig. 9 represents the survey of environmental studies, which considered in design and used for preparation of simulation process.

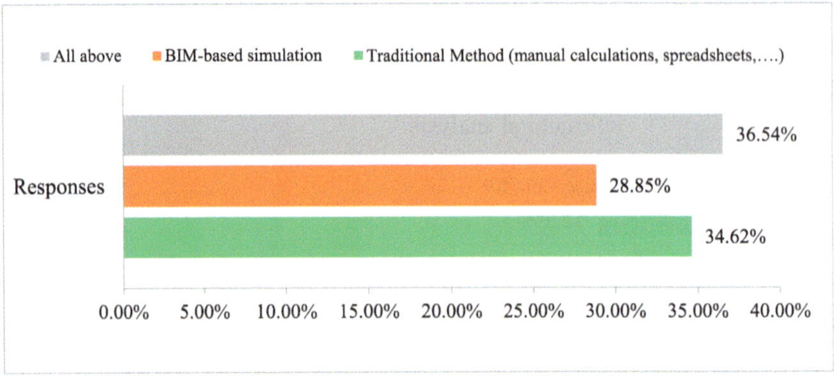

Fig. 4. Using of performance analysis methods-questionnaire result.

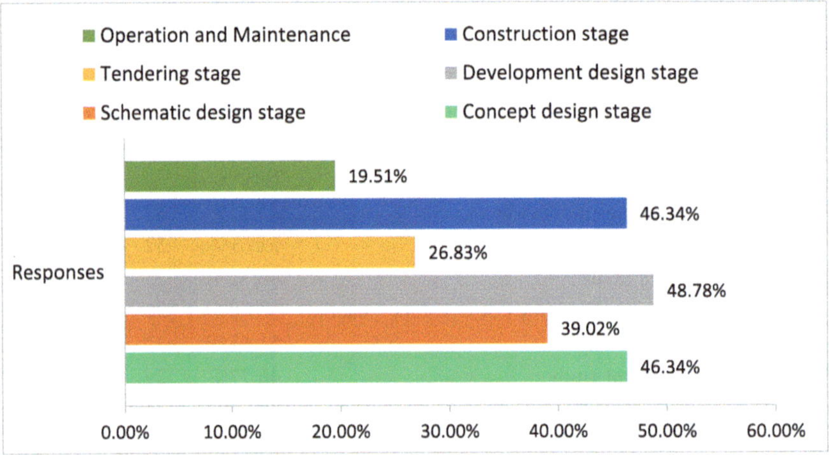

Fig. 5. Building performance analysis during project phases - questionnaire result.

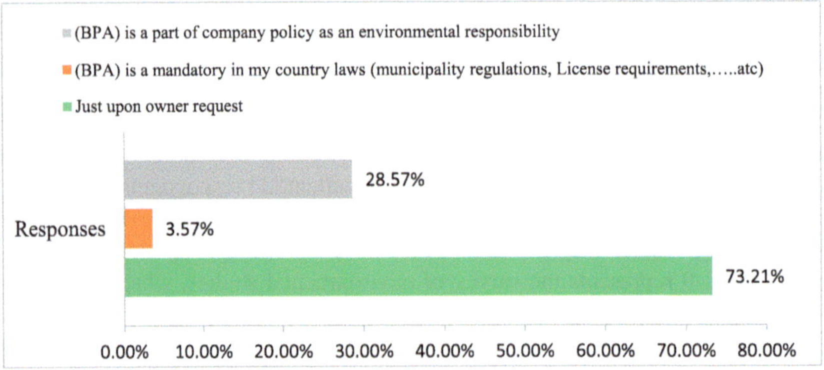

Fig. 6. Reasons of conducting building performance analysis - questionnaire result.

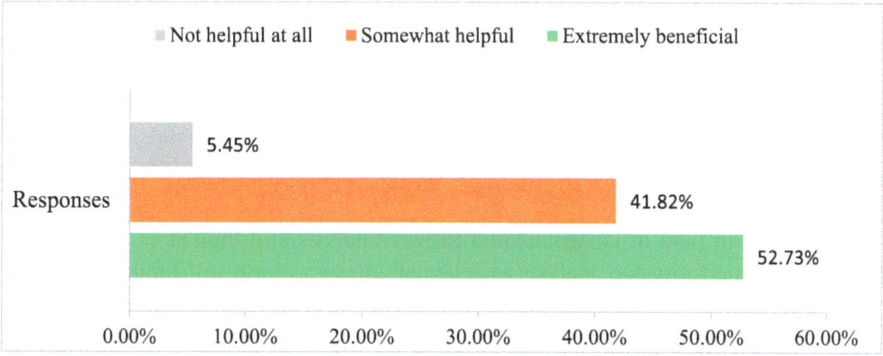

Fig. 7. BIM-based simulation tools benefit for design output - questionnaire result.

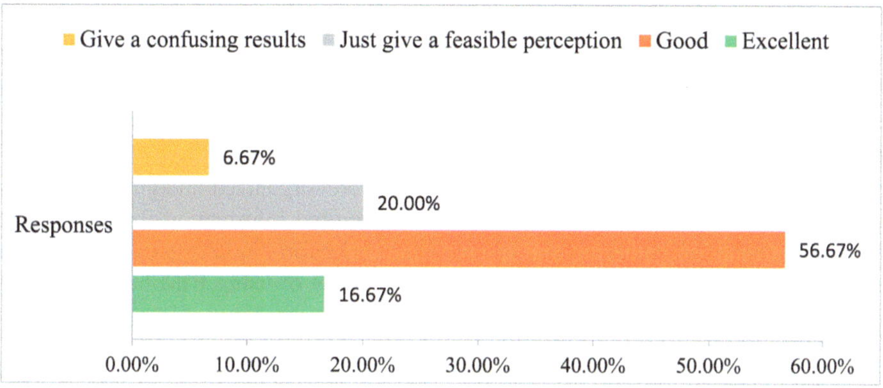

Fig. 8. Accuracy of simulation results - questionnaire result.

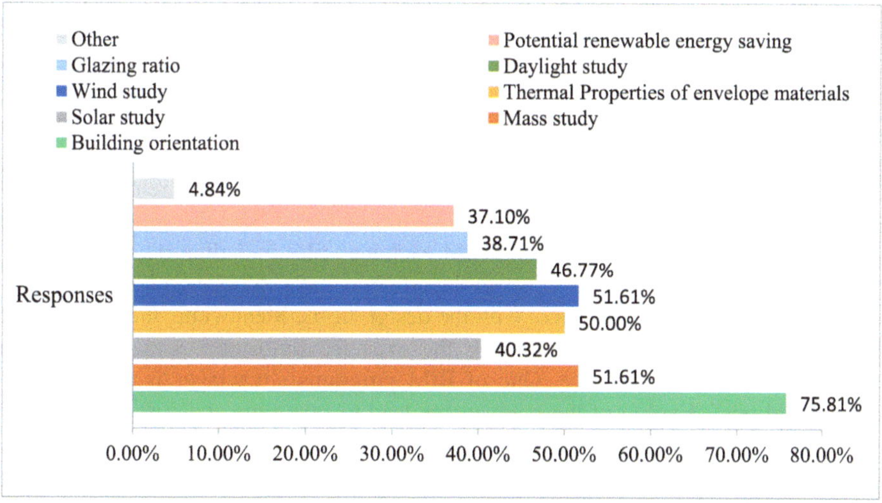

Fig. 9. Considered environmental studies for simulation run - questionnaire result.

6.3 Findings of the Questionnaire

The questionnaire results reflect the following development in the MENA region:

- Increased attention to environmental design during the last decade
- Increased attention to BIM tools during the last five years
- (LEED) is the most used green building system
- Autodesk® Revit is the most used BIM-based software
- Using BIM in field of environmental Energy analysis still limited
- Practice of performance analysis in schematic design phase still limited
- The main reason of practicing performance analysis is up to owner's request not due to commitment toward a government roles or international environmental initiatives
- It was difficult to determine the accuracy of the simulation results because of the divergence of practices. However, there is a tendency for the simulation results are helpful.

7 Proposed Integration Process

Based on the literature review, questionnaire results and applied case study, the research introduces a simplified approach for more organized integration process between the design team and outsources of the required information. In addition, to facilitate information gathering and reserve time and efforts in design phases.

Figure 10 shows the proposed integration workflow of using BIM-based software in sustainable building design. It starts from importing exchange file with extension (.IFC) that contains a 3d geospatial map of project location surrounding area with minimum radius buffer (800 m). The IFC map file considered to be owned by local authorities or city server. Design team has aligned the project conceptual model with the imported IFC map and real geographic location. Furthermore, the 3D-model developed with the required details depends on design Level of Development (LOD100 & 200). In this LOD, an Revit API Add-in were developed to extract information required to calculate points could be achieved related to the following LEED v4 credit categories: (Location & Transportation, *LTc1, LTc2, LTc3, LTc4, LTc5* and, *LTc6*) and (Sustainable Sites, *SSc3*). Then, develop the 3d model through (LOD 300 & 400) in conjunction with other BIM Simulation and Optimization tools like (Green Building Studio, Insight 360 and Revit API Add-ons); in order to extract the remaining information for LEED v4 credit categories. The research proposes that the design team shall be linked with many outsources using BIM exchange files including related GIS data. Accordingly, the concerned government departments have to develop a responsibility assignment matrix (RAM) protocol with the design firms. Meanwhile, the design team should develop a spreadsheet including the expected outsources and classifies the required data from each source. One of BIM programs advantages is the ability to accept "Add-ons". Exploit of these advantage lead to save time and cost of the design team during modeling the project, particularly, if these "Add-ons" are employed to support LEED certification process.

7.1 Outputs of Proposed Framework

The research directed the BIM tools to introduce more specific deliverables by analyzing the required information for particular target results. The outputs of Add-ons, simulation and indirect collected information has been transferred to a predesigned simple spreadsheet contains the maximum potential points that are potential to be obtained as per LEED-v4 requirements. Furthermore, a feedback report and exchange file with extensions (.IFC & gbXML) format exported to the government departments to be included in the city server (and/or) regional planning data base. The research considers the exported (.IFC & gbXML) files are the core of future City Information Modeling (CIM). The final pre-assessment spreadsheet is and feedback files are the final output of the proposed framework.

7.2 Modeling Software

Based on the questionnaire results, (Autodesk® Revit) is the common used BIM-based software for modeling. In addition, the cloud-based (Autodesk® Green Building Studio & Autodesk® Insight360) are recommended for simulation and optimization processes. There are various of BIM software are used for specific sustainable studies like; (Location & Transportation Add-on, Autodesk® CFD, daylighting analysis plugin and EXNO plugin for acoustic analysis).

7.3 Adopting the Green Building Rating System

Each green building council around the world developed a specific green building rating system in order to fit their regional demands. A lot of them established with intent to produce a global rating system. Based on the questionnaire results, LEED® is the common used rating system for green building certificates. Therefore, the credit categories system of LEED v4 is adopted for step-by-step workflow study. The following are the major reasons of selecting LEED v4, 2014:

- The research conducted a questionnaire asking international and regional firms in many countries about the rating system they are using. There are 70 respondants participated in the survey, 62 respondants said they are using LEED which means 88.57% of total respondants, Fig. 11 shows the result of survey about the most common rating system used.
- The availability and easiness of accessibility of all required information related to LEED v4, 2014 credit category using USGBC web site.
- Widely use of LEED v4, 2014 rating system around the world.

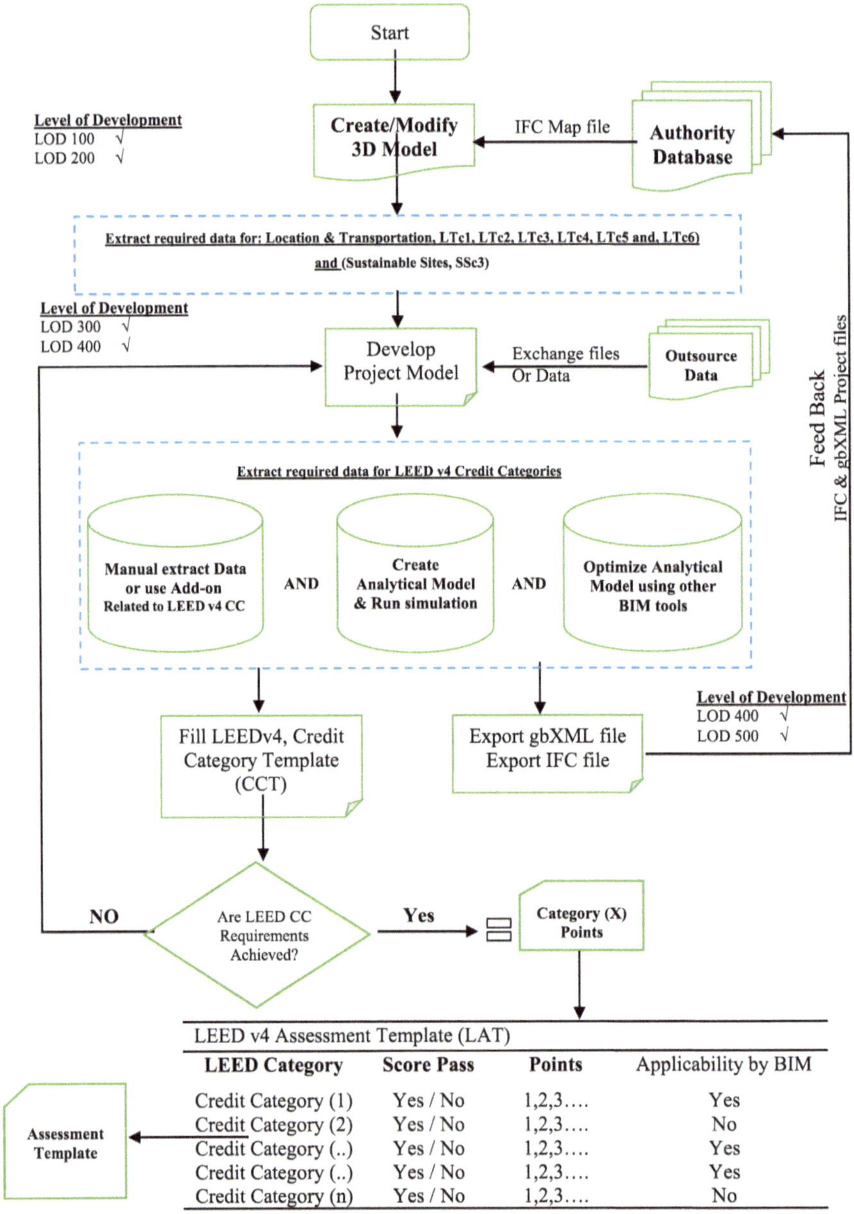

Fig. 10. Proposed integration workflow of using BIM-based software in sustainable design

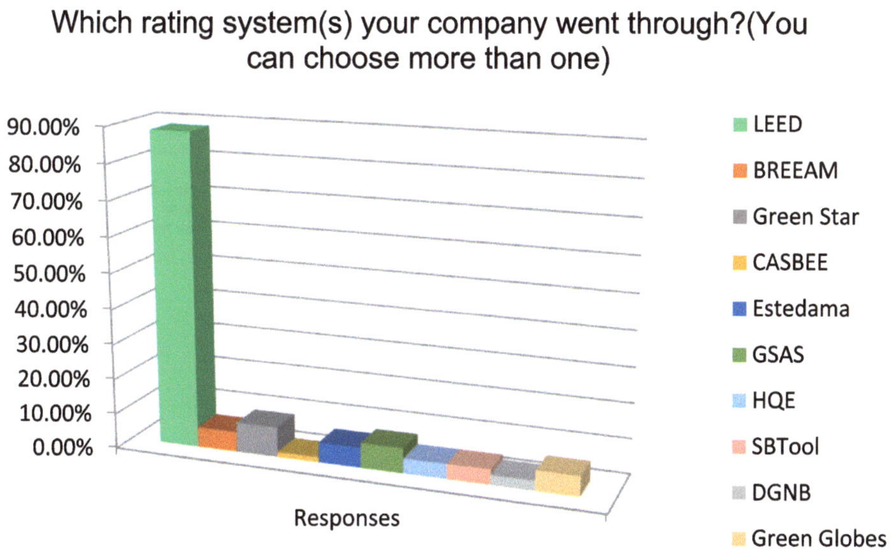

Fig. 11. Green building rating systems used in Middle East countries - questionnaire result

8 Case Study

A case study was carried out as a demonstration to the process of LEED gold certification from USGBC. The case study selected is "4 Stars Hotel" project located at Northern Province of Saudi Arabia. The project land has an area of 4230 m^2, and the building consists of (100 hotel rooms) and other hotel service facilities distributed among seven floors. Each floor has approximate area of 1825 m^2, for a total gross area of 12600 m^2 and a perimeter of 180 m and project boundary 376 m. By applying the previous prepared checklist in the 3d model to measure the potential green building points could be achieved under LEED v4 credit categories. The proposed methodology has been applied to verify the capabilities of the model and investigate its performance among the total potential scores of different credit categories. Table 6 describes the analysis of required information on the Location and Transportation credit category. Other credit categories of LEED rating system like sustainable cities, water efficiency, energy efficiency, materials & resources and indoor environmental quality can be simulated similarly. In addition, Table 7 describes the analysis of required information for all credit categories according to LEED v4 for (NC). The case study results show that, the total potential scores can be assessed using the proposed integration method is 70 points out of the total 100 scores. Table 8 shows the design process & interoperability steps in LOD 100&200 applied on the case study, Table 9 shows the remaining design process & interoperability steps in LOD 300&400. The following figures describe some examples captured during design process; Fig. 12 shows project geographic location and nearest weather station using Revit 2018. Figure 13 shows IFC file assumed to be transferred formally from local authority loaded by geospatial information related to surrounding area with approximately radius 1000 m.

Table 6. Extracted information of the location and transportation credit category

LOCATION and TRANSPORTATION (LT)						Verify Project Data		
Credit Code	**Credit Title**	**Option**	**Options**	**Data Inputs**	**Points**	**Maximum Points**	**Proposed Software**	**Notes**
LTc1	LEED for Neighborhood Development Location		The project location is within the boundary of a development certified under LEED for Neighborhood Development (Stage 2 or Stage 3 under the Pilot or 2009 rating systems, Certified Plan or Certified Project under the LEED v4 rating system).	Gold	0 0 12 0	12	**AutoCAD map 3d** **ArcGIS** **Google Maps**	**If project meet these option, cannot get more points under LT credit category**
LT c2	Sensitive Land Protection?	a	Locate the development footprint on land that has been previously developed	YES	1	1		**Data must be loaded on ArcGIS file provided from local authority**
		b	**(OR)** Locate the development footprint on land that has been previously developed that meet LEED v4 requirement	YES	1			
LT c3	High Priority Site	a	Locate the project on an infill location in a historic district.	YES	1	2		**Data must be loaded on ArcGIS file provided from local authority**
		b	**(OR)** Locate the project on an location meets LEED v4 requirement	YES	1			
		c	**(OR)** Locate on a brownfield where soil or groundwater contamination has been identified, and where the local, state, or national authority (whichever has jurisdiction) requires its remediation. Perform remediation to the satisfaction of that authority.	YES	2			
LT c4	Surrounding Density and Diverse Uses		**Locate on a site whose surrounding existing density within a ¼-mile [400-meter] radius of the project boundary meets the values in Table 1. Use either the "separate residential and nonresidential densities" or the "combined density" values.**			5	**AutoCAD map 3d** **ArcGIS** **Google Maps**	**Need an add-on can calculate density & diverse uses according to specific equation**
		a	Combined density = - (5,050 Square meters per hectare of buildable land) - Residential density (17.5 DU/hectare) - Nonresidential density (0.5 FAR)	YES	2			
		b	Combined density = - (8,035 Square meters per hectare of buildable land) - Residential density (30 DU/hectare) - Nonresidential density (0.8 FAR)	YES	3			
		d	Diverse uses - within ½-mile (800-meter) walking distance of the main entrance of four to seven	YES	1			
		e	Diverse uses - within ½-mile (800-meter) walking distance of the main entrance of ≥ eight to seven	YES	2			
LT c5	Access to Quality Transit		**Locate any functional entry of the project within a ¼-mile (400 meter) walking distance of existing or planned bus, streetcar, or informal transit stops:**			5	**AutoCAD map 3d** **ArcGIS** **Google Maps**	**Need an add-on can calculate Transit options according to specific equation**
		a	Weekday trips (72) & Weekend trips (40)	YES	1			
		b	Weekday trips (144) & Weekend trips (108)	YES	3			
		c	Weekday trips (360) & Weekend trips (216)	YES	5			
			Locate any functional entry of the project within a ½-mile (800-meter) walking distance of existing or planned bus rapid transit stops, light or heavy rail stations, commuter rail stations or ferry terminals:					
		d	Weekday trips (24) & Weekend trips (6)	YES	1			
		e	Weekday trips (40) & Weekend trips (8)	YES	2			
		f	Weekday trips (60) & Weekend trips (12)	YES	3			
LT c6	Bicycle Facilities	a	**Bicycle network**			1		**Need an add-on can calculate density &diverse uses according to specific equation**
			Design or locate the project such that a functional entry and/or bicycle storage is within a 200-yard (180-meter) walking distance or bicycling distance from a bicycle network that connects to at least one of the following: - at least 10 diverse uses; - a school or employment center, if the project total floor area is 50% or more residential; or - a bus rapid transit stop, light or heavy rail station, commuter rail station, or ferry terminal. **All destinations must be within a 3-mile (4800-meter) bicycling distance of the project boundary.**	YES	0.5			
		b	**(AND) Bicycle Storage and Shower Rooms**					

(continued)

Table 6. (*continued*)

		b-1	**Case 1. Commercial or institutional projects:** - Provide short-term bicycle storage ≥ 2.5% of all peak visitors, but ≥ storage spaces per building. - Provide long-term bicycle storage ≥ 5% of all regular building occupants, but ≥ storage spaces per building in addition to the short-term bicycle storage space. - Provide at least one on-site shower with changing facility for the first 100 regular building occupants and one additional shower for every 150 regular building occupants thereafter.	YES	**0.25**			
		b-2	**Case 2. Residential projects:** - Provide short-term bicycle storage ≥ 2.5% of all peak visitors, but ≥ storage spaces per building. - Provide long-term bicycle storage ≥ 30% of all regular building occupants, but ≥ storage spaces per building in addition to the short-term bicycle storage space.	YES	**0.25**			
		b-3	**Case 3. mixed-use projects:** Meet the Case 1 and Case 2 storage requirements for the nonresidential and residential portions of the project, respectively.					
LT c7	Reduced Parking Footprint	a	**Do not exceed the minimum local code requirements for parking capacity.** **Provide parking capacity that is a percentage reduction below the base ratios recommended by the Parking Consultants Council, as shown in the Institute of Transportation Engineers' Transportation Planning Handbook, 3rd edition, Tables 18-2 through 18-4.**			**2**	**REVIT**	**Need information bar to be available in REVIT**
			Case 1.a Baseline location Projects that have not earned points under LT Credit Surrounding Density and Diverse Uses or LT Credit Access to Quality Transit must achieve a 20% reduction from the base ratios.	YES	**1**			
			Case 1.b Baseline location Projects that have not earned points under LT Credit Surrounding Density and Diverse Uses or LT Credit Access to Quality Transit must achieve a 40% reduction from the base ratios.	YES	**2**			
			Case 2. Dense and/or transit-served location Projects earning 1 or more points under either LT Credit Surrounding Density and Diverse Uses or LT Credit Access to Quality Transit must achieve a 40% reduction from the base ratios.	YES	**1**			
			Case 2. Dense and/or transit-served location Projects earning 1 or more points under either LT Credit Surrounding Density and Diverse Uses or LT Credit Access to Quality Transit must achieve a 60% reduction from the base ratios.	YES	**2**			
LT c8	Green Vehicles		**Designate 5% of all parking spaces used by the project as preferred parking for green vehicles. Clearly identify and enforce for sole use by green vehicles. Distribute preferred parking spaces proportionally among various parking sections (e.g. between short-term and long-term spaces).** **Green vehicles must achieve a minimum green score of 45 on the American Council for an Energy Efficient Economy (ACEEE) annual vehicle rating guide (or local equivalent for projects outside the U.S.)**			**1**	**REVIT**	**Need information bar to be available in REVIT**
		a	**Option 1. Electric vehicle charging:** Install electrical vehicle supply equipment (EVSE) in 2% of all parking spaces used by the project.	YES	**1**			
		b	**(OR) Option 2. Liquid, gas, or battery facilities:** Install liquid or gas alternative fuel fueling facilities or a battery switching station capable of refueling a number of vehicles per day equal to at least 2% of all parking spaces.	YES	**1**			
		Possible LT Category Points				**16**		
					YES	Drop list (YES/NO) - User Input		

Table 7. Potential points assessment on the case study using the proposed methodology of integration

Credit	Credit Title	Points - New Construction Buildings	Design/ Construction	PROGRAMMING	CONCEPTUAL	SCHEMATIC	DESIGN DEVELOPMENT	CONSTRCUTION DOCUMENTS	Required Information Source	Nominated Source	Data Feed Method	Proposed Improvement	Potential points after Proposed Methodology
IP c1	Integrative Process	1	D	•	•	•	•	•	BIM Implementation plan	Project Manager	Documentation	Any	1
	Possible IP Category Points Per Rating System:												
TL c1	LEED for Neighborhood Development Location	16	D	•					Outsource	USGBC	Add-on (Drop List)	API Add-on & Export IFC file for Feedback	16
TL c2	Sensitive Land Protection	1	D	•					Outsource	Local Authority	Add-on (Drop List)		1
TL c3	High Priority Site	2	D	•					Outsource	Local Authority	Add-on (Drop List)		2
TL c4	Surrounding Density and Diverse Uses	5	D	•					Calculations on map	Calculations on map	Add-on (Calculation)		5
TL c5	Access to Quality Transit	5	D	•					Calculations on map	Calculations on map	Add-on (Calculation)		5
TL c6	Bicycle Facilities	1	D	•					Calculations on map	Calculations on map	Add-on (Calculation)		1
TL c7	Reduced Parking Footprint	1	D	•					Manual input	Design Team	Add-on (Drop List)		1
TL c8	Green Vehicles	1	D	•					Manual input	Design Team	Add-on (Drop List)		1
	Possible LT Category Points Per Rating System:												
SS c1	Site Assessment	1	D	•					VAR	Design Team	Add-on (Drop List)	API Add-on & Export IFC file for Feedback	1
SS c2	Site Development—Protect or Restore Habitat	2	D	•					Manual input	Design Team	Manual input		2
SS c3	Open Space	1	D	•	•	•			Model Calculation	Model Calculations	Add-on (Calculation)		1
SS c4	Rainwater Management	3	D	•	•	•			Manual input	Design Calculations	Manual input		3
SS c5	Heat Island Reduction	2	D	•	•	•	•		Manual input	Design Calculations	Manual input		2
SS c6	Light Pollution Reduction	1	D			•	•			**Not applicable in design phases**			0
	Possible SS Category Points Per Rating System:												
WE c1	Outdoor Water Use Reduction	2	D	•	•	•	•		Manual input	Design Calculations	Manual input		2
WE c2	Indoor Water Use Reduction	6	D	•	•	•	•			Available in GBS Calculation			6
WE c3	Cooling Tower Water Use	2	D			•	•			**Not applicable in design phases**			0
WE c4	Water Metering	1	D			•	•			**Not applicable in design phases**			0
	Possible WE Category Points Per Rating System:												

(*continued*)

Table 7. *(continued)*

EA c1	Enhanced Commissioning	6	C					•		**Not Applicable in design phases**		0
EA c2	Optimize Energy Performance	18	D	•	•	•	•			**Available in Autodesk Insight**		18
EA c3	Advanced Energy Metering	1	D			•	•			**Not applicable in design phases**		0
EA c4	Demand Response	2	C	•				•		**Not applicable in design phases**		0
EA c5	Renewable Energy Production	3	D	•	•	•	•			**Not applicable in design phases**		0
EA c6	Enhanced Refrigerant Management	1	D		•	•	•		Design manual input	Design Calculations	Manual input	1
EA c7	Green Power and Carbon Offsets	2	C	•				•		**Not Applicable in design phases**		0
Possible SS Category Points Per Rating System:										**Possible SS Category Points Per Rating System:**		
							MATERIALS and RESOURCES (MR)					
MR c1	Building Life Cycle Impact Reduction	5	C			•	•			Available in "One Click LCA" Add-in		5
MR c2	Building Product Disclosure and Optimization---Environmental Product Declarations	2	C			•	•			**Not applicable in design phases**		0
MR c3	Building Product Disclosure and Optimization---Sourcing of Raw Materials	2	C			•	•			**Not applicable in design phases**		0
MR c4	Building Product Disclosure and Optimization---Material Ingredients	2	C			•	•			**Not applicable in design phases**		0
MR c5	Construction and Demolition Waste Management	2	C					•		**Not applicable in design phases**		0
Possible MR Category Points Per Rating System:										**Possible MR Category Points Per Rating System:**		
							INDOOR ENVIRONMENTAL QUALITY (EQ)					
EQ c1	Enhanced Indoor Air Quality Strategies	2	D	•	•	•	•			Available in Autodesk Revit		2
EQ c2	Low-Emitting Materials	3	C			•	•		Material Properties	Design Team	Autodesk Revit	3
EQ c3	Construction Indoor Air Quality Management Plan	1	C	•	•			•		**Not applicable in design phases**		0
EQ c4	Indoor Air Quality Assessment	2	C					•		**Not applicable in design phases**		0
EQ c5	Thermal Comfort	1	D			•	•			Available in Autodesk Revit & Design Builder		1
EQ c6	Interior Lighting	2	D			•	•			Partially Available in Autodesk Revit Elum Tools & DIAlux evo		2
EQ c7	Daylight	3	D	•	•	•	•			Available in Autodesk Revit and Insight & IESVE		3
EQ c8	Quality Views	1	D	•	•	•	•			**Not applicable in design phases**		0
EQ c9	Acoustic Performance	1	D			•	•			Available in Autodesk Revit and EXNO Add-in		1
Possible EQ Category Points Per Rating System:												
Total LEED v4 Points for New Construction (NC)		**100**								**Potential Points project may achieve using BIM programs**		**70**

Table 8. Design process & interoperability steps in LOD 100&200

I.	LOD 100			
No.	Step and Requirement	Recommended Software	Responsible	Notes
1.	Open new project file	Autodesk® Revit 2018	City Server/Local Authority	*The research considered the project has been in well managed area using shared source for City Model which allow design team to import the needed part of 3D map model (as described in literature review)*
2.	Create a conceptual 3d model for the project	Google maps & AutoCAD® map 3d Autodesk® Revit 2018		
3.	Import a map for project surrounding area to Revit file	Google maps & Autodesk® Revit 2018		
4.	Assign project in real geographic location and align the exported map using survey point	Autodesk® Revit 2018		
5.	Create 3d map for 1 km buffer area around the project location	Autodesk® Revit 2018		
6.	Identify the 3D map objects	Autodesk® Revit 2018		
7.	Program a Revit API Add-in to recognize the identified objects on the 3D map	Autodesk® Revit 2018	Design Team & API programmer	
8.	Extract required information for LEED credit categories (Location & Transportation, LTc1, LTc2, LTc3, LTc4, LTc5 and, LTc6) (Sustainable Sites, SSc3)	Need Add-in for generate automatic calculations	Design Team	
9	How many points could be achieved?			

Table 9. Design process & interoperability steps in LOD 300&400

I.	LOD 300 & 400		
No.	Step and requirement	Recommended software	Notes
10	Develop the 3d model elements for the building and major site components (Parking, shades, walkways, hardscape area, soft scape area, including at least envelope components & materials and specify accurate zoning and spaces	Autodesk® Revit 2018	
11	Assign materials for model envelop	Autodesk® Revit 2018 & use trusted vendor families	LEED requirement were considered in material properties, energy settings and thermal properties
12	Assign rooms, spaces and zoning	Autodesk® Revit 2018	This step is very important for a lot of calculations like; (Cooling & heating, Lighting, and Acoustic calculations)
13	Create energy model	Autodesk® Revit 2018	
14	Conduct energy simulation	Autodesk® Revit 2018	
15	Check energy simulation results	Autodesk® Revit 2018	
16	Export gbXML file to GBS cloud	Autodesk® Revit 2018	
17	Optimize Energy consumption and estimating Water demand baseline and improving efficiency, Also, try and compare other alternatives. Then, export gbXML file	Green Building Studio®	
18	Re-Import gbXML file in Revit	Autodesk® Revit 2018	
19	Check and optimize energy behavior. Compare project with Architecture203 and Net Zero target	Insight 360®	
20	Use other BIM tools for more analysis and details	Autodesk CFD, Design Builder, DIAlux evo, IES-VE, eQuest.. and other	
21	Use Revit plug-ins (One click LCA, Elum Tools, EXNO, UL. and other when needed) for related LEED calculations	Autodesk® Revit 2018 Plug-ins/Add-ins	

(continued)

Table 9. (*continued*)

I.	LOD 300 & 400		
No.	Step and requirement	Recommended software	Notes
22	Extract remaining information related to LEED v4 credit categories		
23	In parallel with previous steps, complete required information in the proposed Excel sheets	Proposed Excel Sheet	
24	Excel sheet is pre-programed to calculate the potential LEED v4 point	Proposed Excel Sheet	

Figure 14 illustrates the estimation parameters of water use in Green Building Studio cloud. Figure 15 shows a part of optimization parameters used in Insight 360 cloud with simple manipulators. Figure 16 shows sun path for solar and shadow study in Autodesk® Revit 2018. Figure 17 illustrate Insight lighting analysis in Autodesk® Revit 2018 & in cloud according to LEED v4 EQc7 option 2. Figure 18 shows artificial lighting analysis using specific add-in Elum Tools (and/or) DIAlux evo software.

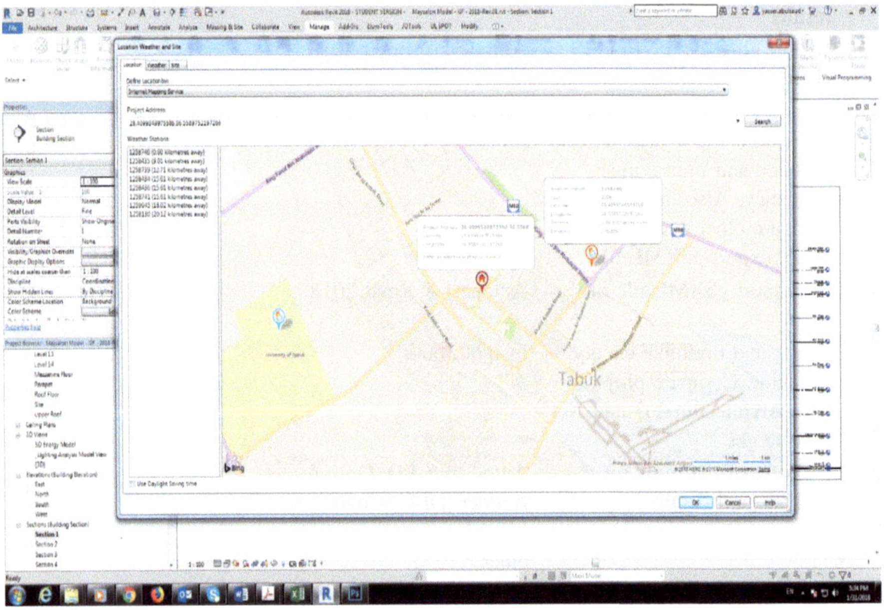

Fig. 12. Assign project location and nearest weather station in Autodesk® Revit 2018

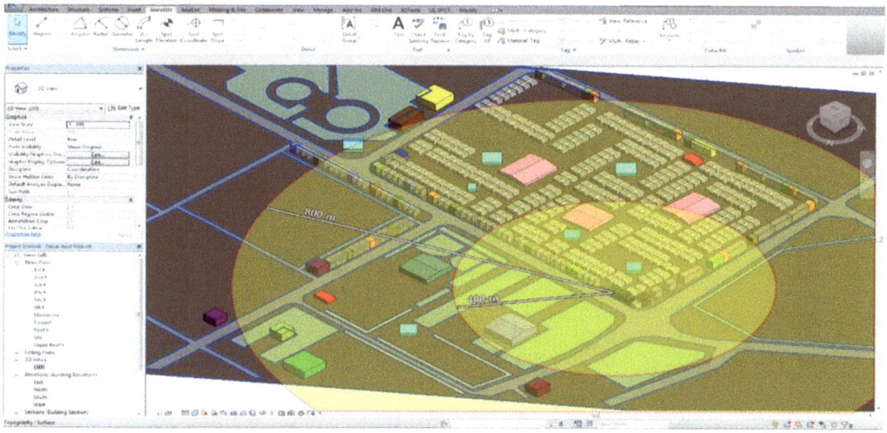

Fig. 13. Location and Transportation calculations on IFC file received from City Server

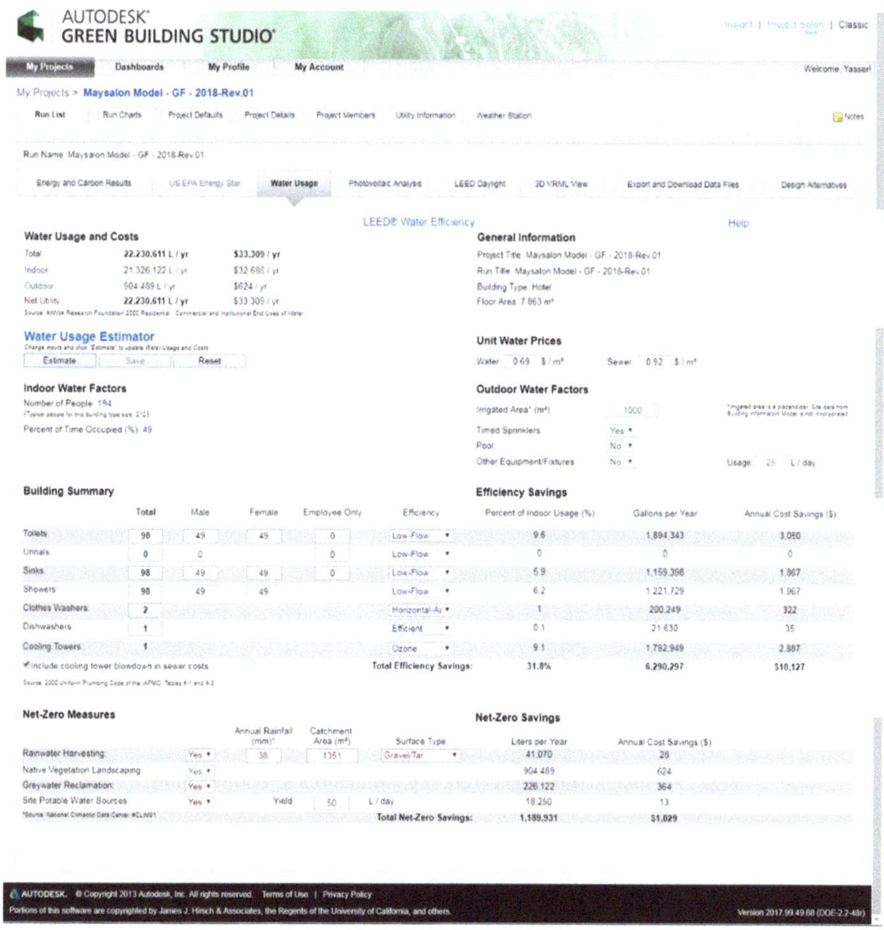

Fig. 14. Green Building Studio® cloud - water use estimation

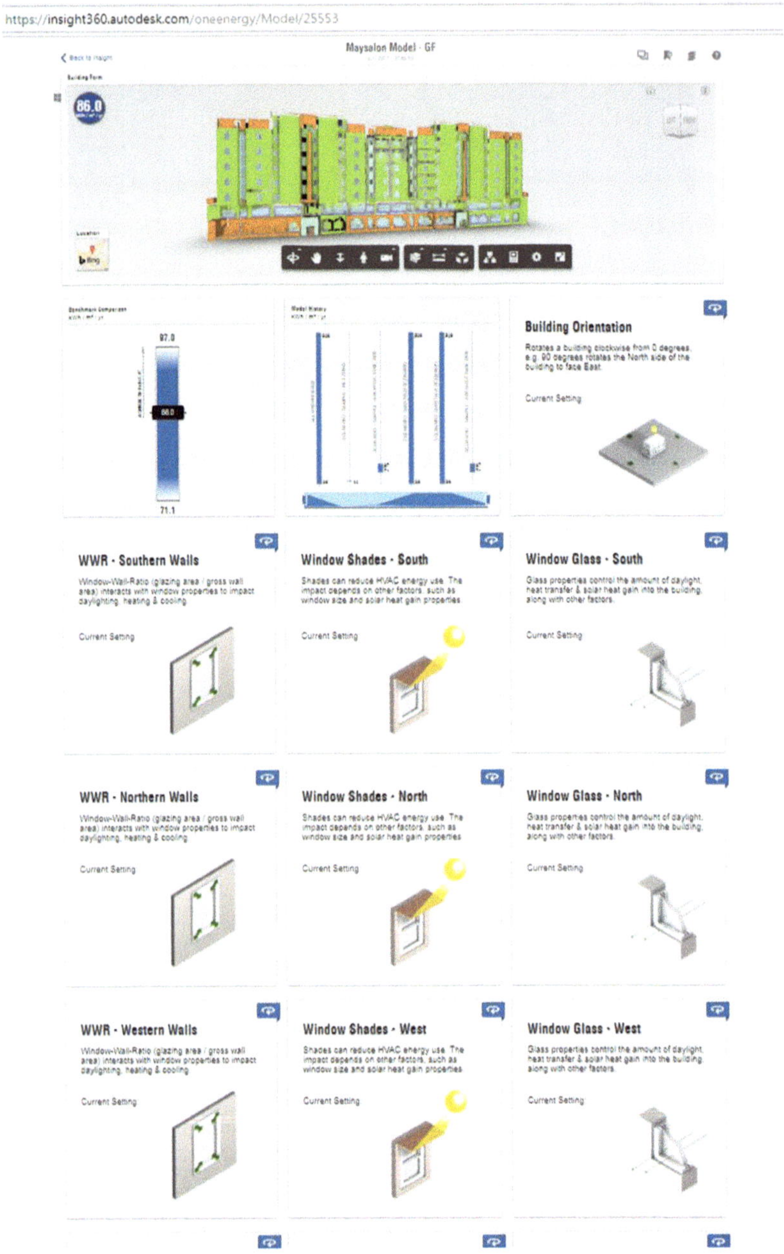

Fig. 15. Insight360 cloud - energy simulation & optimization

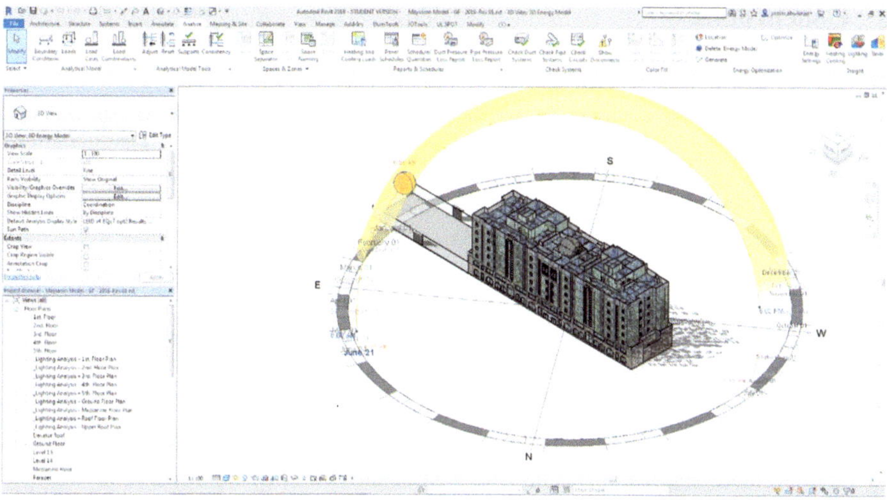

Fig. 16. Solar and shadow study in Autodesk® Revit 2018

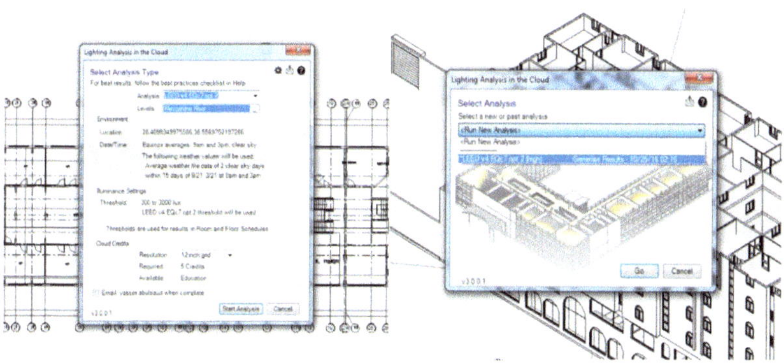

Fig. 17. Insight lighting analysis in Autodesk® Revit 2018 & in cloud

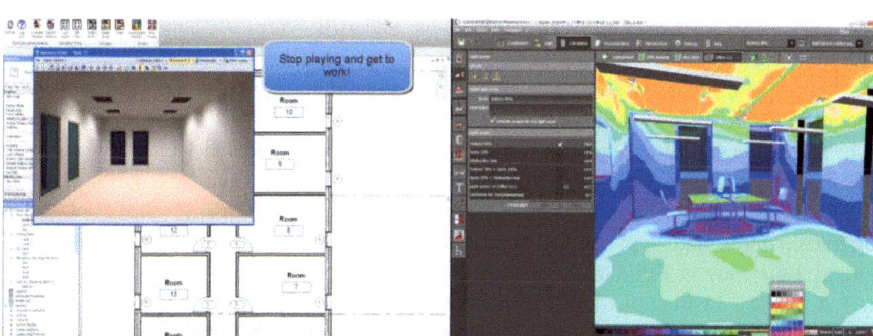

Fig. 18. Artificial lighting analysis using specific add-in Elum Tools (and/or) DIAlux evo

9 Conclusions

The current research is divided into two parallel paths; the first is a literature review of the relevant researches in the past 7 years. The second is a questionnaire distributed to specialists in both sustainable design and BIM fields. In order to investigate the ability of BIM-Based sustainable tools to simulate a particular green building rating system using a simple approach based on suggested framework module. In addition, the research proposed a programmed Excel Sheet to produce a green building assessment template according to a specific green building rating system. The proposed framework applied on a hotel project as a verification case study pursue to achieve green building certification using LEED v4 rating system for new construction (NC) projects. The research findings concluded as follows:

- Improve the database management system (DBMS) between local authorities/outsources and design firms in field of GIS-BIM database became a mandatory for more accurate simulation results and significant saving of time and cost.
- BIM-based sustainable design process needs to be managed and developed properly. Therefore, understanding the level of development (LOD) is essential for successful integration.
- BIM-based sustainable simulation results will be improved when all design/performance analysis team members work collaboratively within an iterative process of design decision-making.
- BIM-based software can be used in simple interoperability framework to simplify the green building certification process.
- BIM-based software can support the design team to achieve (70) points in LEED v4 NC certification process.
- BIM-based software can utilize a simplified approach as a pre-assessment process for LEED v4 NC projects. as well, a similar study can be applied for other sustainable building rating systems.

References

(AIA), A.I.: www.aia.org. http://contractdocs.aia.org:800/!SSL!/Multiscreen_HTML5/desktop/Get_Document_Answers/Document_Instruction_Sheets/By_Series/E-Series/E202-2008.htm (2016)

(EPA), U. E. (n.d.): Green Building. (EPA). https://archive.epa.gov/greenbuilding/web/html/about.html. Accessed 13 July 2016

(EPA), U. E. (n.d.). www.epa.gov. (EPA). https://www.epa.gov/sustainability/learn-about-sustainability#what. Accessed 11 July 2016

Abul Abdullah, B.C.: Whole building energy analysis: a comparative study of different simulation tools and applications in architectural design. In: ACEEE Summer Study on Energy Efficiency in Buildings. Pacific Grove, CA (2014)

AIA, T.A.: Guide, Instructions and Commentary to the 2013 AIA Digital Practice Documents, USA (2013). www.aiacontracts.org. https://www.aiacontracts.org/resources/69541-guide-instructions-and-commentary-to-the-2013-aia-digital-practice-documents

Ajla Aksamija, T.L.: BIM-based building performance analysis: evaluation and simulation of design decisions. In: ACEEE Conference on Energy Efficiency in Buildings, pp. 2–3. American Council for an Energy-Efficient Economy, Pacific Grove, CA (2012)

Aksamija, A.: A strategy for energy performance analysis at the early design stage: predicted vs actual building energy performance. J. Green Build. **10**(3), 161–176 (2015)

Attia, S.: Simulation-based decision support tool for early stages of zero-energy building design. In: Architecture et Climat, Université Catholique de Louvain, 1348 Louvain La Neuve, Belgium, **49**(Energy and Buildings), 2–15 (2012)

Azhar, S.: Building information modeling (BIM): trends, benefits, risks, and challenges for the AEC industry. Leadership Manag. Eng. **11**(3) (2011)

Borong Linn, Q.Y.: Research on parametric design method for energy efficiency of green building in architectural scheme phase. Front. Archit. Res. **2**(1) (2013). (Open Access funded by Higher Education Press Limited Company)

Bringezu, S.: Construction Ecology and Metabolism (Construction Ecology: Nature as the Basis for), pp. 196–215. Spon Press, London (2002)

Dictionary, C.: Cambridge Dictionary. Cambridge University Press (1995). http://dictionary.cambridge.org/dictionary/english/heritage. Accessed 10 Jan 2017

Dictionary.com "sustainability": Dictionary.com Unabridged. Random House, Inc. (2016). http://www.dictionary.com/browse/sustainability. Accessed 11 July 2016

Division, S. o.-O.-B.: Performance Measure Guide. Office of Financial Management (OFM), Washington (2016)

Dowsett, R.M.: Evaluating the benefits of BIM for sustainable design–a review. http://www.reading.ac.uk/web/FILES/tsbe/Dowsett_TSBE_Conference_Paper_2013.pdf. Vancouver (2013, 2014)

Karan, E., Irizarry, J.I.: BIM and GIS integration and interoperability based on semantic web technology. J. Comput. Civ. Eng. (2015)

Executive, T.O. (n.d.): The Federal Commitment to Green Building: Experiences and Expectations. https://www.sustainability.gov/Resources/Guidance_reports/Guidance_reports_archives/fgb_report.pdf

Gerber, S.-H.E.: Designing-in performance: evolutionary energy performance feedback for early stage design. In: 13th Conference of International Building Performance Simulation Association. Chambéry, France (2013)

Green Building Education Services, L: LEED Principles and LEED Green Associate Study Guide, 3rd edn. Green Building Education Services LLC, Atlanta (2014)

Shah, H.H., Ma, Y., Gulliver, S.R.: Selecting Key Performance Indicators for Sustainable. ACM New York, NY, USA ©(IWCSC 2010 First Interdisciplinary Workshop on Communication for Sustainable Communities), Abstract (2010)

Iman Kiani, S.K.: The barriers and implementation of building information modeling (BIM) based on integrated project delivery (IPD) in the construction industry. In: International Graduate Conference on Engineering Science and Humanity (IGCESH), Kuala Lumpur, Malaysia (2013)

Initiative, U.N.-S.: 2009–2010 Public Annual Report & 2010–2011 Work Plan. UNPE-SBCI - The Annual General Meeting of SBCI, Paris (2010)

Jalaei, F.: Integrate building information modeling (BIM) and sustainable design at the conceptual stage of building projects. Sustain. Cities Soc. **18** (2015)

Jim Sinopoli, P.L.:. How Do We Measure the Performance of a Building? (2009)

Kibert, C.J.: Sustainable Construction—Green Building Design and Delivery, 4th edn. Wiley, Hoboken (2016)

Senave, M., Boeykens, S.: Building Information Modelling (BIM) in Design, Construction and Operations, p. 351. WIT Press, Southampton (2015)

Quinn, S.: Performance based design: the design analysis process. In: Sustainable Building 2013 Conference, Urban Density & Sustainability. Hong Kong (2013)
Richard Reed, S.W.-W.: A comparison of international sustainable building tools—an update. In: The 17th Annual Pacific Rim Real Estate Society Conference, Gold Coast, Australia (2011)
Rizal Sebastian, M.B.: BIM and GIS for low-disturbance construction. In: 13th International Conference on Construction Applications of Virtual Reality, London, UK (2013)
Salman, A.: Building information modeling for sustainable design and LEED® rating analysis. In: McWhorter School of Building Science, Auburn University, Auburn, AL, USA, vol. 20 (Automation in Construction), pp. 217–224 (2012)
Salman, A., Rizwan, F.: BIM-based sustainability analysis: an evaluation of building performance analysis software. In: Sulbaran, T., Sterling, C. (eds.) Proceedings of the 45th Annual International Conference on Associated Schools of Construction (ASC) (2009)
Salmon, S.M.: A Comparative Analysis of Energy Modeling Methods for Commercial Buildings (2013)
Sebastiano Maltese, L.C.: Sustainability assessment through green BIM for environmental, social and economic efficiency. In: International High- Performance Built Environment Conference—A Sustainable Built Environment Conference 2016 Series (SBE 2016), iHBE 2016. Elsevier Ltd, Sydney, Australia (2016)
Standard-US, T. N. (n.d.): National BIM Standard. https://www.nationalbimstandard.org/faqs. Accessed 12 July 2016
Stauffer, N.W.: Energy-efficient design: helping architects weigh their options. In: The Magazine of the MIT Energy Initiative (Energy Futures) (2016)
The Arab Republic of Egypt Ministry of Housing, U. A.: The Green Pyramid Rating System. The Egyptian Green Building Council, Cairo (2017)
USGBC.: Green Building 101: What is an integrated process? (USGBC) (2014). http://www.usgbc.org/articles/green-building-101-what-integrated-process. Accessed 02 July 2017
WorldGBC, W.G.: http://www.worldgbc.org. (World Green Building Council) (2017). http://www.worldgbc.org/index.php?cID=220
Yan, H.A. (2008). Benefits and barriers of building information modelling. In: 12th International Conference on Computing in Civil and Building Engineering, vol. 161. Beijing, China

Smart and Connected Infrastructure Through Sustainable Urban Management: Balancing Economic Development and Environmental Protection Goals

Ahmed Abukhater(✉)

Boeing, Seattle, USA
Ahmedabukhater@yahoo.com

Abstract. Geographic Information Systems (GIS) is being deployed in many planning disciplines to combat urban and environmental problems in cities and communities around the world. GIS cutting-edge technology is an effective information management, analytical, and communication tool that is critical in planning for future growth while ensuring smart and connected infrastructure to foster sustainable urban management and growth. Because of its effectiveness in saving time and effort, and its ability to easily share data and provide accurate results, GIS has been increasingly used in almost every field. This paper argues for basic changes in the way GIS software has been commonly used as mapping tools. The widespread misuse and reliance on GIS-based applications as mapping tools should end. The ability to monitor economic, environmental, and social impacts on spatial dimensions provides a comprehensive lens for collaborative community decision making that goes beyond mapping and cartographic purposes to boosting sustainable infrastructure and overall livability of communities.

GIS built-in capabilities allow for multiple criteria evaluation (MCE), which is mainly characterized by allocating various weights to assessment criteria for exploring and ranking different growth scenarios and infrastructure development alternatives. An important advantage of GIS spatial planning support tools is the convenience of changing the valuation criteria to visually illustrate and depict the implications of different spatial decisions and options. The availability of these capabilities needed for decision making in a single system makes GIS an effective planning support system and one that can be easily integrated into the planning processes and workflows.

This paper provides an example of how GIS spatial analytical tools are used as in shaping the decisions we make to foster smart and connected infrastructure conducive for urban growth management, namely identifying and selecting suitable sites for low-density residential development from two contradictory points of view. This paper will demonstrate the value of using GIS to accommodate future growth demand and guide the planning and annexation of cities. It will further provide a framework for sustainable growth management strategies that take into account both the socio-economic and environmental needs.

Keywords: GIS · Multiple criteria evaluation · Land use
Smart and connected infrastructure · Planning · Geo-spatial · Sustainability

M. Shehata and F. Rodrigues (Eds.): GeoMEast 2018, SUCI, pp. 239–248, 2019.
https://doi.org/10.1007/978-3-030-01905-1_13

1 Introduction

The need to make land-use decisions on a national and regional scale in Canada was the impetus for the development of GIS. Roger Tomlinson, the acknowledged father of GIS, led the Agricultural Rehabilitation and Development team that developed what became known as the Canada Geographic Information System. Land use analysis has remained an important GIS application. This paper illustrates how criteria reflecting different planning goals can be incorporated into analysis by modifying the parameters of GIS tools to enable planners and policy makers balance economic development and environmental protection goals.

Today, with the rapid global population growth and its socio-economic and environmental consequences, planners and policy makers are challenged on a whole new biblical scale both locally and globally. As policy makers we aspire to achieve desired future conditions. However, the world is rapidly changing creating new realties and new challenges for planning professionals. The combination of the natural world and human footprint resulted in many complex urban and environmental problems, such as social and environmental conflicts, resource limitation, air pollution, climate change, and increasing desertification, to name a few.

Echoing the trends of other regions around the nation and the state of Illinois, the Champaign-Urbana region will continue to experience rapid population growth in the foreseeable future. In the wake of the proposed multi-modal transportation system, the area is expected to proliferate and attract more residents, suggesting that nearby areas need to be annexed to make room for new housing development including high- and low-density residential projects. To select appropriate areas for sustainable future annexation, city planners need to consider several important criteria for new housing construction, as illustrated in Fig. 1. These criteria reflect three main competing interests: equity, economy, and the environment, or what is known as the "3 Es" of sustainability. By and large, new development cannot be located in environmentally sensitive areas, such as watersheds, or in areas considered environmentally hazardous, such as near landfills. Predicting and managing future growth of the region is a compelling necessity that should be addressed today so that a more environmentally friendly and economically viable image of the future can be attained.

2 Purpose of the Study

The focus of this study is to identify desirable locations for these expected new residential developments, namely low-density residential projects with their associated infrastructure components, based on two almost antithetical approaches. In essence, the sites will be analyzed and evaluated according to two scenarios. The first scenario is based on the developer's point of view, taking into account the purchasers' preferences. This scenario considers the developer's preferences pertaining to economic and marketing factors. To that end, the developer is interested in maximizing revenue and minimizing cost of the development paying little, or no, attention to environmental concerns. The second scenario is based on the environmentalists' point of view, which is quite the opposite of the first scenario. In this case, sites are ranked and evaluated to

Fig. 1. The sustainability triangle of the three conflicting planning goals

foster low impact and environmentally friendly development. Protecting the agricultural and prime forest lands and maintaining the integrity of the environment are the foremost decisive factors for the environmentalists' criteria. Site selection suitability analysis conducted in this study includes weighing the different factors in both scenarios and ranking desirable sites. The outcome of both scenarios will be evaluated and analyzed based on specific parameters discussed below.

3 Target Area

This study covers the Champaign-Urbana region, which is comprised of the municipal boundaries of both the city of Champaign and the city of Urbana in addition to a 1.5-mile buffer zone around the municipal boundaries of both cities, taking into consideration the established urban growth boundary, as shown in Fig. 2. According to the most recent census estimate, Champaign, the larger of the two cities, reached an estimated population of 79,389 in 2008, whereas Urbana has a population of approximately 38,725. The entire region has been growing slowly but steadily with Champaign experiencing rapid growth over the last three decades. The area has substantial potential for improvement and development opportunities with respect to businesses, employment, and in turn housing developments.

4 Site Selection Criteria

On the whole, two fundamental considerations materialize in determining the selection criteria, which are socio-economic and environmental factors. Low-density residential projects are high quality housing units characterized by a less compact pattern and advantageous location and site aesthetic features. More advantageous features mean a higher price. For the developer, ensuring high revenue and minimizing construction costs are of great importance. To that end, the first scenario attempts to rank suitable

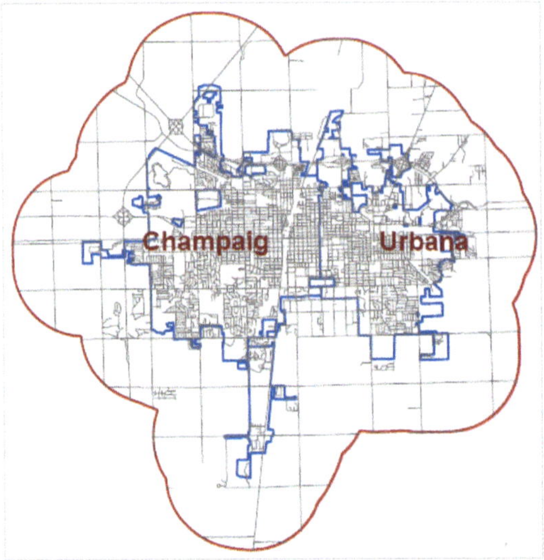

Fig. 2. Boundary of the study area (Urbana-Champaign, IL)

sites based on the developer's preferences. The second scenario represents a more environmentally contentious approach to growth management. This scenario provides a more discerning analysis regarding new developments by fostering environmentally friendly growth patterns and mitigating harm to the region's most precious environmental resources, meanwhile, avoiding locations that are more vulnerable to significant environmental hazards. In essence, the following criteria are used for the analysis:

A. Endogenous Factors (features of the location):

1. 100-year floodplain and wetland areas: the site must not be located in a floodplain designated area or areas that have high run-off rates to prevent and avoid any potential environmental hazards.
2. Soil type: the type of the soil must also be conducive for building housing unit. This means avoiding soil types with low bearing strength or poor drainage. Soil must be strong enough to establish foundations for multi-story buildings, namely two story dwelling units, and permeable enough to allow for the construction of septic systems.
3. Topography: relatively flat locations are preferable. Avoiding sites with steep slopes of more than 6% tends to reduce the cost of site grading which is necessary for adequate drainage and sewage systems.
4. Minimum lot size: the site should be of a minimum size to increase the overall project return on investment (ROI). Assuming 200 dwelling units per site, with a minimum lot size of 600 sq. ft./DU, the minimum size of the entire site should be 120,000 sq. ft.

B. Exogenous Factors (features of the surrounding locations):

5. Proximity to existing urbanized areas: the site should be connected to the existing residential areas with high growth potential. This condition will ensure that the site is adequately served by the existing facilities and infrastructure, such as utilities, fire and police departments, medical facilities, educational facilities able to service with adequate capacity, and retail and commercial districts. It also ensures the availability of community's assets, such as parks and recreational facilities, in the vicinity of the site. Parks and other essential community facilities should be centrally located within and easily accessible from surrounding neighborhoods so that they can adequately serve a maximum range of a half-mile radius, which reflects the smart growth standard of a short 5-min walking distance. Therefore, a desirable distance from the nearest built-up areas should not exceed a half mile.
6. Environmental compliance: to preserve the environmental integrity and natural heritage of the area, the site should avoid prime agricultural land and open spaces.
7. Accessibility: the site must have access to transportation network. It should be well-connected by roads, railroads, interchanges, trials, and highways to ensure that the commute time required for work, entertainment, or shopping trips does not exceed 30 min. For simplification, a 300-m buffer zone from major roads is considered an acceptable range for accessibility.
8. Proximity to industrial and landfill sites: a minimum distance of one mile from landfill and industrial locations, and other noxious land uses must be ensured in order to prevent noise and eliminate immediate threat of harmful chemical emissions to public health. There are two major industrial sites located within the study area and another site located to the northeast of the region.
9. Airport location: there is only one major airport, the University of Illinois Willard airport, located to the south of the Champaign-Urbana region. A minimum distance of one mile is deemed sufficient to avoid adverse impacts of the airport noise.

5 Geospatial Methods, Analysis, and Results

The analysis followed three major steps. These include data acquisition, data management and analysis. Data were collected from various sources such as the US Department of agriculture, USGS, Illinois State Geographical Survey, Illinois Department of Natural Resources, and the Inventory of Land based Disposal Sites to name a few. Data management included preparing the data for analysis, creating other derivative GIS layers and defining and delineating the data to fit within the boundary of the target area. Figure 3 shows the basic steps carried out to accomplish the analysis which is composed of two major categories: vector analysis and raster analysis.

A. Vector-based analysis:
Vector-based analysis (using geo-processing tools) was performed on the first set of development criteria including floodplain, wetlands, road networks, industrial sites, airport location, built-up areas, and minimum site size. The final layers produced from the vector analysis are juxtaposed into one map shown in Fig. 4, which displays the

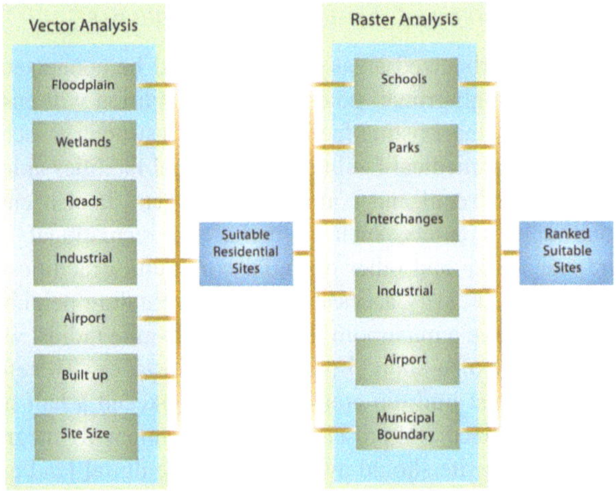

Fig. 3. Analytical procedure

first-cut of suitable sites for the low-density residential development. At this stage of the analysis, the output obtained does not reflect either the developer's or the environmentalist's concerns. Interestingly, most of these suitable sites are located outside the municipal boundaries of both cities, yet within the 1.5 mile boundary. This is due in part to the lack of sites of suitable substantial sizes available closer to the central parts of either city.

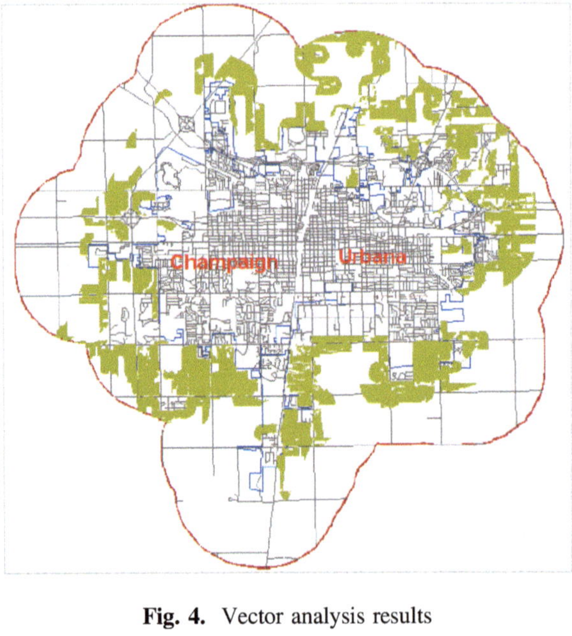

Fig. 4. Vector analysis results

B. Raster-based analysis:
The raster-based analysis (using geospatial analysis, model builder, and density, proximity, and propensity tools) considers the two scenarios introduced in the beginning of the study (the developer's point of view and the environmentalists' point of view). Raster-based analytical steps and outcomes are illustrated in Fig. 5. All layers were classified according to a ten-category color scale for consistency purposes. In this analysis, the closer the location is to schools, interchanges, municipal boundary, and parks, the more desirable it is for development (designated compatible sites), while the closer the site to industrial locations and the airport, the less desirable it is for development (designated incompatible sites).

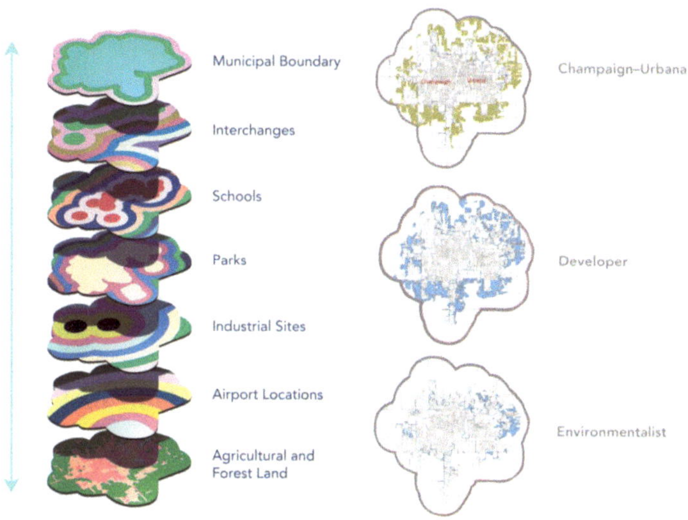

Fig. 5. Layers and outcome of raster analysis

6 First Scenario (The Developer's Point of View)

This scenario assumes that a private developer intends to initiate a new low-density residential subdivision in the Champaign-Urbana region. The developer knows that the key to achieving a successful and profitable project is to meet the preferences of the buyers. By doing so, the price of the housing units will be high and more revenue will be generated. To cater to the buyers' preferences and ensure project success, the best locations where facilities, amenities, and infrastructure are available must be identified as priority sites. To achieve these goals, different weights are assigned to each of the factors identified above based on these considerations. Proximity to schools, interchanges, and municipal boundary are assigned higher weights, while environmental factors regarding preserving the prime agricultural and forest land are given no or insignificant weight. After overlaying all the layers obtained from both analyses, a preference map that shows rated locations is obtained as shown in Fig. 6. The red color represents the most desirable locations from the developer's perspective. It is clear that the most desirable sites for development are located in west Champaign and northeast Urbana.

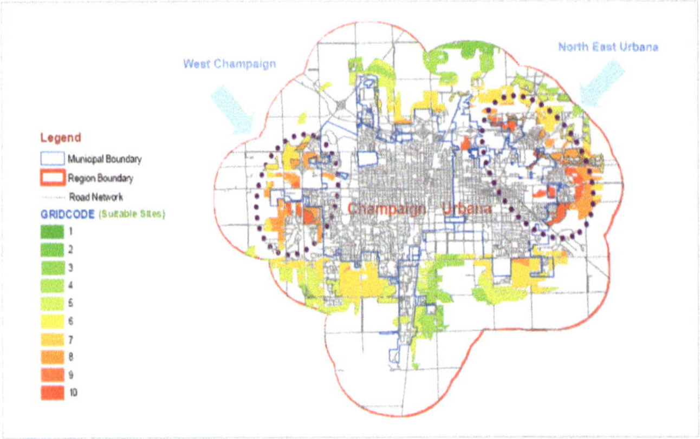

Fig. 6. Results from the first scenario (developer)

7 Second Scenario (The Environmentalist's Point of View)

In this scenario, the environment is the main focus of the analysis and evaluation process. This approach aims to maintain the integrity of the agricultural and forest land of the region, which are assigned high weight values. By combining and weighting the different raster layers, a preference map that shows desirable locations is obtained for the entire area. The map shows that the most desirable sites, presented in red, are centrally located within the region avoiding the forest and agricultural land, which primarily exist on the periphery. Coinciding with the results obtained from the first scenario, the most desirable locations, according to this scenario, are those located in west Champaign and northeast Urbana. The difference is that this scenario is more conservative in allocating and land and recourses to accommodate future population growth and land use demand (Fig. 7).

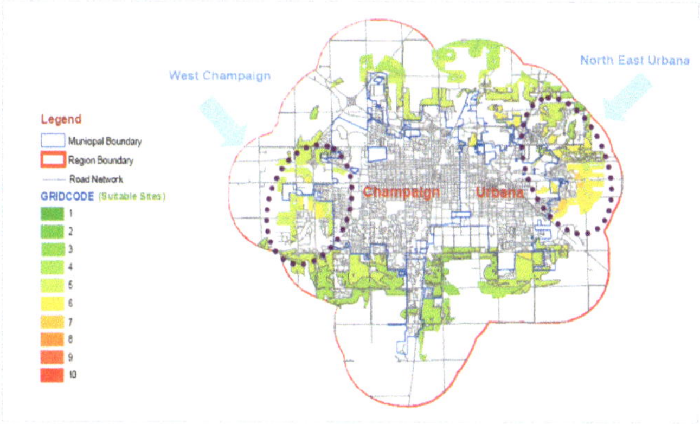

Fig. 7. Results from second scenario (environmentalists)

8 Conclusions

The findings from this study are summarized in Fig. 8, which provides a comparison of the results obtained from both scenarios showing locations with weight values higher than 5. It can be observed that the suitable areas for the low-density residential development in the first scenario cover a significantly larger portion of region than that of the second more conservative scenario. Because developers pay little attention to environmental factors, considerably large areas are proposed for development in the first scenario. In contrast, the fact that the second scenario relies on more conscientious efforts to preserve the natural resources of the region, greatly limits the amount of land perceived suitable for the proposed development. Knowing the potential sites for such development helps in determining future growth strategies including roads, infrastructure, schools, parks, shopping centers and other amenities. Knowing how and where cities tend to expand in the future helps in identifying locations and capacities of these amenities and assets needed for future expansion.

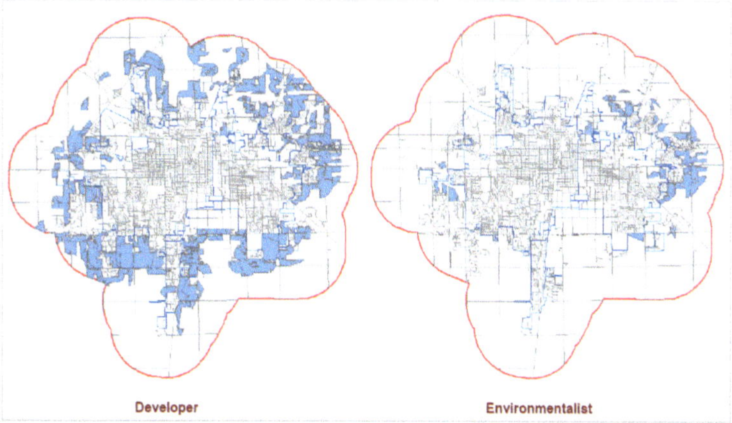

Fig. 8. Suitable sites whose value is greater than 5

This study provided an example of how GIS is used to support planning tasks and help make better decisions regarding real-world planning issues and community development more effectively for the betterment of urban growth management practices. The previous example is used as a vehicle to shed light on the role of GIS in land use planning and decision-making. This study argues for the full utilization of GIS in order to maximize its contribution to the planning practice and refrain from limiting its applications to only mapmaking and cartographic purposes. Beyond mapmaking, GIS can deliver insight and actionable intelligence that helps planners accomplish their goals of livable and sustainable cities. The sole purpose of GIS based-technology is to provide state of the art analytical and management tools to study and explore spatial patterns and relationships in order make informed decisions regarding how to strike the right balance between socioeconomic development and environmental protection goals.

Understanding Marine Geo-Technical Engineering, Gas Hydrate Energy Release and the Role of External Stimuli

J. Rajaraman[1(✉)] and S. Narasimha Rao[2]

[1] Department of Harbour and Ocean Engineering, AMET University, Kanathur, Chennai 603112, India
rajaraman.usha44@gmail.com
[2] Dredging Corporation of India (Government of India), Visakhapatnam, India
kuttyharirao@gmail.com

Abstract. In the modern world globalization is good because when trade stops war begins. In the Geo environment methane is non toxic, yet it is extremely flammable and may form explosive mixtures with air. Methane is violently reactive with oxidizers, halogen and some halogen containing compounds. Methane is also an asphyxiant and may displace oxygen in an enclosed space. Methane will burn between lower explosive limit (LEL) and upper explosive limit (UEL). LEL is 5% UEL is 15% by percentage volume. In this paper the application of marine geo technical engineering related to soil shear strength is linked to the chemical reactions of methane hydrates. Methane clathrate compounds buried in marine sediments is a powerful green house gas. By increasing the temperature leads to further temperature rise and further methane clathrate destabilization. Initiating a runaway process as irreversible once firing of a gun is started. Runaway methane clathrate may cause drastic alteration of the ocean environment (ocean acidification and stratification). In Permian – Triassic event (P-T event) (great dying)/Great Permian extinction, runaway method was also responsible. Marine invertebrates suffered the greatest losses but fortunately life recovered quickly after P-T extinction. Methane clathrate is released as gas into the surrounding water column or soils when ambient temperature increases. By controlled bombing the methane can be burnt continuously in the marine environment. In this method the marine temperature is raised and accelerated runaway chemical reaction is created artificially. In the coastal low pressure region reasonably high temperature can be created by bombing and the gas hydrates takes responsibility for the rest of the destruction. The marine soil slump will further help the flammable methane. This type of non nuclear bombing supported by methane hydrate will keep two warring nations to think about future peace and existence. Since it is non nuclear and temporary shock treatment the nations will spend more time on peace and avoid war. Before resorting to this method all nations should concentrate on mutually benefiting trade practices because when globalization trade stops war begins.

1 Introduction

Clathrate Gun Hypothesis

The clathrate gun hypothesis is the popular name given to the idea that increases in sea temperatures or drops in sea levels can trigger a catastrophic positive feedback

M. Shehata and F. Rodrigues (Eds.): GeoMEast 2018, SUCI, pp. 249–267, 2019.
https://doi.org/10.1007/978-3-030-01905-1_14

effect on climate: first warming causes a sudden release of methane from methane clathrate compounds buried in seabeds and seabed permafrost; second, because methane itself is a powerful greenhouse gas, temperatures rise further, and the cycle repeats. This runaway process, once started, could be as irreversible as the firing of a gun.

In its original form, the hypothesis proposed that the "clathrate gun" could cause abrupt runaway warming on a time scale less than a human lifetime (Kennett et al. 2003). It was thought to be responsible for warming events in and at the end of the Last Glacial Maximum (Kennett et al. 2000), but this is now thought to be unlikely (Sowers 2006; Severinghaus 2006). However, there is stronger evidence that runaway methane clathrate breakdown may have caused drastic alteration of the ocean environment (such as ocean acidification and ocean stratification) and of the atmosphere on a number of occasions in the past, over timescales of tens of thousands of years. These events include the Paleocene-Ecocene Thermal Maximum 56 million years ago, and most notably the Permian-Triassic extinction event, when up to 96% of all marine species became extinct, 252 million years ago (The Day The Earth….. BBC 2002).

Mechanism

Methane clathrate, also known commonly as methane hydrate is a form of water ice that contains a large amount of methane within its crystal structure. Potentially large deposits of methane clathrate have been found under sediments on the ocean floors of the Earth, although the estimates of total resource size given by various experts differ by many orders of magnitude, leaving doubt as to the size of methane clathrate deposits (particularly in the viability of extracting them as a fuel resource). The sudden release of large amounts of natural gas from methane clathrate deposits in runaway climate change could be a cause of past, future, and present climate changes. The release of this trapped methane is a potential major outcome of a rise in temperature; some have suggested that this was a main factor in the planet warming 6 °C, which happened during the end-Permian extinction, as methane is much more powerful as a greenhouse gas than carbon dioxide.

Subsea Permafrost

Subsea permafrost occurs beneath the seabed and exists in the continental shelves of the polar regions. This source of methane is different from methane clathrates, but contributes to the overall outcome and feedbacks.

Metastable Methane Clathrates

Another kind of exception is in clathrates associated with the Arctic ocean where clathrates can exist in shallower water stabilized by lower temperatures rather than higher pressures; these may potentially be marginally stable much closer to the surface of the sea - bed, stabilized by a frozen 'lid' of permafrost preventing methane escape.

Current Outlook

Most deposits of methane clathrate are in sediments too deep to respond rapidly, and modelling by (Archer 2007) suggests the methane forcing should remain a minor component of the overall greenhouse effect. Clathrate deposits destabilize from the deepest part of their stability zone, which is typically hundreds of metres below the seabed. A sustained increase in sea temperature will warm its way through the sediment eventually, and cause the shallowest, most marginal clathrate to start to break down; but it will typically take on the order of a thousand years or more for the temperature signal to get through. However, there is also a possibility for the formation of gas migration pathways within fault zones in the East Siberian Arctic Shelf, through the process of talik formation or pingo-like features (Climate Hydrate Interactions USGS 2013; Shakova and Semiletov 2010; Methane bubbling ...Montery Bay Aquarium Research Institute 2008).

Continental Slopes
A trapped gas deposit on the continental slope off Canada in the Beaufort Sea, located in an area of small conical hills on the ocean floor is just 290 m below sea level and considered the shallowest known deposit of methane hydrate.

2 The Dangers of Gas Hydrates

Hazardous Nature of Gas Hydrates
The recovery of gas hydrates is where the hazards begin. The natural gas is frozen but bringing it to the surface risks a sudden dissociation into gas and water, which would cause it to expand exponentially. Another risk is the fact that oil pipelines present ideal conditions for gas hydrates to form. This can clog up the pipeline leading to reduced flow or a complete blockage which would result in damage to the pipe itself. Gas hydrates can be removed after the formation but it is a difficult and time-consuming process.

A simplified explanation of the removal process would be raising the temperature within the pipeline coupled with lowering the pressure. Despite the proven methods, the removal process is time consuming and anything that slows down the flow, ends up costing money in the long run. A far more effective method is to block the formation of the clathrates before they have a chance to form.

Depending on the likelihood of gas hydrate formation, the location of the pipe itself, and the cost involved, various mitigation techniques have been formed. Various chemical processes, such as glycol dehydration, have proven effective in decreasing the optimal temperature for hydrate formation. Other benign chemical additives have shown promise in shifting the flow to a lower temperature while at the same time increasing pressure. Still another technique involves chemicals used as inhibitors, lengthening the amount of time it takes for gas hydrates to form.

Standard off shore oil drilling stations face hazards from gas hydrates as well. Submerged in deep water, oil reservoirs can contain gas pockets that can get sucked into the well bore and form gas hydrates. The formations can then be intermingled with the oil, mud, and other fluids. The upward trajectory, combined with rising temperatures, cause the hydrates to dissociate as mentioned previously. This leads to a cycle of fluid ejection, increased pressure, and more potential hydrate formation. The ferocious ejection of fluid can lead to a scenario known as a "kick". This type of situation was what exacerbated the BP Deepwater Horizon oil spill in 2010. As you can see, from collecting gas hydrates as a potential energy source, to avoiding the potential dangers involved in collecting other energy sources nearby, very real hazards are presented in oil and natural gas collection due to gas hydrates.

Methane Hydrate: A Surprising Compound
What do you get when you combine water and swamp gas under low temperatures and high pressures? You get a frozen lattice like substance called methane hydrate., huge amounts of which underlie our oceans and polar permafrost. This crystalline combination of a natural gas and water (known technically as a clathrate) looks remarkably like ice but burns if it meets a lit match. Methane hydrate was discovered only a few decades ago, and little research has been done on it until recently. By some estimates, the energy locked up in methane hydrate deposits is more than twice the global reserves of all conventional gas, oil and coal deposits combined. But no one has yet figured out how to pull out the gas inexpensively, and no one knows how much is actually recoverable. Because methane is also a greenhouse gas, release of even a small percentage of total deposits could have a serious effect on Earth's atmosphere.

Research on methane hydrate has increased in the last few years, particularly in countries such as Japan that have few native energy resources. As scientists around the world learn more about this material, new concerns surface. For example, ocean-based oil-drilling operations sometime encounter methane hydrate deposits. As a drill spins through the hydrate, the process can cause it to dissociate. The freed gas may explode, causing the drilling crew to lose control of the well. Another concern is that unstable hydrate layers could give way beneath oil platforms or, on a larger scale, even cause tsunamis.

3 Deepwater Methane Hydrates and the BP Gulf Explosion

The vast deepwater methane hydrate deposits of the Gulf of Mexico are an open secret in big energy circles. They represent the most tantalizing new frontier of unconventional energy- a potential source of hydrocarbon fuel thought to be twice as large as all the petroleum deposits ever known. For the oil and gas industry, the substances are also known to be the primary hazard when drilling for deepwater oil.

Methane hydrates are volatile compounds- natural gas compressed into molecular cages of ice. They are stable in the extreme cold and crushing weight of deepwater, but are extremely dangerous when they build up inside the drill column of a well. If destabilized by heat or a decrease in pressure, methane hydrates can quickly expand to 164 times their volume.

Survivors of the BP rig explosion told interviewers that right before the blast, workers had decreased the pressure in the drill column and applied heat to set the cement seal around the wellhead. Then a quickly expanding bubble of methane gas shot up the drill column before exploding on the platform on the ocean's surface. Even a solid steel pipe has little chance against a 164 –fold expansion of volume – something that would render a man six feet six inches tall suddenly the height of the Eiffel Tower.

Scientists are well aware of the awesome power of those strange hydrocarbons. A sudden large scale release of methane hydrates is believed to have caused a mass extinction 55 million years ago. Among planners concerned with mega-disasters, their sudden escape is considered to be a threat comparable to an asteroid strike or nuclear war. The Lawrence Livermore National Laboratory, a Livermore, Ca-based weapons design centre, reports that when released on a large scale, methane hydrates can even cause tsunamis.

Oil and gas industry has a lot of experience with methane hydrates, because they have to be kept from forming in pipes or they will clog the lines, stop the flow of oil, and pose a danger. Drillers use inhibitors such as methanol to keep the hydrates from crystallizing inside drill rigs operating at great depths, where conditions for methane hydrate formation are ideal.

Vast Deposits in Ocean Sediments

Methane Hydrates are believed to exist in vast deposits underneath the ocean floor, trapped by nature in ocean sediments. Deepwater drillers could find themselves drilling through these natural hydrate deposits. According to the National Academy of Sciences, which published a bullish report on the energy potential of methane hydrates, "Industry practice is to avoid methane-bearing areas during drilling for conventional oil and gas resources for safety reasons".

Professor Sum explained that because "with oil there is usually gas present", it is possible for methane hydrates to form in the pipe even when not drilling through hydrate bearing sediments. The pressure and cold of the deepwater create conditions that encourage gas flowing into the pipe to form hydrates, and if the rate of crystallization is rapid enough, the hydrates can clog the pipe.

The cofferdam that BP lowered over the broken pipe gushing oil to contain the spill was almost immediately clogged by methane hydrates, which formed spontaneously. Gas escaping with the oil from the well, when trapped in the steel structure with cold water under great pressure, rapidly accumulated into an ice like matrix.

Documented Explosive Hazard

In a book about methane hydrates, which Professor Koh co-authored, brief mention is made of a case in which methane hydrates caused a gas pipe to rupture on land, leading to loss of life. Two workers were attempting to clear a line in which a hydrate plug had formed. The authors say that "the impact of a moving hydrate mass" caused the pipe to fail. The explosion caused a large piece of pipe to strike the foreman, killing him. The book then quotes from the *Canadian Association of Petroleum Producers Hydrate Guidelines* to describe proper procedures for safely removing a hydrate plug in a pipe on land.

Solve- Climate was not able to find more detailed public documentation of this incident in Alberta, but mention is made in an article in a publication of the Oak Ridge National Laboratory, a federal research centre associated with the Department of Energy, of a different unspecified incident on a drilling rig.

"Forces from methane hydrate dissociation have been blamed for a damaging shift in a drilling rig's foundation, causing a loss of $100 million," the article reports.

Although public discussion of damage from methane hydrate accidents appears to be minimal, the danger is well-recognized within the industry.

One Trade Ministry is targeting to start commercial production, even as scientists warned against causing a massive undersea landslide that suddenly trigger a massive methane hydrate release.

4 Mechanisms Leading to Co-existence of Gas and Hydrate in Ocean Sediments

The understanding of the influence of Geochemistry of Gas Hydrates in the shear strength and stability of Marine Sediment movement is necessary (Rajaraman et al. 2017; Satyanarayana et al. 2011). The explanation for the Figs. 1, 2, 3, 4, 5, 6, 7, 8 and 9 involves the fundamentals of Geotechnical Engineering, such as effecting pressure, total pressure, Cohesion and angle of internal friction, Capillary forces, gram-sizes, void ratios, shear strength ($\tau = c + \sigma \tan \varphi$) (1). These figures are included in this paper to understand simple fundamentals of Geotechnical Engineering in earth processes in marine environment. Accelerated destabilization of Gas hydrates only need simple explosions through sediments and dropping bombs from air.

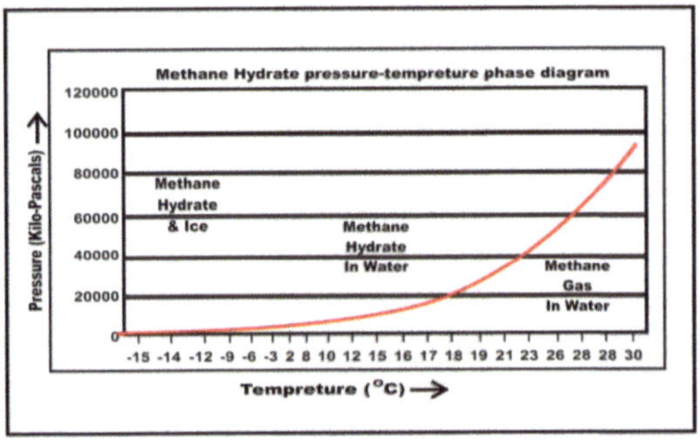

Fig. 1. Methane clathrate is released as gas into the surrounding water column or soils when ambient temperature increases

The temporary "Firewall" created will last till the entire volume of methane gas burns. It will not last for long because these "Firewalls" are self controlling processes.

Figure 1: Explains the situation when methane clathrate is released as gas into the surrounding water column or soils when ambient temperature increases.

Figure 2: Idealized cross section of the hydrate stability zone (HSZ) along a continental shelf. The thickness of the HSZ increases with the depth of the sea floor (After Dickens 2003).

Figures 3, 4, 5, 6, 7, 8 and 9 are after Juanes and Bryant (2006).

Figure 3: The Conceptual model of methane hydrates in ocean sediments involves the following key physical processes. (1) Focussed and possible episodic upwards flow of gas through faults, fractures and high-permeability conduits. (2) Presence of a laterally-extended free mobile gas zone beneath the HSZ, whose height responds to gradual build-up from deep methane gas and episodic release into the HSZ. (3) Vertical invasion of methane gas from below the HSZ, either by exceeding the capillary entry pressure or the fracturing pressure: (4) Lateral invasion of methane gas into sediment beds, also by the two models of invasion (Remark: the thresholds for capillary pressure and fracture opening: invasions are dependent upon pre-existing hydrate in the pore space). (5) Increase in pore water salinity by ion exclusion from the hydrate crystalline

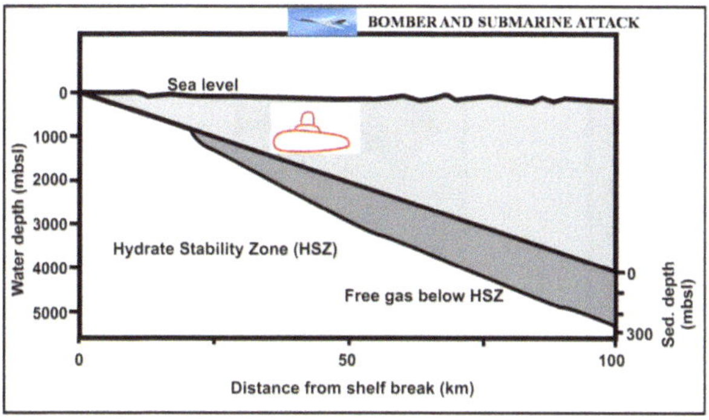

Fig. 2. Hydrate stability zone (HSZ) along a continental shelf.

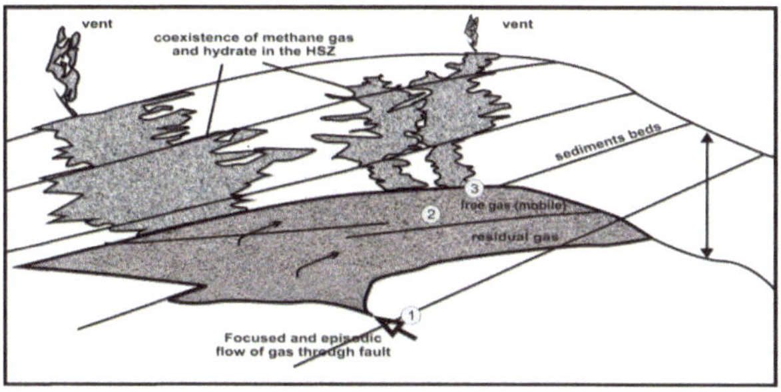

Fig. 3. Methane hydrates in ocean sediments.

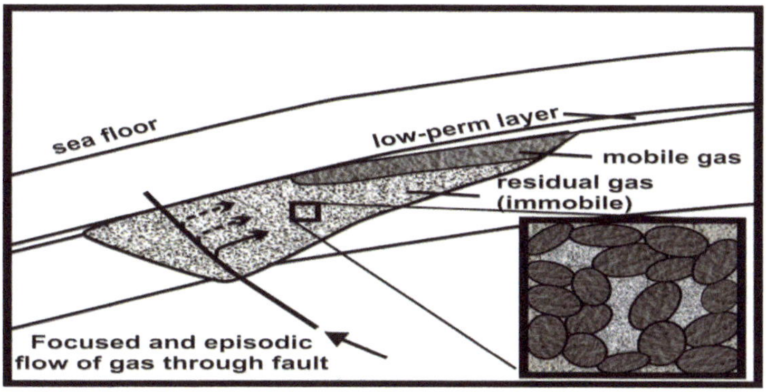

Fig. 4. Residual (immobile) gas saturation results from capillary trapping during an imbibitions process.

structure –ion diffusion out of macroscopic drainage areas can be very slow; (Remark processes 5–7 occur at the grain scale and are not explicitly depicted). (6) Imbibition events (water saturation increases locally) that are driven by the inability to maintain

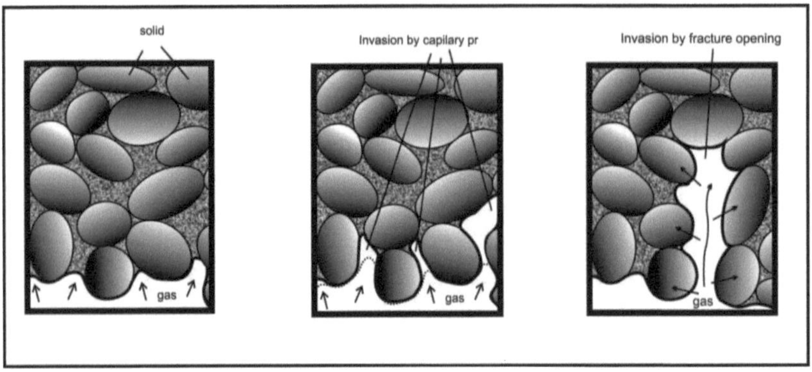

Fig. 5. Schematic diagram of the two modes of methane gas invading a segment.

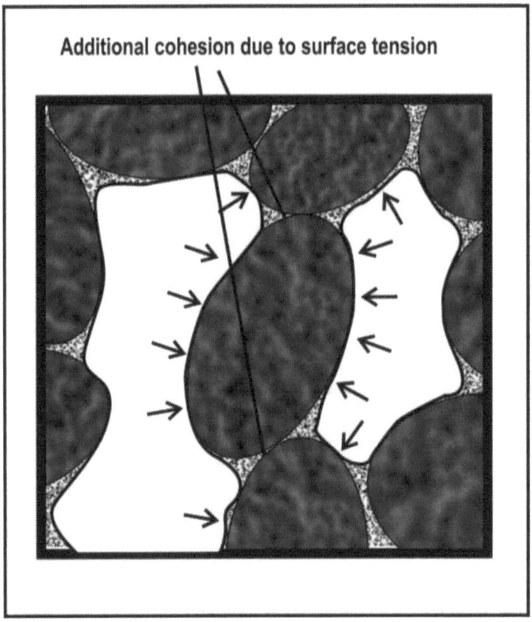

Fig. 6. Meniscus pinning in the presence of two fluid phases in the sediment.

gas pressure due to the finite volume of free gas beneath the HSZ – imbibition leads to capillary trapping and a complex distribution of methane gas and methane hydrate. (7) Possible rupture of hydrate shell around disconnected volumes of gas, responsible for increased mobility of methane and formation of additional gas-water interface.

Figure 4: Residual (immobile) gas saturation results from capillary trapping during an imbibition process (increasing water saturation). Trapping can occur below and within the HSZ, a consequence of the irreversible character of multiphase displacement in porous media: capillary pressure and relative permeability functions exhibit

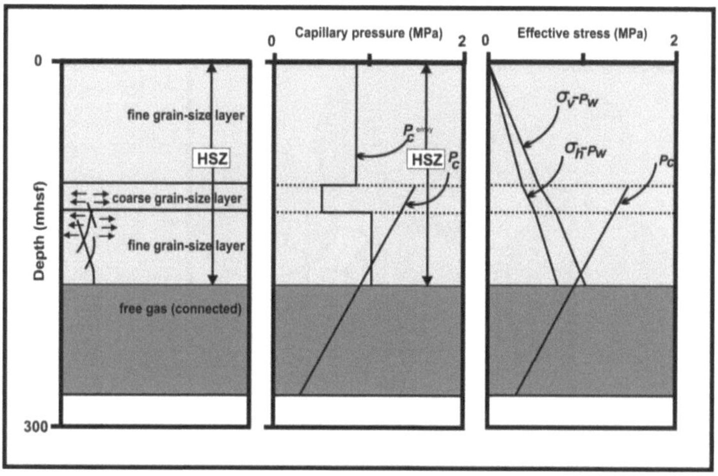

Fig. 7. Conceptual profiles of capillary entry pressure and minimum principal stress vs sediment depth.

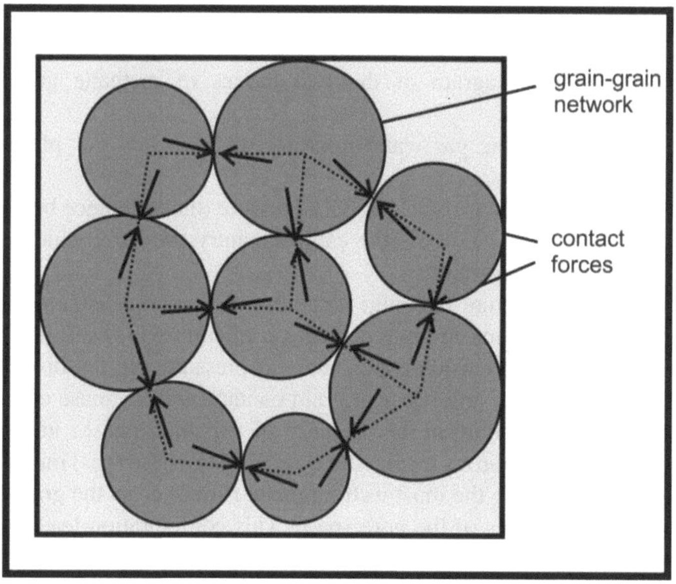

Fig. 8. Schematic diagram of the network of grain-grain contact forces.

hysteresis. Imbibition takes place naturally during the upwards migration of methane through the sediment whenever the methane is driven by episodic events in which the pressure is not held constant.

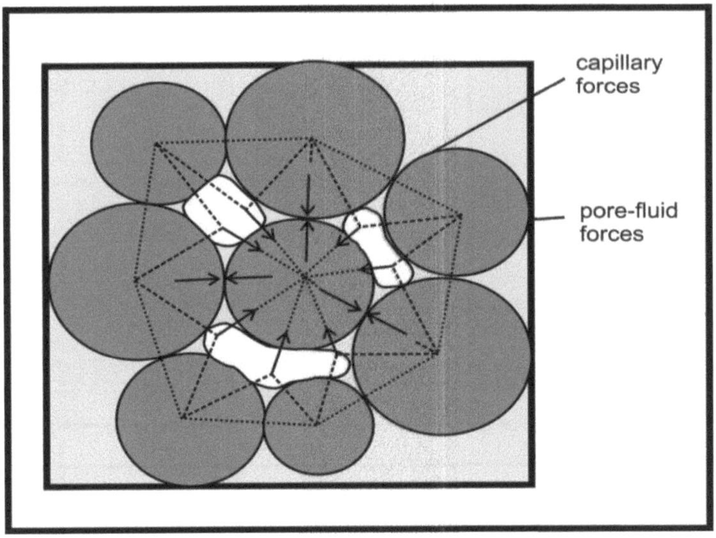

Fig. 9. Schematic diagram of the networks of grain-pore body fluid forces and grain-grain capillary forces.

Figure 5: Schematic diagram of the two modes of methane gas invading a sediment.

Left: before invasion, the gas water interface of a buoyant gas plume underlies water filled sediment.

Centre: invasion will occur if the capillary pressure (the difference between the gas pressure and water pressure) exceeds the capillary entry pressure, which is inversely proportional to the pore diameter.

Right: invasion by fracture opening: if the exerted pressure is sufficient to overcome compression and friction at grain contacts, a fracture will form. In a multiphase environment, due to surface tension effects, the gas pressure will not dissipate quickly through the porous medium, and water at grain contacts will increase cohesion.

Figure 6: Meniscus pinning in the presence of two fluid phases in the sediment. During multiphase flow in porous media, the least wetting phase (gas) migrates through the center of the pores, while the most wetting phase (brine) coats the grains and forms filaments around the crevices of the pore space. This configuration leads to gas-water menisci around the grain contacts. Due to interfacial tension, these menisci are responsible for an attraction force between grains. At the macroscopic level it can be interpreted as, an increment in the cohesion of the material. This is a purely multiphase flow effect, not present in single phase poromechanics.

Figure 7: Conceptual profiles of capillary entry pressure and minimum principal stress vs sediment depth.

Left: A connected mobile free gas zone exist underneath the HSZ. Gas will migrate upwards if the capillary pressure is sufficient to overcome the capillary entry pressure or if the gas pressure is sufficient to fracture the sediment.

Centre: The capillary entry pressure P_c^{entry} is higher in fine grain sediments, because it is inversely proportional to the pore diameter. If the gas phase is connected, the capillary pressure P increases with elevation. The situation depicted is such that $P_c < P_c^{entry}$ at the bottom of the HSZ, so gas cannot invade. However, if gas were to invade by opening a vertical fracture (right panel), the capillary pressure would exceed P_c^{entry} at higher elevations, leading to lateral invasion of sediment beds above the bottom of the HSZ.

Right: The vertical (lithostatic) stress increases with depth in excess of hydrostatic stress. In this figure we plot the difference between the stress and pore water pressure for both vertical and horizontal stresses. If the capillary pressure exceeds the minimum horizontal stress, a vertical fracture will propagate, serving as a conduit for upwards gas migration. This is the case depicted here. These fractures will open in response to episodic releases of methane and can transport methane gas to shallower beds, from which gas may flow laterally by capillary or fracture invasion.

Figure 8: Schematic diagram of the network of grain-grain contact forces. Even though the contact forces generally involve a normal and tangential component, their action can be associated with a network that connects the centers of grains that are in contact. This network of forces is sufficient to characterize "dry samples".

Figure 9: Schematic diagram of the networks of grain-pore body fluid forces (in blue) and grain-grain capillary forces (in black) in a DEM model. These force networks are required to model grain-scale mechanics in the presence of multiphase fluid displacements, and must be supplemented to the grain-grain network of contact forces shown in Fig. 8.

Methane Transport and Hydrate Formation at the Bed Scale: A Threshold/ Leakage Model

We will integrate the process models developed above into a threshold/leakage model. This model will describe methane movement vertically and laterally within the HSZ from a postulated source below the HSZ from a postulated source below the HSZ. The "threshold" refers to the criteria for vertical gas phase movement (fracture initiation or drainage) and for lateral gas movement from a fracture into a sediment bed. These criteria will be determined in the models discussed above. The "leakage" refers to the movement of methane laterally into sediment beds satisfying the threshold criteria. Should vertical movement continue all the way to the sea floor, the leakage term will include the flux of methane into the ocean.

The thresholds in this model depend on capillary pressure $P_c = P_g - P_w$ as depicted in (Fig. 7) and so P_c will be the primary variable. The dynamic of this model derive from the following observations.

- Buoyancy causes P_c at the top of the rising gas to increase as it rises, as long as it remains connected to the source below the HSZ.
- Vertical variation in sediment properties makes the evaluation of thresholds non-trivial, even though P_c increases with elevation.
- Lateral movement of methane into a sediment bed will be driven by capillary forces, establishing a water saturation S_w that depends only on P_c and hence upon elevation S_w may therefore vary within a bed that is invaded. The rate at which methane enters the bed will depend on the gas relative permeability at the relevant values of S_w.

- If the Methane source below the HSZ is finite and is being charged slowly, then leakage of methane into sediment beds will result in water rising into the bottom of the methane source. This will reduce P_c uniformly in the column of rising methane, causing hysteresis and requiring re-evaluation of thresholds.

5 Favourable Geotechnical Properties/Parameters for Submarine and Air Attack to Explode Methane and Methane Hydrates

Cluster Munitions

A Cluster munitions is a form of air-dropped or ground –launched explosive weapon that releases or ejects smaller sub-munitions. Commonly, this is a cluster bomb that ejects explosive bomblets that are designed to kill personnel and destroy vehicles. Other cluster munitions are designed to destroy runways or electric power transmission lines, disperse chemical or biological weapons, or to scatter land mines. Because cluster bombs release many small bomb-lets over a wide area, they pose risks to civilians both during attacks and afterwards. Unexploded bomb-lets can kill or wound civilians and/or unintended targets long after a conflict has ended, and are costly to locate and remove.

Threat to civilians

While all weapons are dangerous, cluster bombs pose a particular threat to civilians for two reasons: they have a wide area of effect, and they have consistently left behind a large number of unexploded bomb-lets. The un-exploded bomb-lets can remain dangerous for decades after the end of a conflict.

Cluster munitions are opposed by many individuals and hundreds of groups, such as the Red Cross, the Cluster Munition Coalition and the United Nations, because of the high number of civilians that have fallen victim to the weapon. Since February 2005, Handicap international called for cluster munitions to be prohibited and collected hundreds of thousands of signatures to support its call. 98% of 13,306 recorded cluster munitions casualties that are registered with Handicap International are civilians, while 27% are children.

The area affected by a single cluster munition, known as its footprint, can be very large; a single unguided M26 MLRS rocket can effectively cover an area of 0.23 km^2.

Arguments favoring the use of cluster munitions are that their use reduces the number of aircraft and artillery systems needed to support military operations and if they were eliminated, significantly more money would have to be spent on new weapons, ammunitions, and logistical resources. Also, militaries would need to increase their use of massed artillery and rocket barrages to get the same coverage, which would destroy or damage more key infrastructures.

Bombing Top of continental slope will release Methane gas because the temperature will be increased and pressure will be released simultaneously. The shear strength decreases down the slope and the chain reaction will propagate into the ocean and not towards coastal area or coastal community. The unexploded bomb-lets will sink into ocean (water) and will not explode in the future after the conflict. When the temperature lowers and pressure increases down the slope the burning of methane gas automatically stop, since the Geotechnical conditions (shear strength) will limit damage. This burning process of methane is self limiting process and stops after sometime. Special efforts are not needed to stop it.

The Basic Shear strength Equation:

$\tau = c + \sigma \tan \varphi$, as far as methane hydrate is concerned is a function of σ (vertical pressure) and temperature and therefore deep in the ocean sediments the methane hydrate will be only in solid form and change in (cohesion) c and φ angle of internal friction value will not affect the formation of gas hydrates.

Surgical strike

A surgical strike is a military attack which results in, was intended to result in, or is claimed to have resulted in only damage to the intended legitimate military target, and no or minimal collateral damage to surrounding structures, vehicles, buildings, or the general public infrastructure and utilities.

A swift and targeted attack with the aim of minimum collateral damage to the nearby areas and civilians is a surgical strike. Neutralization of targets with surgical strikes also prevents escalation to a full blown war. Surgical strike attacks can be carried out via air strike, airdropping special ops teams or a swift ground operation or by sending special troops.

Precision bombing is another example of a surgical strike carried out by aircraft – it can be contrasted against carpet bombing, the latter which results in high collateral damage and a wide range of destruction over an affected area which may or may not include high civilian casualties.

In Gas hydrate explosion used as a surgical strike, the collateral damage is totally avoided. All war operations are confined to marine environment away from coast. This artificial and accelerated explosion is only a warning between two nations separated by a part of ocean. This prevents escalation to a full war. Precision bombing is not necessary because it is operated only in ocean by creating a "Firewall" by simultaneously bombing from aircrafts and through submarine operations from ocean bottom in a restricted, selected area.

6 War and Energy War (Gas Hydrates)

In the future Wars will be fought on the basis of:

(1) **War for the Energy:** A country on a war-footing basis engaged to utilize available energy sources within their territorial limits.
(2) **War of the Energy:** Energy as the cause of the war. Trying to utilize another nation's energy sources.
(3) **War by the Energy:** Burning other nation's Gas Hydrates to create and utilize the opportunities of energy costs and consequences.

In Modern Wars, Resource-Related objectives have been generally determined by broader strategic aims and not vice-versa.

War for the Energy (Type 1 War)

Until recently, methane hydrates had never been tapped as a source of energy to most increasing global demands. It has generally been considered that other sources of fossil fuels, notably conventional oil and gas (and more recently shale oil and gas), have been easier and cheaper to access. But in March 2013, Japan became the first country to successfully flow gas from methane hydrate deposits under the Pacific Ocean. The Experimental data not available to the academic community (Spalding and Fox 2014).

Japan kick-starts exploration

The Japanese energy organization, estimates that there is some 1.1 trillion cubic meters of methane held in methane hydrate deposits in marine sediments in the Nankai Trough off the Pacific coast of central Japan. This is equivalent to around 11 years of the amount of LNG that is currently imported into Japan. A further 120,000 cubic meters of gas from methane hydrates has been discovered some 50 miles off the coast of Aichi Prefecture in central Japan following recent tests. These resources could represent just the tip of the iceberg. (Experimental Data not available.)

Such discoveries are even more significant for Japan once the repercussions of the 2011 Fukushima Daiichi disaster are taken into account. Indeed, all but two of Japan's 54 nuclear power plants (which once provided 30% of the nation's energy) have ceased operations. With virtually no exploitable domestic fuel sources to take the place of nuclear, Japan is therefore increasingly reliant on expensive imports of oil and LNG. In 2011 Japan consumed 123 billion cubic meters of natural gas, of which 117 billion cubic meters were imported.

The implications of methane hydrates as a new energy source for countries lacking access to conventional resources are therefore profound. It is thought that methane hydrates within Japan's territorial waters may be able to supply the nation's natural gas needs for a century. For Japan, historically resource-poor, domestically produced methane hydrate gas has the potential to reduce the dependence on imports that has defined its energy system throughout the modern era.

Potential environmental impacts

The role that methane hydrates play in stabilizing the seafloor should not be underestimated. For example, drilling deep into oceanic deposits could impact both marine life and the seabed, potentially causing sediment to slide down the continental slope. Some evidence suggests that such underwater landslides have already occurred. Geologists consider that the movement of so much sediment could also trigger tsunamis.

Further, methane is a greenhouse gas and is considered to be 25 times more potent than carbon dioxide in trapping solar radiation in the atmosphere. Scientific studies have revealed that the gas has already started to leak from oceans and soils in the Arctic. As a result, there is concern over any uncontrolled release of methane from hydrate formations.

Even modest leakage rates could nix any potential climate benefit of burning hydrates instead of coal. Potential ecological and environmental risks are being flagged. The considerable rewards of releasing methane from gas hydrate fields must therefore be balanced with risks, and more research may be required to determine the likelihood of such risks materialising and whether there are ways of mitigating them. Some researchers say that they ultimately expect the issues to be comparable to those of offshore conventional natural gas production.

Technology plays its part

The high cost of LNG imports has given Japan greater motivation to press ahead with its methane hydrate research program, the value of which is currently in the region of US$700 million. Researchers in Japan hope to develop production technology that achieves controlled release of the methane from the ice into the production well, thereby minimizing the risk of gas escaping into the atmosphere. But despite Japan's ground-breaking success in producing gas from undersea hydrates, there is still much work to be done to develop a commercially profitable set of technologies for efficient methane hydrate extraction.

Yielding commercial quantities of natural gas from hydrates at an affordable price presents numerous challenges. Such challenges stem in part from the difficult environments in which hydrates are found and where would-be hydrate producers must drill: the frozen expanses of the Arctic and the deep sea abyss.

A variety of technologies could be developed to produce methane hydrates using pressure reduction, ion exchange and other processes that take advantage of their unique chemical and physical properties. The US, Canada, Japan, and India all have vigorous research programs working to discover viable technologies. Most of these research programs remain in the exploratory, experimental, and laboratory phases.

Despite recent advances, commercial production is still unlikely for at least 10 to 15 years. Japan believes that commercial production will be possible by 2018, while the US Geological Survey estimates that countries with the "political will" to pursue methane hydrates could see production by around 2025.

The cost of developing any new energy is high and methane hydrates are no different. The current cost of gas produced from methane hydrates is estimated to be US$30 to US$50 per million British thermal units (MMBTUs), compared to the current Henry Hub price of approximately US$6 per MMBTU. The International Energy Agency estimates that once efficient practices and processes are developed, natural gas produced from methane hydrates will cost between US$4.70 and US$8.60 per MMBTU.

The future

The world's demand for energy continues to increase rapidly and global energy consumption is expected to rise by 41% between 2012 and 2035. Among fossil fuels, gas demand is growing fastest and is increasingly being used as a cleaner alternative to coal for power generation as well as in other sectors. Any potential new source of gas production is therefore likely to be given considerable attention by both industry and government.

War of the Energy (Type 2 War)

Worries are mounting that methane hydrate, touted as a dream energy source, could spark a new energy war in Asia.

U.S.-based magazine "Foreign Policy" recently pointed out that, "The fact that a bulk of methane hydrate is buried under the center of Asia's territorial dispute is a big misfortune for the surrounding nations," meaning that the new energy source turns out to be a new factor that could exacerbate territorial conflicts among major energy-importing countries like Korea, China, and Japan.

Methane hydrate is a solid crystal in the form of ice that is formed when water and gas meet at high pressure and low temperature. It is compressed gas, which in gaseous form would be 160–170 times that of its solid mass, making it an ideal future energy source.

However, some data points to a rosier outlook in that its massive burial, up to 700,000 trillion cubic feet, is distributed evenly all around the world, dispelling concerns about the possible international dispute (Mathew 2014).

However, the energy source, nicknamed fire ice, could fuel the conflict especially in Asia.

This is because Korea, China, and Japan top the list of countries that import the greatest amount of energy sources. According to the U.S. Energy Information Administration (EIA), as of 2012 China is the number 2 crude oil importer, followed by number 3 Japan and number 5 Korea. In natural gas imports, Japan and Korea top the list as numbers 1 and 2.

The 3 energy consumption giants are trying to be less dependent on energy imports by actively mining methane hydrate, but the big three and even other South East Asian countries are entangled in a territorial dispute.

As for the major methane hydrate deposit sites, experts cite the Senkaku Islands, the southern part of the South China Sea, and the East Sea as the epicenter of intense territorial tension.

In fact, according to the Nihon Keizai newspaper, China's marine department and geological survey department launched a full-fledged methane hydrate energy probe last month in the South China Sea, causing a big backlash from territorial disputing countries such as Vietnam and the Philippines.

The Chinese government is set on reinforcing the probe and research system to commercialize the energy source by 2030.

The Japanese media quoted a Chinese expert as saying that the methane hydrate burial amount in the South China Sea is estimated to be enough to sustain China for the next 130 years.

Japan also joined in by sending its probe vessel to the East Sea at the beginning of April.

It has been already confirmed that the methane hydrate that is buried under the East Sea, near Dokdo Island, could feed Korea for 200 years.

Experts predict that it will take 10–20 years for methane hydrate to be commercialized. However, on the back of Japan's success in the first mining of the energy source at the beginning of last year, 2017 its commercial viability is looming closer on the horizon.

War by the Energy Type 3 (burning gas hydrates on the continental slope areas)
Figure 10: There are methane hydrates within the permafrost in Arctic areas, where cold temperatures year round keep them frozen. The sediments of the deeper parts of the continental shelves also contain methane hydrates, where they are stable under the cold, high-pressure conditions found there. These areas could be attacked by

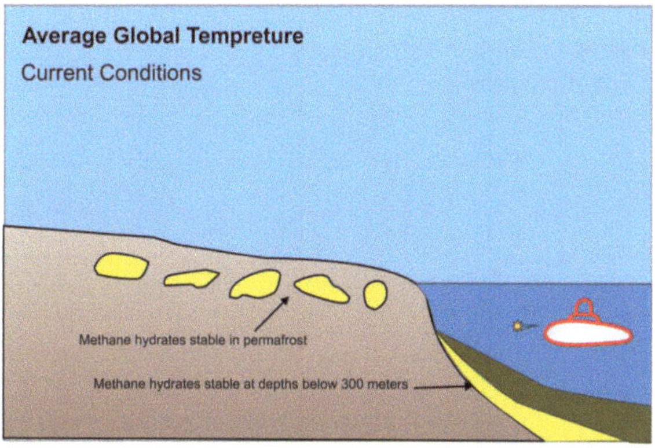

Fig. 10. Methane hydrates within the permafrost in arctic areas current conditions.

Submarines amply supported by air strikes and bombing. The Geotechnical properties are already discussed at length in the initial pages of the paper.

Figure 11: When average temperature rises significantly, sea level goes up because glacial ice and snow melts, and because warm water expands. As the oceans rise, they flood low-lying lands. Even in high latitudes, this water is above freezing, and any permafrost it covers will start to thaw. As the ground warms up, methane hydrates within it will break down - the ice will melt, allowing the trapped gas to escape into the atmosphere. Marine hydrates will remain stable, because rising sea level will only increase the pressure on them. The Geotechnical conditions are favorable for the release of methane and submarine attacks supported by air strike will create a Fire Wall temporarily in a war situation.

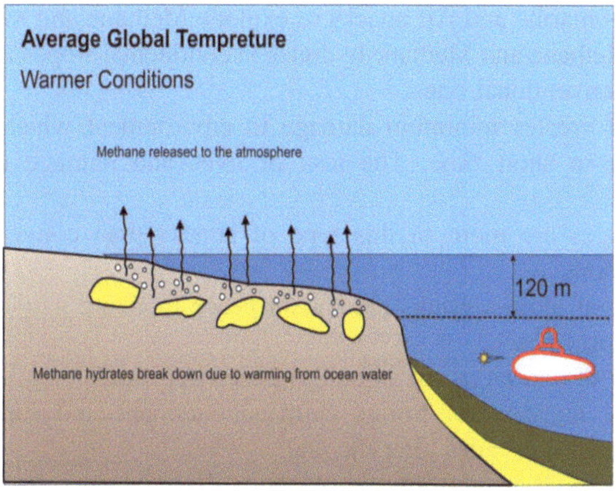

Fig. 11. Gas hydrates in warmer conditions.

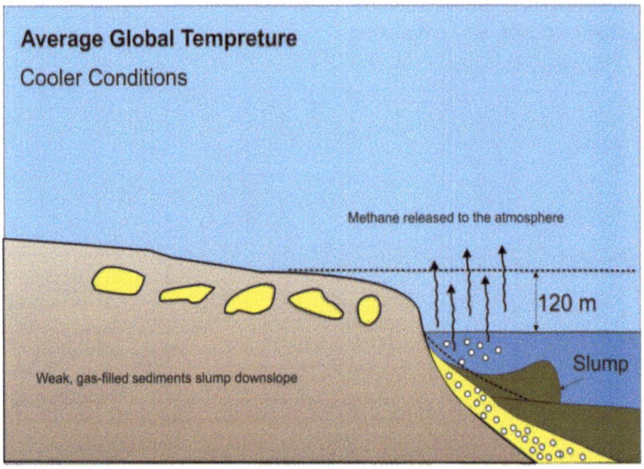

Fig. 12. Gas hydrates in cooler conditions.

Figure 12: When temperature drops, so does sea level, as more and more water is locked into ice and snow on the continents. Falling sea level will reduce the water pressure on ocean floor deposits, so hydrates will start to break down, beginning in the shallowest areas. Methane gas will bubble up from the bottom, rise through the sea, and eventually diffuse into the air. Terrestrial hydrates are not affected, because colder air preserves the permafrost that protects them. The submarine slump will further help to release methane and the disaster will be more.

7 Conclusions

(1) Favourable Geotechnical properties/parameters in continental slope will augur well with Submarine and Air attacks to explode Methane and Methane Hydrates.
(2) Exploding Methane and Methane hydrates in continental slopes is a better surgical strike than conventional ones.
(3) This method creates minimum damage to environment which will recover to normal state in short time. The loss of lives and damage to properties are minimum.
(4) The advantages are more in this type of attacks than conventional war with nuclear weapons.
(5) This method of attacks brings permanent peace to erring nations, which in turn will bring world peace.
(6) This method of attacks is not an end to peace but means to peace.
(7) Conventional nuclear war brings permanent damage and sorrow to mankind lasting for several decades after the war.
(8) In Modern Wars, Resource-Related objectives have been generally determined by broader strategic aims and not vice-versa.

References

Archer, D.: Methane Hydrate stability and anthropogenic climate change (PDF). Biogeosciences **4**(4), 521–544 (2007)

Climate- Hydrate interactions USGS, January 2013

Spalding, D., Fox, L.: Challenges of methane hydrates. Oil Gas Financ. J. (2014)

Dickens, G.R.: Rethinking the global carbon cycle with a large dynamic and microbially mediated gas hydrate capacitor. Earth Planet Sci. Lett. **213**, 169–183 (2003)

Kennet, J.P., Cannariato, K.G., Hendy, I.L., Behl, R.J.: Methane Hydrates in Quaternary Climate Change: The Clathrate Gun Hypothesis (2003)

Kennet, J.P., Cannariato, K.G., Hendy, I.L., Behl, R.J.: Carbon isotopic evidence for methane hydrate instability during quaternary interstadials. Science **288**(5463), 128–133 (2000)

Mathew: Methane hydrates may ignite new energy war in Asia. Business Korea News Portal Dream Energy Source, 2 May 2014 (2014)

Methane Bubbling Through Seafloor Creates Undersea Hills: Monterey Bay Aquarian Research Institute, October 2008

Rajaraman, J., Thiruvenkataswamy, K., Narasimha Rao, S.: The Influence of Geo Chemistry of Gas Hydrates on the Shear Strength and Stability of Marine Sediment Movement. Springer (2017)

Juanes, R., Bryant, S.L.: Mechanisms Leading to Co-existence of Gas and Hydrate in Ocean Sediments. Massachusetts Institute of Technology, 30 November 2006 (2006)

Satyanarayana, D.: Petroleum Geochemistry, pp. 342–362. Daya Publishing House, New Delhi (2011)

Severinghaus, J.P., Whiticar, M.J., Brook, E.J., Petrenko, W., Ferreti, D.F., Severinghaus, J.P.: Ice record of ^{13}C for atmospheric CH_4 across the younger dryas-pre-boreal transition. Science **313**(5790), 1109–1112 (2006)

Shakova, N., Semiletov, I.P.: Methane Release from the East Siberian Arctic Shelf and the Potential Abrupt Climate Change, 30 November 2010, April 2014

Sowers, T.: Late quaternary atmospheric CH_4 isotope record suggests marine clathrates are stable. Science **311**(5762), 838–840 (2006)

The Day The Earth Nearly Died. BBC. http://www.bbc.co.uk/science/horizon/2002. Accessed 5 Dec 2002

Author Index

M. Shehata and F. Rodrigues (Eds.): GeoMEast 2018, SUCI, p. 269, 2019.
https://doi.org/10.1007/978-3-030-01905-1

MIX
Papier aus verantwortungsvollen Quellen
Paper from responsible sources
FSC® C105338
FSC
www.fsc.org

If you have any concerns about our products,
you can contact us on
ProductSafety@springernature.com

In case Publisher is established outside the EU,
the EU authorized representative is:
Springer Nature Customer Service Center GmbH
Europaplatz 3, 69115 Heidelberg, Germany

Printed by Libri Plureos GmbH
in Hamburg, Germany